Fossil Free Fuels

Fossil Free Fuels
Trends in Renewable Energy

Edited by
Maniruzzaman A. Aziz, Khairul Anuar Kassim,
Wan Azelee Wan Abu Bakar, Aminaton Marto, and
Syed Anuar Faua'ad Syed Muhammad

CRC Press
Taylor & Francis Group
Boca Raton London New York

CRC Press is an imprint of the
Taylor & Francis Group, an **informa** business

CRC Press
Taylor & Francis Group
6000 Broken Sound Parkway NW, Suite 300
Boca Raton, FL 33487-2742

© 2020 by Taylor & Francis Group, LLC
CRC Press is an imprint of Taylor & Francis Group, an Informa business

International Standard Book Number-13: 978-0-367-34762-8 (Hardback)

Library of Congress Cataloging-in-Publication Data

Names: Aziz, Maniruzzaman Bin A., editor.
Title: Fossil free fuels : trends in renewable energy / edited by
Maniruzzaman Bin A. Aziz, Khairul Anuar Kassim, Wan Azelee Wan Abu
Bakar, Amination Marto, Syed Anuar Faua ad Syed Muhammad.
Description: Boca Raton : CRC Press, 2020. I Includes bibliographical
references and index.
Identifiers: LCCN 2019028555 (print) I LCCN 2019028556 (ebook) I ISBN
9780367347628 (hardback : acid-free paper) I ISBN 9780429327773 (ebook)
Subjects: LCSH: Renewable energy sources. I Biodiesel fuels. I Biomass
energy--Environmental aspects. I Fossil fuels--Environmental aspects.
Classification: LCC TJ808 .F68 2020 (print) I LCC TJ808 (ebook) I DDC
333.79/4--dc23
LC record available at https://lccn.loc.gov/2019028555
LC ebook record available at https://lccn.loc.gov/2019028556

Visit the Taylor & Francis Web site at
http://www.taylorandfrancis.com

and the CRC Press Web site at
http://www.crcpress.com

Contents

Foreword

Energy is everything around us and nothing is possible without energy. Predominantly, mankind has consumed fossil fuels for millions of years and this finite resource will be depleted in the future. Over decades, researchers and industrial practitioners have been looking out for ways to reduce our reliance on fossil fuels. Other than the scarce-resources issue, the use of fossil fuels also raises many serious environmental problems. The burning of fossil fuels releases huge amounts of carbon dioxide (CO_2) and other air pollutants into the atmosphere. Notably, these activities are associated with the environmental and public health risks global warming and global climate change.

According to the International Energy Agency (IEA), the world's CO_2 levels grew from 17.78 billion tons in 1980 to 32.10 billion tons in 2015. If no action is taken to mitigate greenhouse gases (GHGs), the concentration of GHGs in the atmosphere would double its preindustrial level by as early as 2035. The statistical data from the Global Report on Human Settlement (GRHS), a source studying cities and climate change, revealed that about 75% of CO_2 in the atmosphere originated from industrial power plants, especially those that were involved in the burning of fossil fuels. Thus, the use of renewable energy must be explored for sustainable energy. Globally, nations have made many efforts to provide new alternative energy sources in every industry, including transportation, construction and service providers, as well as the household and commercial sectors.

This book, *Fossil Free Fuels: Trends in Renewable Energy*, written by three renowned professors together with associate professors, lecturers and PhD students, addresses the present problems of the world as well as the future generation of ideas around sustainability. It also covers the following sectors: (i.) green construction—roads and highways; (ii.) green economy; (iii.) environmental pollution, greenhouse gases and solutions to climate change and traffic pollution; (iv.) fuel and energy: clean hydrogen fuel; (v.) biofuel for future generations; (vi.) proper utilization of waste and recycled materials; and (vii.) renewable energy.

In addition, a comprehensive analysis of the current world energy situation, solutions to sustainable problems and recommendations for future assessment of energy are discussed and evaluated in this book. These contents are covered in 12 chapters, around 95,192 words and 1,054 references from around the world which intend to reach out and share information to students, researchers, scientists, practicing engineers and the general public who endlessly support renewable energy initiatives.

Last but not least, it is hoped that this book could expand readers' horizons and serve as a reference for policy makers who may be able to bring about a revolution in renewable energy and shape our future.

Best wishes,

Prof. Datuk Ir. Dr. Wahid Bin Omar
Vice-Chancellor
Universiti Teknologi Malaysia

Preface

Fossil Free Fuels: Trends in Renewable Energy provides abatement strategies on the usage of fossil fuels and the emission of carbon dioxide into the environment. As climate change and global warming have become alarming global scenarios over the past few decades, there have been many approaches to overcome and mitigate this recurring environmental problem. This book covers critical topics such as the exploration of biofuels as the next sustainable renewable resources, untapped waste to wealth products, the green economy and carbon-neutral strategies in various industries such as the transportation, construction and manufacturing sectors. In this book, various novel agricultural waste products have been explored as potent substrates for the production of biobutanol and the recycling of construction waste as an eco-friendly material, as well as numerous other green technologies for the production of clean hydrogen. Other than the emission of greenhouse gases, municipal solid waste management through landfill sites have been discussed intensively, as leachate, a product of landfill, leaves a ripple effect on the contamination of ground and surface water. The authors also take a multidisciplinary approach and share their findings by reviewing the most current strategies and solutions to reduce environmental concerns, such as the implementation of carbon capture and sequestration (CCS) techniques, the modeling method, critical analysis, advanced technologies and operations, as well as a case study to deliver present alternatives and updates to readers in great detail. This book attempts to provide resources and information to readers such as students, researchers, scientists, non-specialist associates with renewable energy industries, the general public and those who actively support renewable energy strategies for the sustainable development of future generations.

Prof. Datuk Ir. Dr. Wahid bin Omar
Vice-Chancellor
Universiti Teknologi Malaysia

Editors

Maniruzzaman A. Aziz received his MSc and PhD from University Putra Malaysia (UPM) and joined as a senior lecturer in the University Technology Malaysia (UTM) in 2012. Prior to that, he worked 15 years in Malaysian highways and airport runways, including the North–South Interurban Toll Expressway (PLUS Bhd), Kuala Lumpur International Airport (KLIA), XVI Commonwealth game, Public Works Department (JKR) projects, Malaysian Highways Authority (MHA) projects and others. He received the Excellent Service Award (Anugerah Perkhidmatan Cemerlang Tahun) for the year 2017, University Technology Malaysia (UTM). He has also received many awards from institutions all over the world, such as the *Journal of Traffic and Transportation Engineering* (JTTE), 10th Malaysian Road Conference (MRC) & Exhibition 2018 and PIARC International Seminar on Asset Management. He was a member of Technical Committee of 8th Innovation, Arabia Conference, Hamdan Bin Mohammed Smart University, United Arab Emirates (UAE), February 2015. He has published many articles in the *Journal of Renewable and Sustainable Energy Reviews* (RSER) and given a number of talks as the invited speaker/presenter/facilitator on the topics of how to get articles published in ISI journals and high-impact journal publication.

Khairul Anuar Kassim is a civil engineering lecturer and the Chairman of School of Civil Engineering, University Teknologi Malaysia. He received his B.Eng in civil engineering from the University of Middlesex, London, and his MSc and PhD from the University of Newcastle upon Tyne, United Kingdom. Professor Khairul's research includes work on ground improvement using active additives and biomediated soil stabilization. He has published more than 100 journal papers on related areas.

Wan Azelee Wan Abu Bakar was born in 1959, and he has worked at Universiti Teknologi Malaysia since 1983. He was promoted to full professorship in 2000. His expertise is in heterogeneous catalysis chemistry. He has published more than 100 papers in national and international journals and has supervised 23 MSc and 20 PhD students. He is very active in consultation work and has owned three patents with two commercializable products.

Aminaton Marto started her career in Universiti Teknologi Malaysia (UTM) as an assistant lecturer 'A' in July 1983 and was promoted to full professor in 2008. She has been appointed as a visiting scholar in Japanese and American universities and a visiting professor at Tokyo City University, Japan. Currently, she is affiliated with the Malaysia-Japan International Institute of Technology (MJIIT), UTM Kuala Lumpur. In MJIIT, she belongs to the Chemical and Environmental Engineering Department and a Member of Disaster Preparedness and Prevention Centre. In addition, she is also an associate fellow at the Centre of Tropical Geoengineering, UTM. Since joining UTM, Professor Marto has established numerous successful research programs in the field of geotechnical engineering. Her specializations are in soft soil engineering, geotechnical investigation and geotechnical earthquake engineering. She pioneered the research on bamboo-geotextile composite for soil reinforcement and the use of coal ash as an alternative material for soft soil improvement. Her other research includes chemical and recycled blended tiles for soil stabilization, tunneling, liquefaction and geothermal energy pile. Currently she is focusing on sustainable development and disaster risk reduction. This includes exploring the use of screw driving sounding test for determining fast and accurate sub-surface soil profiles and the use of industrial waste for carbon sequestration. Throughout Professor Marto's 35 years of service in UTM, she has successfully secured more than 30 research projects from national and international agencies. She had published numerous journal papers and supervised more than 50 local and international postgraduate students.

Syed Anuar Faua'ad Syed Muhammad was born in 1971 and has worked at Universiti Teknologi Malaysia since 1997. He was appointed senior lecturer in 2015. His expertise is in the crystallisation of pharmaceutical compounds and aerobic wastewater treatment plant for oleochemical industries. He has published more than 50 papers in national and international journals and conferences. He has supervised ten MSc and two PhD students. He is very active in consultation works, especially in the areas of an aerobic and anaerobic for wastewater treatment plant.

Contributors

Bawadi Abdullah
Department of Chemical Engineering
and
CO_2 Utilization Group, Institute of
    Contaminant Management for
    Oil and Gas
Chemical Engineering Department
Universiti Teknologi PETRONAS
Seri Iskandar, Malaysia

Kamarudin Ahmad
Associate Professor, School of
    Civil Engineering
Universiti Teknologi Malaysia
Johor Bahru, Malaysia

**Mohammed Ali Mohammed
Al-Bared**
Department of Civil and
    Environmental Engineering
Universiti Teknologi PETRONAS
Bandar Seri Iskandar, Malaysia

A. B. M. Amimul Ahsan
Department of Civil Engineering
Uttara University
Dhaka, Bangladesh

Yusuf Babangida Attahiru
School of Civil Engineering, Faculty
    of Engineering
Universiti Teknologi Malaysia
Johor Bahru, Malaysia

Ali Awad
Chemical Engineering Department
Universiti Teknologi PETRONAS
Seri Iskandar, Malaysia

Indra Sati Hamonangan Harahap
Department of Civil and Environmental
    Engineering
Universiti Teknologi PETRONAS
Seri Iskandar, Malaysia

Nurhamieza Md. Huzir
Department of Bioprocess and Polymer
    Engineering, School of Chemical
    and Energy Engineering
Universiti Teknologi Malaysia
Johor Bahru, Malaysia

Shahrul Ismail
Eastern Corridor Renewable Energy,
    School of Ocean Engineering
Universiti Malaysia Terengganu
Terengganu, Malaysia

Fauzan Mohd Jakarmi
Faculty of Engineering
Universiti Putra Malaysia, UPM
    Serdang
Selangor, Malaysia

M. Ehsan Jorat
Division of Natural and Built
    Environment
School of Science Engineering and
    Technology
Abertay University
Dundee, United Kingdom

Nik Azmi Nik Mahmood
Department of Bioprocess and Polymer
    Engineering, School of Chemical
    and Energy Engineering
Universiti Teknologi Malaysia,
    UTM Skudai
Skudai, Malaysia

Zulkepli Majid
Faculty of Geoinformation and
    Real Estate
Universiti Teknologi Malaysia
Johor Bahru, Malaysia

Azman Mohamed
School of Civil Engineering, Faculty
    of Engineering
Universiti Teknologi Malaysia
Johor Bahru, Malaysia

Habiba Ibrahim Mohammed
Faculty of Geoinformation and
    Real Estate
Universiti Teknologi Malaysia
Johor Bahru, Malaysia

George Moses
Department of Civil Engineering
Nigerian Defense Academy
Kaduna, Nigeria

Nurul Hidayah Muslim
School of Civil Engineering, Faculty
    of Engineering
Universiti Teknologi Malaysia
Johor Bahru, Malaysia

Thanwa Filza Nashruddin
School of Civil Engineering, Faculty
    of Engineering
Universiti Teknologi Malaysia
Johor Bahru, Malaysia

Farahiyah Abdul Rahman
School of Civil Engineering, Faculty
    of Engineering
Universiti Teknologi Malaysia
Johor Bahru, Malaysia
and
School of Environmental Engineering
Universiti Malaysia Perlis
Jejawi, Malaysia

Salmiah Jamal Mat Rosid
Unisza Science and Medicine
    Foundation Centre
Universiti Sultan Zainal Abidin
Terengganu, Malaysia

Hoofar Shokravi
School of Civil Engineering
Faculty of Engineering
Universiti Teknologi Malaysia
Johor Bahru, Malaysia

Hooman Shokravi
Department of Civil Engineering
Islamic Azad University
Tabriz, Iran

Zahra Shokravi
Department of Microbiology
Faculty of Basic Science
Islamic Azad University, Science and
    Research Branch of Tehran
Arak, Iran

Susilawati Toemen
School of Science
Universiti Teknologi Malaysia
Johor Bahru, Malaysia

Noor Azrimi Umor
Department of Microbiology, Faculty of
    Applied Science
Universiti Teknologi MARA
Negeri Sembilan, Malaysia

Yamusa Bello Yamusa
School of Civil Engineering
Universiti Teknologi Malaysia
Johor Bahru, Malaysia
and
Department of Civil Engineering
Nuhu Bamalli Polytechnic
Zaria, Nigeria

1 Next Generation of Agro-Industrial Lignocellulosic Residues to Eco-Friendly Biobutanol

Nurhamieza Md. Huzir, Md. Maniruzzaman A. Aziz, Shahrul Ismail, Bawadi Abdullah, Nik Azmi Nik Mahmood, Noor Azrimi Umor, and Syed Anuar Faua'ad Syed Muhammad

CONTENTS

1.0 INTRODUCTION

In our lives, everything around us is surrounded by energy which is in a state of flux. As the demand for global energy increases rapidly, existing resources come to their shrinking phases and require alternatives to mitigate this scenario. Due to this, sustainable energy has become a prime concern around the world. As quoted by Tim Wirth, *"energy is essential for development and sustainable energy is essential for sustainable development."* According to the World Energy Council, demand for energy will continuously increase and peak before 2030 [1]. This forecast has triggered the production of sustainable biofuels to fulfill the required demand, reducing the global environmental footprint caused by climate change and the emission of greenhouse gas (GHG) into the atmosphere. Figure 1.1 shows the total world biofuel productions from 2006 to 2016. Notably, bioalcohols such as bioethanol are

1

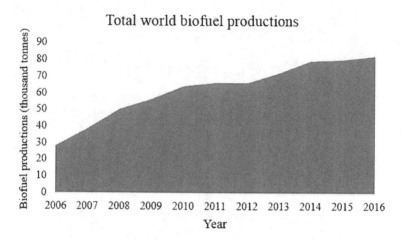

FIGURE 1.1   Global biofuel productions from 2006 to 2016 [2].

widely used as biofuels in internal combustion engines in several countries such as the United States and Brazil. However, to fulfill the extensive demands of energy in biofuel productions, biobutanol is a viable option as the next generation of biofuels since it has advanced and superior fuel properties.

Intense research and development (R&D) in biofuel studies has led to the different generation of biofuels based on the feedstock being used. First-generation biofuels are produced from edible biomass, which are food crops such as sugarcane, maize and cereal grains, while second-generation biofuels are generated from lignocellulosic materials that can be obtained from agricultural and forestry residues. Meanwhile, there are also several studies using algae for the third generation of biofuels [3, 4]. Besides, current trends in biofuel production have also introduced the fourth generation of biofuels by using photobiological solar fuels and electrofuels [5].

Generally, there is a growing consensus that second-generation biofuels outweigh the others as they are capable of reducing tons of waste generated by the agricultural sector, providing green energy and environmental sustainability, as well as improving the economy and development in rural areas [6]. Figure 1.2 illustrates top world agricultural production in 2017. As there is a rising demand in the agricultural sector, the agricultural by-product seems to produce tremendous quantities of waste. To address this issue, agricultural waste can be utilized as a resource to produce biofuels and it also provides a win-win solution for the agricultural and transportation sectors. Therefore, this chapter will update the potential uses of agricultural residue for the production of biobutanol.

1.1   CLASSIFICATION OF BIOFUELS

Biofuels that are produced from biomass can be categorized into primary (solid fuels) and secondary biofuels (liquid or gaseous fuels). Any types of solid materials that produce energy through combustion are referred to as solid fuels, such as coal, charcoal, wood and pellets, while secondary biofuels are commonly used for transportation,

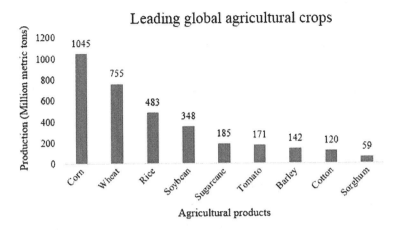

FIGURE 1.2   Total agricultural products in 2017 [7].

which can be classified into four generations as listed in Table 1.1. Typically, biofuels that can be produced from biomass are bioethanol, biobutanol, biomethanol, biodiesel, syngas (hydrogen, carbon monoxide and some carbon dioxide) and biomethane. Among the renewable fuels, biobutanol is regarded among the future green biofuels due to its superior properties which do not require any modification to be used in engines. Biobutanol, sometimes referred to as n-butanol, is a product of the acetone-butanol-ethanol (ABE) fermentation process, where the sugar is converted using the genus *Clostridium* into butanol, acetone and ethanol in a ratio of 6:3:1, respectively. This four-carbon alcohol is capable of replacing gasoline and can be used as fuel in

TABLE 1.1
First to fourth biofuel generations [3, 12, 13]

Biofuel generations	Advantages	Disadvantages
First biofuel generation (Sugar and starch feedstocks)	• High yield of biofuel can be achieved • Requires simple pretreatment process	• Occupied large land for cultivation of crops • Food insecurity issues
Second biofuel generation (Lignocellulosic biomass)	• Massive amount of resources • Does not compete with food supply	• Complex and recalcitrant feedstock • Requires monotonous pretreatment
Third biofuel generation (Algal biomass and seaweeds)	• Low or no lignin and fermentation inhibitors • Devoid arable land for Cultivation	• Costly in downstream process • Requires intense research for further development
Fourth biofuel generation (Biologically engineered crop to fix carbon dioxide)	• More CO_2 capture ability • High production rate	• Research still at its primary stage

TABLE 1.2

Comparison of bioalcohols with gasoline and diesel fuels [14, 15]

Properties	Butanol	Ethanol	Methanol	Gasoline	Diesel
Molecular formula	C_4H_9OH	C_2H_5OH	CH_3OH	$C_4 - C_{12}$	$C_{12} - C_{25}$
Cetane number	~25	5–8	3.8	5–20	40–55
Motor octane number	78	102	104	90	81–89
Rating octane number	96	129	136	91–99	15–25
Higher heating value (MJ/kg)	36.6	29.7	22.7	47.3	44.8
Lower heating value (MJ/kg)	33.1	26.9	20.1	43.1	42.8
Viscosity (10^{-3} Pa s)	2.593	1.078	0.545	0.24–0.32	–
Flash point (°C)	35	13	12	42	74
Oxygen content (%)	22	34.7	49.9	–	–
Solubility in 100 g water	Immiscible	Miscible	Miscible	Immiscible	Immiscible

vehicles without altering their internal combustion (CI) engines. It is superior to ethanol since it has greater intersolubility with gasoline and diesel compared to ethanol [8, 9]. Other than that, several studies have also proved that biobutanol produces a cleaner burn than bioethanol since it has a lower oxygen content, and, when it is used in internal combustion engines, it produces less carbon monoxide, low nitrogen oxide (Nox) and near-zero smoke emissions, which make it an eco-friendly biofuel [10, 11]. Other physico-chemical properties of liquid fuels are listed in Table 1.2.

1.2   AGRO-INDUSTRIAL LIGNOCELLULOSIC BIOMASS

"We can get fuel from fruit, from that shrub by the roadside, or from apples, weeds, saw-dust—almost anything! There is fuel in every bit of vegetable matter that can be fermented."

Henry Ford, 1925.

As predicted by Henry Ford in 1925, lignocellulosic biomass, which includes agricultural waste, forestry residue, industrial waste, grasses and woody materials, is currently being utilized to produce biofuels. The efficiency of plant materials to be used as feedstock in the production of biofuels depends on the biomass composition, where most of it contains 40–50% of cellulose, 20–30% of hemicellulose and 10–25% of lignin [16]. As pointed out by Baral et al. [17], high cellulose and hemicellulose content with a low lignin content is more preferable to produce high amounts of monomeric sugar with minimal inhibitory effects during the fermentation process. A high amount of fermentable sugar is capable of producing a high yield of biobutanol. However, since even the biomass itself stores a high amount of lignocellulosic composition, the role of pretreatment is the key to break down the lignin in order to recover the cellulose prior to enzymatic hydrolysis and the fermentation process. Figure 1.3 illustrates the overview of lignocellulosic biomass to biobutanol production, while Table 1.3 depicts the lignocellulosic composition in agro-industrial residues.

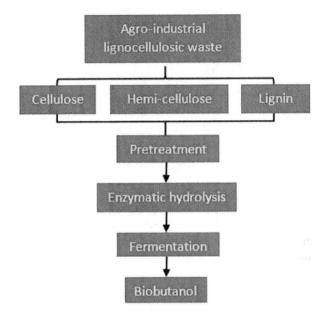

FIGURE 1.3 Overview of biobutanol production from agricultural residue.

All of the viable agro-waste used in biobutanol production are discussed below:

1.2.1 Corn Residues

Zea mays, corn or maize is one of the top agricultural crops in the world. In 2016, the major production of corn was dominated by the United States (US) and China, with 345 and 225 million tons, respectively [42]. Prior to the debate in "food versus fuel," corn itself was widely being used as the predominant feedstock to produce biofuels. To avoid this issue, another potential source with a low cost of production is being selected from cornfields, which is corn stover. Corn stover consists of stalks, leaves and cobs of the corn that are generated after the corn grain is harvested. When the grain yields increase, the corn stover yields also increase. It is estimated that 1.0 dry kg corn stover is produced per 1.0 dry kg of corn grain harvested [43].

As for biobutanol production, several researchers use corn stover as a substrate with different kinds of pretreatments. Alkaline and acid pretreatment are the most employed pretreatments used by researchers. From the previous work, alkaline-pretreated corn stover (NaOH) is capable of producing as much as 10.2 g/L of biobutanol, which is slightly higher than acidic pretreatment's (H_2SO_4) 8.98 g/L [44, 45]. Another study by Cai et al. [46] used alkaline pretreatment on various parts of corn, such as the flower, leaf, cob, husk and stem of corn, and the biobutanol produced from each part was as much as 7.5 g/L, 7.6 g/L, 9.4 g/L, 7 g/L and 7.6 g/L, respectively. Other than that, corn fiber is also utilized to its potential as a substrate for biobutanol production. Corn fiber consists of corn kernel hulls and residual starch that is not extracted during the corn wet-milling process. Guo et al. [47] have extensively studied the use of corn fiber as a

TABLE 1.3

Lignocellulosic compositions of agro-industrial wastes

Agro-waste	Cellulose (%)	Hemicellulose (%)	Lignin (%)	References
Corn stover	37	31.3	17.8	[17]
Corn fiber	14	39	5.7	[18]
Sugarcane bagasse	40–45	30–35	20–30	[19]
Wheat straw	30	50	15	[20]
Wheat bran	40–50	25–35	15–20	[21]
Rice straw	26.87	26.66	13.27	[22]
De-oiled rice bran	39	31	4	[23]
Rice husk	34.4	24.3	19.2	[24]
Barley straw	35	33	24	[25]
Sweet sorghum bagasse	36.9	17.8	19.7	[26]
Cassava bagasse	43	25	9	[27]
Sugar beet pulp	29.50	27.51	3.82	[28]
Banana stem	40.8	29.9	6.57	[29]
Mango peels	25	8.1	–	[30]
Orange peels	11.97	14.4	2.15	[31]
Pineapple peels	35	19	16	[32]
White grape pomace	9.2	4	11.6	[33]
Apple pomace	36	11	19	[34]
Oil palm frond	30.4	40.4	21.7	[35]
Oil palm trunk	45.81	17.74	24.49	[36]
Empty fruit bunch	50.49	28.46	17.84	[37]
Palm kernel cake	35.7	30.3	15.7	[38]
Palm oil mill effluent	39.56	23.33	25.02	[39]
Cauliflower waste	17.32	9.12	5.94	[40]
Pea pod waste	26	20.5	3.92	[41]

substrate by using *Clostridium beijerinckii* mutant RT66 as a microorganism involved in ABE fermentation, which eventually produced 12.9 g/L total solvents, consisting of 3.1 g/L of acetone, 9.3 g/L of butanol and 0.5 g/L of ethanol.

1.2.2 SUGARCANE WASTE

Sugarcane (*Saccharum officinarum*) belongs to the grass family Gramineae, and it is a major crop in Brazil. This makes Brazil renowned as the largest sugar-producing country in the world. When it comes to the use of biofuels, Brazil is one of the leading countries in the world to use biofuels derived from sugarcane in the transportation industry. Apart from utilizing the sugarcane itself, the crop residues can also be maximized for biobutanol production. One of the largest sugarcane residues is sugarcane bagasse. It is a fiber residue that remains after the sugarcane stalks are crushed to extract the juice from the sugarcane. Sugarcane bagasse has gained so much interest in the biofuel industry and served as a feasible substrate for microbial to produce value-added products such as enzymes, amino acids and organic acids [48].

Recent research conducted by Li et al. [49] used *Clostridium acetobutylicum* ATCC 824 as sources of inoculum with two different pretreatments on sugarcane bagasse. From the experiments, sugarcane bagasse that was pretreated with oxidate ammonolysis and a combination of oxidate ammonolysis with dilute acid pretreatment (DA-OA) produced 4.62 g/L and 7.68 g/L biobutanol, respectively. In biobutanol production, the fermentation process also plays an important role other than the pretreatment process. Previously, there have been studies that used similar pretreatment toward sugarcane bagasse with different modes of fermentation operation, which were batch and fed-batch fermentation (FBF). Alkaline-pretreated sugarcane bagasse in batch fermentation was capable of producing 9.3 g/L biobutanol, whereas in FBF, 14.17 g/L of biobutanol was produced [50, 51].

1.2.3 WHEAT AND RICE RESIDUES

Wheat (*Triticum aestivum*) and rice (*Oryza sativa*) are included in the top three leading food crops in the world, and they are the staple foods for most of the world's population. Higher production of these industrial food crops generates a voluminous amount of waste, which triggers environmental problems and health risks to the population if the disposal of the by-product is not managed properly. The main by-product of wheat and rice is straw, husk and bran. Straw is produced after the harvesting process where the grains are threshed while husk and bran are generated during the milling process. The amount of straw yield varies depending on the country and region. For instance, in northern Europe, 7.5 tons of grain are produced after harvesting, which eventually leads to higher yields of wheat straw, whereas in southern Europe, only 3 tons of wheat grain are generated per harvest, leading to a lower quantity of wheat straw [52]. For rice straw, it is estimated that 1.35 tons of rice straw are produced per ton of harvested rice grain, while in the milling industry, 200 kg of rice husk and 100 kg of rice bran are generated per ton of processing paddy rice [53]. Major residues, such as that of wheat and rice straw, are capable of producing a high concentration of biobutanol, while bran and husk are still at the primary stage of research. Table 1.4 shows the biobutanol production from different by-products of wheat and rice.

1.2.4 BARLEY AND SORGHUM RESIDUES

For the past few decades, the annual world harvest of barley, or its botanical name *Hordeum vulgare*, was approximately 140 million tons, which contributed to the high generation of barley straw. According to Kim and Dale [43], 1.2 kg of barley straw is produced per kg of barley grain. Conventionally, the barley straw can be used as animal feedstock in addition to being used in the field to prevent soil erosion. Besides, barley straw has also been utilized as a viable raw material to produce biobutanol for the past few years. Yang et al. [59] have compared the effect of using pure glucose and barley grain as a substrate by implementing dilute-sulfuric acid

TABLE 1.4

Biobutanol production from wheat and rice residue

Feedstock	Microorganisms	Pretreatment	Biobutanol titer (g/L)	Reference
Wheat straw	*Clostridium beijerinckii* DSM 6422	Steam explosion	7.21	[54]
Wheat bran	*Clostridium beijerinckii* ATCC 55025	Dilute sulfuric acid pretreatment	8.80	[55]
Rice straw	*Clostridium acetobutylicum* MTCC 481	Steam explosion	1.72	[56]
De-oiled rice bran	*Clostridium acetobutylicum* YM 1	Diluted acetyl chloride	6.48	[57]
Rice husk	*Clostridium acetobutylicum* MTCC 481	Acid pretreatment	0.018	[58]

pretreatment prior to ABE fermentation. From the process, the biobutanol production using barley straw has similar concentrations with pure glucose, which are 7.9 g/L and 8.0 g/L, respectively. This indicates that barley straw is another promising feedstock that can be exploited for fuel and energy in the future, where extensive research is required to maximize its benefit to the biofuel industry.

Another prominent feedstock in biobutanol production is sweet sorghum bagasse (SSB). Sweet sorghum (*Sorghum bicolor*) has a higher content of carbohydrates and a high content of monomeric sugar in its stalk, which make it possible to be used as raw material in ABE fermentation [60]. Once the stalk has been crushed after the extraction of its juice, the bagasse is left on the field. Cai et al. [61] have discussed the effect of SSB pretreated by acid pretreatment and detoxification process, as well as implementing the recovery process in the fermentation broth by using pervaporation separation. After the fermentation process, 12.3 g/L butanol concentration is produced, while after the recovery process, the pinnacle of biobutanol titer is produced at 201.9 g/L. Generally, a significant increase in biobutanol production can be obtained by operating recovery techniques from a fermentation broth, such as adsorption, distillation, gas stripping, liquid-liquid extraction, membrane distillation, pervaporation, pertraction (membrane extraction) and reverse osmosis [62]. It is reported that a hybrid of gas stripping with pervaporation is capable of obtaining a maximum of biobutanol at 482.55 g/L from a fermentation broth of SSB [63].

1.2.5 CASSAVA AND SUGAR BEET RESIDUES

Cassava (*Manihot esculenta*), also known as tapioca, belongs to the Euphorbiaceae family, where it is a perennial and sustainable crop that can tolerate prolonged drought and has a resistance to nutrient deprivation in the soil. Furthermore, cassava root has a high starch content, which can be up to 90% on a dry weight basis [64]. These characteristics mean that cassava can be easily cultivated, placing the cassava-based industry in its blooming phase. The industry produces a huge generation of by-products, and a major residue of this is cassava bagasse, also called cassava pomace, cassava fiber and cassava pulp. Cassava bagasse is a fibrous material that remains after the extraction of flour or starch content from the tuber. Lu et al. [27] reported that cassava bagasse used as a feedstock in fed-batch fermentation of ABE integrated with gas stripping is capable of obtaining 76.4 g/L of biobutanol titer. Notably, when glucose is used as a feedstock, the biobutanol concentration produced after gas stripping is 113.3 g/L, which indicates a 32.57% increment over cassava bagasse [65]. From batch fermentation without the recovery and separation process, cassava bagasse and glucose produced as much as 9.71 g/L and 19.1 g/L of biobutanol, respectively.

Twenty percent of the global supply of sugar is derived from sugar beet (*Beta vulgaris*), while the remainder is dominated by sugarcane. In 2016, the world's leading producer of sugar beet crop is the European Union (EU), constituting 50% of the global production of sugar beet, which is as much as 111.7 million tons [66]. From the manufacturing of sugar beet, valuable by-products that can be used as a source of energy are sugar beet pulp and sugar beet molasses. Sugar beet pulp is highly recommended for use as a feedstock since sugar beet can be harvested daily throughout the year and the pulps contain low lignin, making it suitable for use in the biofuel industry [67]. The use of acid-pretreated sugar beet pulp for biobutanol production shows higher results with an increment of solid loading. For 5%, 7.5% and 10% solid loading, the biobutanol obtained in the fermentation broth are 4.5 g/L, 6.6 g/L and 7.8 g/L, respectively [68]. Meanwhile, *Clostridium acetobutylicum* 2N used to ferment sugar beet molasses within 48 hours produced 14.15 g/L biobutanol titer, which is higher than the 13.2 g/L biobutanol produced after a 60-hour fermentation of glucose medium [69, 70].

1.2.6   FRUIT WASTE

The expansion of the biofuel industry has led to the discovery of several novel fruit wastes that may be used as substrates. Moreover, this holistic approach eventually assists in the dwindling of waste production, turning the valuable untapped waste into energy. For the past few years, banana (*Musa sapientum*), mango (*Mangifera indica*), citrus (*Citrus sp.*), acorn (*Quercus brantii*) and pineapple (*Ananas comosus*) wastes have been used as substrates to produce biobutanol.

Bananas are one of those fruits with an enormous demand that also generate a huge amount of waste in the form of rotten bananas, peels, leaves, stalk and pseudostems. Currently, the leading global producers of bananas are India, China and Uganda, where India alone constitutes 27% of the world's banana production [71]. Numerous studies have been done to maximize the by-product of bananas, such as for the production of biogas [72], in wastewater treatment [73], as an animal feed [74], biofertilizer [75] and bioethanol production [76]. Recently, there have been several studies that use bananas as a source of producing biobutanol. Vidhya et al. [77] discussed the effect of different pretreatment methods for the production of biobutanol from banana stems. Three different pretreatment methods, which are alkaline pretreatment, acid pretreatment, and microwave digestion have been done where the biobutanol production was 8.1 g/L, 5.0 g/L, and 6.9 g/L, respectively. When a banana stem was used solely as a substrate without any addition of glucose, 9.1 g/L biobutanol was obtained, whereas when 2% of glucose was added, 18.0 g/L biobutanol titer was produced [78].

Another emerging substrate for biobutanol is mango waste. To date, there have been many studies regarding the use of mango waste, such as mango peels, pulps [79] and stems [80] in bioethanol production, whereas there has only been one study using mango peels in biobutanol production. A study by Avula et al. [30] imposed three kinds of conditions to the mango peel, which were the fermentation of the mango peel in a serum bottle with and without the supplementation of nutrients, and the fermentation of mango peels in a bioreactor with the addition of nutrients. By using *Clostridium acetobutylicum* 2878 throughout the experiments, biobutanol titer from the fermentation of mango peels solely produced 10.5 g/L, while the addition of supplements increased the concentration to 13.3 g/L. On the other side, the fermentation of mango peels with the addition of nutrients in 3L bioreactor was capable of producing 15.42 g/L biobutanol, where the concentration obtained was higher than in the fermentation of rice straw using 2L and 5L bioreactor, which were 12.17 g/L and 12.22 g/L, respectively [81].

Citrus belong to the Rutaceae family, which contains many species such as Mandarin orange (*Citrus reticulata*), pomelo (*Citrus maxima*), citron (*Citrus medica*) and hybrid species such as grapefruit (*Citrus paradisi*), tangerine (*Citrus tangerina*), lemon (*Citrus limon*) and sweet orange (*Citrus sinensis*). In citrus-processing industries, more than 61% of the world's citrus production is dominated by sweet orange where, after juice extraction, an abundance of leaves, fruit peels, seeds and other by-products are also generated [82]. From past research, the utilization of orange peels and by-products in bioethanol production shows that there is a good potential for energy generation. For biobutanol production, one study has been conducted with the application of steam explosion pretreatment toward sweet orange peels. *Clostridium acetobutylicum* NCIM 2877 was used to ferment the pretreated orange peels and produced a 19.5 g/L concentration of biobutanol, which is higher compared to the 9.71 g/L of biobutanol obtained from cassava bagasse [27, 83].

Recent research has also investigated the use of pineapple peel waste and acorn waste and their capability of being utilized as substrates for biobutanol production. Both studies only showed that the total solvent produced after ABE fermentation, which was 5.23 g/L for pineapple waste and 5.81 g/L total ABE for acorn waste [32, 84]. Further study concerned with pretreatments and separation techniques for this substrate must be explored to attain maximum titer of biobutanol from each source.

As for grape (*Vitis vinifera*) and apple (*Malus sp.*), both substrates have long been identified as being involved with fermentation to produce wine and cider, respectively. After the fruits have been pressed to extract their juice, the pomace left aside is the solid residue, consisting of the skin, pulp and seed of the fruit. In the cider and winemaking industries, the pomace is used as animal feed or fertilizer, but much of the waste is sent to landfill [85]. These pomaces have a high moisture content of up to 70–75% and they are also high in biochemical oxygen demand (BOD) and chemical oxygen demand (COD) values, which require the proper handling of the waste as it can cause environmental problems [86]. Hence, several studies suggested the used of pomace primarily in bioethanol and, recently, the use of it in the production of biobutanol. Law and Gutierrez [87] used white grape pomace and the addition of yeast in submerged fermentation with the help of *Clostridium saccharobutylicum* P262, where 6 g/L biobutanol was produced. Meanwhile, apple pomace is capable of obtaining a higher concentration of biobutanol than white grape pomace, which is as much as 9.11 g/L [88]. Even though only a slight amount was produced, these pomaces have already proved their possibility to be used as biobutanol sources.

1.2.7 OTHERS

Elaeis guineensis, or palm oil, is another booming agricultural sector, especially in Indonesia, Malaysia and Thailand. Rapid expansion in this industry contributes to the abundant production of waste in its plantations, such as oil palm frond (OPF) and oil palm trunk (OPT). From the palm oil milling industry, empty fruit bunches (EFB), palm kernel cake (PKC) and palm oil mill effluent (POME) are the major residues being generated. Mainly, OPF and EFB are used as fertilizer while OPT is involved with the manufacturing of plywood [89]. Meanwhile, the majority of PKC is used as animal feedstock and POME is used as a source of energy generation, typically biogas, where the energy produced is recycled back to the milling plant. Apart from that, several researchers have also tested these oil palm residues as feasible substrates in biobutanol production. From the results obtained in Table 1.5, oil palm biomass is capable of becoming one of the other sources for the generation of biofuels in the future. OPT is capable of producing a higher production of biobutanol due to the higher amount of sugar composed in the trunk. It is reported that the

TABLE 1.5
Compilation of viable oil palm waste as substrate in biobutanol productions

Feedstock	Microorganisms	Pretreatment	Biobutanol titer (g/L)	Reference
OPF juice	*Clostridium acetobutylicum* ATCC 824	Mechanical	9.24	[91]
OPT S : Sap H : Hydrolysate	S: *Clostridium acetobutylicum* DSM 1731 H: *Clostridium beijerinckii* TISTR 1461	• S: Mechanical treatment • H: Acid hydrolysis	S: 12.76 H: 10.03	[92]
EFB	*Clostridium acetobutylicum* ATCC 824	Alkaline pretreatment	2.75	[93]
PKC	*Clostridium acetobutylicum* YM1	• N: None • AC: Activated charcoal • O: Overliming	N: 3.55 AC: 4.27 O: 3.26	[94]
POME	*Clostridium saccharoperbutyl-acetonicum* N1-4	No pretreatment	0.90	[95]

amount of sugar stored in the sap is higher than the sugar stored in sugarcane juice [90]. Meanwhile, a little concentration of biobutanol from POME could occur due to no pretreatment being implemented prior to the fermentation processes.

In addition, vegetable waste is also not excluded from the search for a suitable biobutanol substrate. According to Plazzotta et al. [96], vegetable waste can be defined as the disposing of any inedible part of the vegetable that occurs throughout the collection, handling, transportation, processing activities, food supply chain or, in other words, any loss of vegetable materials from farm to fork. Due to the bulk generation of vegetable waste, researchers have been tackling this problem by transforming waste into wealth. For example, this perishable waste can be used as antioxidants, a source of dietary fiber and the generation of bioenergy in the forms of biomethane, biohydrogen, bioethanol and biodiesel [97, 98]. Latterly, biobutanol production from vegetable waste has also been gaining much attention in the research and development prospect. Latest research utilizes several samples of vegetable waste such as pea pod, cauliflower and lettuce waste. From the research, pea pod waste obtains the highest amount of biobutanol followed by cauliflower and lettuce waste, which are 3.82 g/L, 2.99 g/L and 1.10 g/L, respectively [40, 41, 99]. Even if only a slight amount of biobutanol is produced, these wastes have proved their viability for use as substrates.

Ideally, the use of biofuels from agro-wastes complies with the three pillars of sustainability, which benefit the economy, the environment and society. As the world is trying to secure energy away from fossil fuels, the role of feedstocks in the production of biofuels will increase for sustainable energy, especially biobutanol it is regarded as being the fuel of the future. Other than the feedstocks, genetically modified butanol-producing microorganisms must be explored to make the strain able to withstand high butanol titer. Besides, integration of pretreatment, hydrolysis and the fermentation with recovery process are also required to ensure the system efficiency. Conventionally, sugarcane and corn are prioritized and exploited for biobutanol production. However, there are ample and feasible resources that can be used as raw materials without depending on single sources only. Additionally, the plentiful generation of agro-wastes has become a great challenge faced by the majority of the agricultural sector. Hence, to tackle these hurdles, the application of the 3R policies, which are reduce, reuse and recycle, the agricultural residues as raw material in biobutanol production lead to waste minimization. Notably, the direct and continuous interaction between agriculture industries and energy industries is important as this will provide a platform to mitigate environmental problems, as well as generate sustainable energy that benefits the society, economy and the planet.

REFERENCES

1. World Energy Council, "World Energy Scenarios", 2016.
2. BP, "BP Statistical Review of World Energy", 2017.
3. Y.-S. Jang, A. Malaviya, C. Cho, J. Lee, and S. Y. Lee, "Butanol production from renewable biomass by clostridia", *Bioresource Technology*, vol. 123, pp. 653–663, 2012.
4. T. Potts, J. Du, M. Paul, P. May, R. Beitle, and J. Hestekin, "The production of butanol from Jamaica bay macro algae", *Environmental Progress & Sustainable Energy*, vol. 31, pp. 29–36, 2012.

5. M. Rastogi, and S. Shrivastava, "Recent advances in second generation bioethanol production: An insight to pretreatment, saccharification and fermentation processes", *Renewable and Sustainable Energy Reviews*, vol. 80, pp. 330–340, 2017.

6. N. Sharma, B. Bohra, N. Pragya, R. Ciannella, P. Dobie, and S. Lehmann, "Bioenergy from agroforestry can lead to improved food security, climate change, soil quality, and rural development", *Food and Energy Security*, vol. 5, pp. 165–183, 2016.

7. USDA, "World Agricultural Production", 2018.

8. P. Dürre, "Biobutanol: An attractive biofuel", *Biotechnology Journal*, vol. 2, pp. 1525–1534, 2007.

9. M. K. Mahapatra, and A. Kumar, "A short review on biobutanol, a second generation biofuel production from lignocellulosic biomass", *Journal of Clean Energy Technologies*, vol. 5, pp. 27–30, 2017.

10. X. Han, Z. Yang, M. Wang, J. Tjong, and M. Zheng, "Clean combustion of *n*-butanol as a next generation biofuel for diesel engines", *Applied Energy*, vol. 198, pp. 347–359, 2017.

11. L. Siwale, L. Kristóf, T. Adam, A. Bereczky, M. Mbarawa, A. Penninger, et al., "Combustion and emission characteristics of *n*-butanol/diesel fuel blend in a turbocharged compression ignition engine", *Fuel*, vol. 107, pp. 409–418, 2013.

12. K. Dutta, A. Daverey, and J.-G. Lin, "Evolution retrospective for alternative fuels: First to fourth generation", *Renewable Energy*, vol. 69, pp. 114–122, 2014.

13. K. Srirangan, L. Akawi, M. Moo-Young, and C. P. Chou, "Towards sustainable production of clean energy carriers from biomass resources", *Applied Energy*, vol. 100, pp. 172–186, 2012.

14. A. Elfasakhany, "Investigations on performance and pollutant emissions of spark-ignition engines fueled with *n*-butanol–, isobutanol–, ethanol–, methanol–, and acetone–gasoline blends: A comparative study", *Renewable and Sustainable Energy Reviews*, vol. 71, pp. 404–413, 2017.

15. S. Kumar, J. H. Cho, J. Park, and I. Moon, "Advances in diesel–alcohol blends and their effects on the performance and emissions of diesel engines", *Renewable and Sustainable Energy Reviews*, vol. 22, pp. 46–72, 2013.

16. Z. Anwar, M. Gulfraz, and M. Irshad, "Agro-industrial lignocellulosic biomass a key to unlock the future bio-energy: A brief review", *Journal of Radiation Research and Applied Sciences*, vol. 7, pp. 163–173, 2014.

17. N. R. Baral, J. Li, and A. K. Jha, "Perspective and prospective of pretreatment of corn straw for butanol production", *Applied Biochemistry and Biotechnology*, vol. 172, pp. 840–853, 2014.

18. D. Van Eylen, F. Van Dongen, M. Kabel, and J. De Bont, "Corn fiber, cobs and stover: Enzyme-aided saccharification and co-fermentation after dilute acid pretreatment", *Bioresource Technology*, vol. 102, pp. 5995–6004, 2011.

19. T. L. Bezerra, and A. J. Ragauskas, "A review of sugarcane bagasse for second-generation bioethanol and biopower production", *Biofuels, Bioproducts and Biorefining*, vol. 10, pp. 634–647, 2016.

20. B. Bharathiraja, J. Jayamuthunagai, T. Sudharsanaa, A. Bharghavi, R. Praveenkumar, M. Chakravarthy, et al., "Biobutanol–An impending biofuel for future: A review on upstream and downstream processing tecniques", *Renewable and Sustainable Energy Reviews*, vol. 68, pp. 788–807, 2017.

21. M. S. Celiktas, C. Kirsch, and I. Smirnova, "Cascade processing of wheat bran through a biorefinery approach", *Energy Conversion and Management*, vol. 84, pp. 633–639, 2014.

22. P. Amnuaycheewa, R. Hengaroonprasan, K. Rattanaporn, S. Kirdponpattara, K. Cheenkachorn, and M. Sriariyanun, "Enhancing enzymatic hydrolysis and biogas production from rice straw by pretreatment with organic acids", *Industrial Crops and Products*, vol. 87, pp. 247–254, 2016.

23. N. K. N. Al-Shorgani, M. S. Kalil, and W. M. W. Yusoff, "Biobutanol production from rice bran and de-oiled rice bran by *Clostridium saccharoperbutylacetonicum* N1-4", *Bioprocess and Biosystems Engineering*, vol. 35, pp. 817–826, 2012.
24. N. Soltani, A. Bahrami, M. Pech-Canul, and L. González, "Review on the physico-chemical treatments of rice husk for production of advanced materials", *Chemical Engineering Journal*, vol. 264, pp. 899–935, 2015.
25. P. M. Martínez, R. Bakker, P. Harmsen, H. Gruppen, and M. Kabel, "Importance of acid or alkali concentration on the removal of xylan and lignin for enzymatic cellulose hydrolysis", *Industrial Crops and Products*, vol. 64, pp. 88–96, 2015.
26. A. L. Umagiliyage, R. Choudhary, Y. Liang, J. Haddock, and D. G. Watson, "Laboratory scale optimization of alkali pretreatment for improving enzymatic hydrolysis of sweet sorghum bagasse", *Industrial Crops and Products*, vol. 74, pp. 977–986, 2015.
27. C. Lu, J. Zhao, S.-T. Yang, and D. Wei, "Fed-batch fermentation for *n*-butanol production from cassava bagasse hydrolysate in a fibrous bed bioreactor with continuous gas stripping", *Bioresource Technology*, vol. 104, pp. 380–387, 2012.
28. K. Ziemiński, and M. Kowalska-Wentel, "Effect of different sugar beet pulp pretreatments on biogas production efficiency", *Applied Biochemistry and Biotechnology*, vol. 181, pp. 1211–1227, 2017.
29. S. Thakur, B. Shrivastava, S. Ingale, R. C. Kuhad, and A. Gupte, "Degradation and selective ligninolysis of wheat straw and banana stem for an efficient bioethanol production using fungal and chemical pretreatment", *3 Biotech*, vol. 3, pp. 365–372, 2013.
30. S. V. Avula, S. Reddy, and L. V. Reddy, "The feasibility of mango (*Mangifera indica* L.) peel as an alternative substrate for butanol production", *BioResources*, vol. 10, pp. 4453–4459, 2015.
31. R. Sánchez Orozco, P. Balderas Hernández, N. Flores Ramírez, G. Roa Morales, J. Saucedo Luna, and A. J. Castro Montoya, "Gamma irradiation induced degradation of orange peels", *Energies*, vol. 5, pp. 3051–3063, 2012.
32. M. A. Khedkar, P. R. Nimbalkar, S. G. Gaikwad, P. V. Chavan, and S. B. Bankar, "Sustainable biobutanol production from pineapple waste by using *Clostridium acetobutylicum* B 527: Drying kinetics study", *Bioresource Technology*, vol. 225, pp. 359–366, 2017.
33. Y. Zheng, C. Lee, C. Yu, Y.-S. Cheng, C. W. Simmons, R. Zhang, et al., "Ensilage and bioconversion of grape pomace into fuel ethanol", *Journal of Agricultural and Food Chemistry*, vol. 60, pp. 11128–11134, 2012.
34. S. Pathania, N. Sharma, and S. Handa, "Utilization of horticultural waste (Apple Pomace) for multiple carbohydrase production from Rhizopus delemar F 2 under solid state fermentation", *Journal of Genetic Engineering and Biotechnology*, vol. 16, pp. 181–189, 2017.
35. H. Aditiya, W. Chong, T. Mahlia, A. Sebayang, M. Berawi, and H. Nur, "Second generation bioethanol potential from selected Malaysia's biodiversity biomasses: A review", *Waste Management*, vol. 47, pp. 46–61, 2016.
36. S. Ang, E. Shaza, Y. Adibah, A. Suraini, and M. Madihah, "Production of cellulases and xylanase by Aspergillus fumigatus SK1 using untreated oil palm trunk through solid state fermentation", *Process Biochemistry*, vol. 48, pp. 1293–1302, 2013.
37. H. Abdul Khalil, M. Siti Alwani, R. Ridzuan, H. Kamarudin, and A. Khairul, "Chemical composition, morphological characteristics, and cell wall structure of Malaysian oil palm fibers", *Polymer-Plastics Technology and Engineering*, vol. 47, pp. 273–280, 2008.
38. S. Saka, M. Munusamy, M. Shibata, Y. Tono, and H. Miyafuji, "Chemical constituents of the different anatomical parts of the oil palm (*Elaeis guineensis*) for their sustainable utilization", In: *Seminar Proceedings – Natural Resources & Energy Environment JSPS-VCC Program on Environmental Science, Engineering and Ethics (Group IX)*, 24–25 November 2008, Kyoto, Japan, 2008.

39. K. Wong, N. A. Rahman, S. Aziz, V. Sabaratnam, and M. A. Hassan, "Enzymatic hydrolysis of palm oil mill effluent solid using mixed cellulases from locally isolated fungi", *Research Journal of Microbiology*, vol. 3, pp. 474–481, 2008.

40. M. A. Khedkar, P. R. Nimbalkar, P. V. Chavan, Y. J. Chendake, and S. B. Bankar, "Cauliflower waste utilization for sustainable biobutanol production: Revelation of drying kinetics and bioprocess development", *Bioprocess and Biosystems Engineering*, vol. 40, pp. 1493–1506, 2017.

41. P. R. Nimbalkar, M. A. Khedkar, P. V. Chavan, and S. B. Bankar, "Biobutanol production using pea pod waste as substrate: Impact of drying on saccharification and fermentation", *Renewable Energy*, vol. 117, pp. 520–529, 2018.

42. V. Wolf, J. Dehoust, and M. Banse, "World markets for cereal crops", In: Kaltschmitt M., Neuling U. (eds) *Biokerosene*. Springer, pp. 123–145, 2018.

43. S. Kim, and B. E. Dale, "Global potential bioethanol production from wasted crops and crop residues", *Biomass and Bioenergy*, vol. 26, pp. 361–375, 2004.

44. J.-J. Dong, J.-C. Ding, Y. Zhang, L. Ma, G.-C. Xu, R.-Z. Han, et al., "Simultaneous saccharification and fermentation of dilute alkaline-pretreated corn stover for enhanced butanol production by *Clostridium saccharobutylicum* DSM 13864", *FEMS Microbiology Letters*, vol. 363, 2016.

45. N. Qureshi, V. Singh, S. Liu, T. C. Ezeji, B. Saha, and M. Cotta, "Process integration for simultaneous saccharification, fermentation, and recovery (SSFR): Production of butanol from corn stover using *Clostridium beijerinckii* P260", *Bioresource Technology*, vol. 154, pp. 222–228, 2014.

46. D. Cai, P. Li, Z. Luo, P. Qin, C. Chen, Y. Wang, et al., "Effect of dilute alkaline pretreatment on the conversion of different parts of corn stalk to fermentable sugars and its application in acetone–butanol–ethanol fermentation", *Bioresource Technology*, vol. 211, pp. 117–124, 2016.

47. T. Guo, A.-Y. He, T.-F. Du, D.-W. Zhu, D.-F. Liang, M. Jiang, et al., "Butanol production from hemicellulosic hydrolysate of corn fiber by a *Clostridium beijerinckii* mutant with high inhibitor-tolerance", *Bioresource Technology*, vol. 135, pp. 379–385, 2013.

48. R. Sindhu, E. Gnansounou, P. Binod, and A. Pandey, "Bioconversion of sugarcane crop residue for value added products–An overview", *Renewable Energy*, vol. 98, pp. 203–215, 2016.

49. H. Li, L. Xiong, X. Chen, C. Wang, G. Qi, C. Huang, et al., "Enhanced enzymatic hydrolysis and acetone-butanol-ethanol fermentation of sugarcane bagasse by combined diluted acid with oxidate ammonolysis pretreatment", *Bioresource Technology*, vol. 228, pp. 257–263, 2017.

50. X. Kong, H. Xu, H. Wu, C. Wang, A. He, J. Ma, et al., "Biobutanol production from sugarcane bagasse hydrolysate generated with the assistance of gamma-valerolactone", *Process Biochemistry*, vol. 51, pp. 1538–1543, 2016.

51. Z.-W. Pang, W. Lu, H. Zhang, Z.-W. Liang, J.-J. Liang, L.-W. Du, et al., "Butanol production employing fed-batch fermentation by *Clostridium acetobutylicum* GX01 using alkali-pretreated sugarcane bagasse hydrolysed by enzymes from *Thermoascus aurantiacus* QS 7-2-4", *Bioresource Technology*, vol. 212, pp. 82–91, 2016.

52. NL Agency, "Rice Straw and Wheat Straw", Netherlands 2013.

53. C. A. Moraes, I. J. Fernandes, D. Calheiro, A. G. Kieling, F. A. Brehm, M. R. Rigon, et al., "Review of the rice production cycle: By-products and the main applications focusing on rice husk combustion and ash recycling", *Waste Management & Research*, vol. 32, pp. 1034–1048, 2014.

54. C. Bellido, M. L. Pinto, M. Coca, G. González-Benito, and M. T. García-Cubero, "Acetone–butanol–ethanol (ABE) production by *Clostridium beijerinckii* from wheat straw hydrolysates: Efficient use of penta and hexa carbohydrates", *Bioresource Technology*, vol. 167, pp. 198–205, 2014.

55. Z. Liu, Y. Ying, F. Li, C. Ma, and P. Xu, "Butanol production by *Clostridium beijerinckii* ATCC 55025 from wheat bran", *Journal of Industrial Microbiology & Biotechnology*, vol. 37, pp. 495–501, 2010.

56. A. Ranjan, and V. S. Moholkar, "Comparative study of various pretreatment techniques for rice straw saccharification for the production of alcoholic biofuels", *Fuel*, vol. 112, pp. 567–571, 2013.

57. N. K. N. Al-Shorgani, A. I. Al-Tabib, and M. S. Kalil, "Production of butanol from acetyl chloride-treated deoiled rice bran by *Clostridium acetobutylicum* YM1", *BioResources*, vol. 12, pp. 8505–8518, 2017.

58. A. Ranjan, "Butanol production from rice straw: Process development and optimization", Doctor of Philosophy, Center of Energy, Indian Institute of Technology Guwahati, India, 2012.

59. M. Yang, J. Zhang, S. Kuittinen, J. Vepsäläinen, P. Soininen, M. Keinänen, et al., "Enhanced sugar production from pretreated barley straw by additive xylanase and surfactants in enzymatic hydrolysis for acetone–butanol–ethanol fermentation", *Bioresource Technology*, vol. 189, pp. 131–137, 2015.

60. C. Ratnavathi, S. K. Chakravarthy, V. Komala, U. Chavan, and J. Patil, "Sweet sorghum as feedstock for biofuel production: A review", *Sugar Tech*, vol. 13, pp. 399–407, 2011.

61. D. Cai, T. Zhang, J. Zheng, Z. Chang, Z. Wang, P.-Y. Qin, et al., "Biobutanol from sweet sorghum bagasse hydrolysate by a hybrid pervaporation process", *Bioresource Technology*, vol. 145, pp. 97–102, 2013.

62. A. Kujawska, J. Kujawski, M. Bryjak, and W. Kujawski, "ABE fermentation products recovery methods—A review", *Renewable and Sustainable Energy Reviews*, vol. 48, pp. 648–661, 2015.

63. D. Cai, H. Chen, C. Chen, S. Hu, Y. Wang, Z. Chang, et al., "Gas stripping–pervaporation hybrid process for energy-saving product recovery from acetone–butanol–ethanol (ABE) fermentation broth", *Chemical Engineering Journal*, vol. 287, pp. 1–10, 2016.

64. M. Zhang, L. Xie, Z. Yin, S. K. Khanal, and Q. Zhou, "Biorefinery approach for cassava-based industrial wastes: Current status and opportunities", *Bioresource Technology*, vol. 215, pp. 50–62, 2016.

65. C. Xue, J. Zhao, C. Lu, S. T. Yang, F. Bai, and I. Tang, "High-titer *n*-butanol production by *Clostridium acetobutylicum* JB200 in fed-batch fermentation with intermittent gas stripping", *Biotechnology and Bioengineering*, vol. 109, pp. 2746–2756, 2012.

66. Eurostat, "Agricultural Productions – Crops", 2017.

67. L. Panella, "Sugar beet as an energy crop", *Sugar Tech*, vol. 12, pp. 288–293, 2010.

68. C. Bellido, C. Infante, M. Coca, G. González-Benito, S. Lucas, and M. T. García-Cubero, "Efficient acetone–butanol–ethanol production by *Clostridium beijerinckii* from sugar beet pulp", *Bioresource Technology*, vol. 190, pp. 332–338, 2015.

69. J. Fan, W. Feng, H. Yan, J. Jia, and Z. Wang, "Production of butanol from sugar beet molasses by fed-batch fermentation", *Chinese Journal of Bioprocess Engineering*, vol. 8, pp. 6–9, 2010.

70. N. Qureshi, T. C. Ezeji, J. Ebener, B. S. Dien, M. A. Cotta, and H. P. Blaschek, "Butanol production by *Clostridium beijerinckii*. Part I: Use of acid and enzyme hydrolyzed corn fiber", *Bioresource Technology*, vol. 99, pp. 5915–5922, 2008.

71. D. Mohapatra, S. Mishra, and N. Sutar, "Banana and its by-product utilisation: An overview", *Journal of Scientific & Industrial Research*, vol. 69, pp. 323–329, 2010.

72. J. Y. Tock, C. L. Lai, K. T. Lee, K. T. Tan, and S. Bhatia, "Banana biomass as potential renewable energy resource: A Malaysian case study", *Renewable and Sustainable Energy Reviews*, vol. 14, pp. 798–805, 2010.

73. T. Ahmad, and M. Danish, "Prospects of banana waste utilization in wastewater treatment: A review", *Journal of Environmental Management*, vol. 206, pp. 330–348, 2018.

74. M. Wadhwa, M. P. Bakshi, and H. P. Makkar, "Waste to worth: Fruit wastes and by-products as animal feed", *CAB Reviews*, vol. 10, pp. 1–26, 2015.
75. M. del Carmen Rivera-Cruz, A. T. Narcía, G. C. Ballona, J. Kohler, F. Caravaca, and A. Roldan, "Poultry manure and banana waste are effective biofertilizer carriers for promoting plant growth and soil sustainability in banana crops", *Soil Biology and Biochemistry*, vol. 40, pp. 3092–3095, 2008.
76. A. B. Guerrero, I. Ballesteros, and M. Ballesteros, "The potential of agricultural banana waste for bioethanol production", *Fuel*, vol. 213, pp. 176–185, 2018.
77. R. Vidhya, I. Jasmine, and V. Pradeep, "Process for production and quantitation of high yield of biobutanol", Google Patents, 2015.
78. V. Rangaswamy, J. Isar, and H. Joshi, "Butanol fermentation using acid pretreated bio-mass", Google Patents, 2016.
79. N. Boyce, "Bioethanol production from mango waste (*Mangifera indica* L. cv cho-kanan): Biomass as renewable energy", *Australian Journal of Basic and Applied Sciences*, vol. 8, pp. 229–237, 2014.
80. D. Carrillo-Nieves, H. A. Ruiz, C. N. Aguilar, A. Ilyina, R. Parra-Saldivar, J. A. Torres, et al., "Process alternatives for bioethanol production from mango stem bark residues", *Bioresource Technology*, vol. 239, pp. 430–436, 2017.
81. A. Ranjan, R. Mayank, and V. S. Moholkar, "Process optimization for butanol production from developed rice straw hydrolysate using *Clostridium acetobutylicum* MTCC 481 strain", *Biomass Conversion and Biorefinery*, vol. 3, pp. 143–155, 2013.
82. B. Satari, and K. Karimi, "Citrus processing wastes: Environmental impacts, recent advances, and future perspectives in total valorization", *Resources, Conservation and Recycling*, vol. 129, pp. 153–167, 2018.
83. S. Joshi, J. Waghmare, K. Sonawane, and S. Waghmare, "Bio-ethanol and bio-butanol production from orange peel waste", *Biofuels*, vol. 6, pp. 55–61, 2015.
84. F. Heidari, M. A. Asadollahi, A. Jeihanipour, M. Kheyrandish, H. Rismani-Yazdi, and K. Karimi, "Biobutanol production using unhydrolyzed waste acorn as a novel sub-strate", *RSC Advances*, vol. 6, pp. 9254–9260, 2016.
85. M. Kosseva, V. Joshi, and P. Panesar, *Science and Technology of Fruit Wine Production*. Academic Press, 2016.
86. F. Gassara, S. Brar, F. Pelletier, M. Verma, S. Godbout, and R. Tyagi, "Pomace waste management scenarios in Québec—Impact on greenhouse gas emissions", *Journal of Hazardous Materials*, vol. 192, pp. 1178–1185, 2011.
87. L. Law, and N. Gutierrez, "Butanol production by submerged fermentation of white grape pomace", *Current Biotechnology*, vol. 2, pp. 114–116, 2013.
88. M. Hijosa-Valsero, A. I. Paniagua-García, and R. Díez-Antolínez, "Biobutanol production from apple pomace: The importance of pretreatment methods on the fermentability of lignocellulosic agro-food wastes", *Applied Microbiology and Biotechnology*, vol. 101, pp. 8041–8052, 2017.
89. M. A. Sukiran, F. Abnisa, W. M. A. W. Daud, N. A. Bakar, and S. K. Loh, "A review of torrefaction of oil palm solid wastes for biofuel production", *Energy Conversion and Management*, vol. 149, pp. 101–120, 2017.
90. H. Yamada, R. Tanaka, O. Sulaiman, R. Hashim, Z. Hamid, M. Yahya, et al., "Old oil palm trunk: A promising source of sugars for bioethanol production", *Biomass and Bioenergy*, vol. 34, pp. 1608–1613, 2010.
91. N. S. M. Nasrah, M. A. K. M. Zahari, and N. Masngut, "Biobutanol production by *Clostridium acetobutylicum* ATCC 824 using Oil Palm Frond (OPF) juice", In: *The National Conference for Postgraduate Research* 2016, Universiti Malaysia Pahang, Pahang, Malaysia.

92. I. Komonkiat, and B. Cheirsilp, "Felled oil palm trunk as a renewable source for biobutanol production by *Clostridium* spp.", *Bioresource Technology*, vol. 146, pp. 200–207, 2013.

93. M. F. Ibrahim, S. Abd-Aziz, M. E. M. Yusoff, L. Y. Phang, and M. A. Hassan, "Simultaneous enzymatic saccharification and ABE fermentation using pretreated oil palm empty fruit bunch as substrate to produce butanol and hydrogen as biofuel", *Renewable Energy*, vol. 77, pp. 447–455, 2015.

94. A. I. Al-Tabib, N. K. N. Al-Shorgani, H. A. Hasan, A. A. Hamid, and M. S. Kalil, "Assessment of the detoxification of palm kernel cake hydrolysate for butanol production by *Clostridium acetobutylicum* YM1", *Biocatalysis and Agricultural Biotechnology*, vol. 13, pp. 105–109, 2018.

95. N. K. N. Al-Shorgani, H. Shukor, P. Abdeshahian, M. Y. M. Nazir, M. S. Kalil, A. A. Hamid, et al., "Process optimization of butanol production by *Clostridium saccharoperbutylacetonicum* N1-4 (ATCC 13564) using palm oil mill effluent in acetone–butanol–ethanol fermentation", *Biocatalysis and Agricultural Biotechnology*, vol. 4, pp. 244–249, 2015.

96. S. Plazzotta, L. Manzocco, and M. C. Nicoli, "Fruit and vegetable waste management and the challenge of fresh-cut salad", *Trends in Food Science & Technology*, vol. 63, pp. 51–59, 2017.

97. C. Gowe, "Review on potential use of fruit and vegetables by-products as a valuable source of natural food additives", *Food Science and Quality Management*, vol. 45, 2015.

98. A. Singh, A. Kuila, S. Adak, M. Bishai, and R. Banerjee, "Utilization of vegetable wastes for bioenergy generation", *Agricultural Research*, vol. 1, pp. 213–222, 2012.

99. A. Procentese, F. Raganati, G. Olivieri, M. E. Russo, and A. Marzocchella, "Pretreatment and enzymatic hydrolysis of lettuce residues as feedstock for bio-butanol production", *Biomass and Bioenergy*, vol. 96, pp. 172–179, 2017.

2 Carbon Neutral Strategies to Reduce CO_2 Emissions and Mitigate Climate Change

Farahiyah Abdul Rahman and
Md. Maniruzzaman A. Aziz

CONTENTS

2.0   CLIMATE CHANGE MITIGATION

The Intergovernmental Panel on Climate Change (IPCC) estimates that global greenhouse gas reductions of 50–85% will be needed by 2050 to avoid dangerous climate change, representing a radical shift away from today's fossil fuel-derived economy [1]. Climate change mitigation consists of actions to limit the magnitude or rate of long-term climate change [2]. Climate change mitigation generally involves reductions in human activities that create emissions of greenhouse gases (GHGs). Some examples of these human activities are fuel usage, building operations, construction activity, and goods production from industry. These are among the major contributors to the emission of GHGs.

Examples of mitigation include the phasing out of fossil fuels by switching to low-carbon energy sources such as renewable and nuclear energy, and expanding forests and other "sinks" to remove greater amounts of carbon dioxide (CO_2) from the atmosphere [3]. Energy efficiency may also play a role; for example, through improving the insulation of buildings [4]. Mitigation may also be achieved by increasing the capacity of carbon sinks, e.g., through reforestation. Mitigation policies can substantially reduce the risks associated with human-induced global warming [5]. Another approach to climate change mitigation is climate engineering [6].

The methods of mitigating CO_2 emissions can be categorized into carbon source-based, carbon emission minimization-based, and carbon sink-based methods [7]. The carbon neutralization in this study is a compilation of all these methods; however, it will only discuss fuel usage emissions, building emissions (which include the construction phase) and industrial production emissions. Whether or not the mitigation is a success will depend on how quickly and profoundly we can change the way we live our lives, both collectively and individually. There is an urgent need to change from the "business as usual" scenario of a fossil fuels-based, car-centered, throw-away economy to a scenario that practically reduces our emissions levels [8].

In consequence, technology plays a crucial role in the necessary transition to low-carbon development, in terms of making our energy use more efficient, and in developing greater infrastructure for small-scale renewable energy. However, even a massive deployment of all of the available renewable energy technologies could still only generate a fraction of our current energy demand [8]. To help bridge this gap, technological advances in climate change, with the implementation of eco-friendly technologies that take place alongside dramatic cuts in energy consumption levels, need to be implemented. This implies a wider cultural transformation so that society could react with high esteem to energy and climate-conscious behavior, discouraging waste and extravagance.

Hence, a necessary starting point for bringing about this change in society is to acknowledge that the "business as usual" scenario cannot continue in order to make a change. Therefore, the culture of offsetting some types of emission is corrosive to the climate change debate. It presents itself as a way for people to effectively deal with climate change while largely maintaining their levels of energy consumption [8].

2.1 CARBON-NEUTRAL MATERIAL (CNM)

"Carbon neutrality" has at least four terms to describe the concept, including "carbon neutral," "climate neutral," "carbon-free," and "carbon clean" [9]. Carbon neutrality, or having a net zero carbon footprint, refers to achieving net zero carbon emissions by balancing a measured amount of carbon released with an equivalent amount sequestered or offset, or buying enough carbon credits to make up the difference. It is used in the context of carbon dioxide releasing processes associated with transportation, energy production, and industrial processes such as the production of carbon-neutral fuel [10]. In other words, this means taking action to reduce their greenhouse gas emissions to zero, and then "offsetting" an equivalent amount of any remaining emissions [11].

The United Kingdom's Department of Energy and Climate Change (DECC) has developed a series of guidelines surrounding the use of the term carbon neutrality, to discourage "greenwashing" and provide greater clarity around the concept.

According to their definition, "Carbon neutral means that, through a transparent process of calculating emissions, reducing those emissions and offsetting residual emissions, so that the net carbon emissions equal zero" [12]. To achieve carbon neutrality, energy consumption must first be reduced as much as possible [13].

A net zero carbon footprint is closely related to carbon neutrality. In most cases, the total carbon footprint cannot be exactly calculated because of inadequate knowledge of and data about the complex interactions between contributing processes, especially those which influence natural processes storing or releasing carbon dioxide. However, research has been done that defines the carbon footprint as "a measure of the total amount of carbon dioxide (CO$_2$) and methane (CH$_4$) emissions of a defined population, system, or activity, considering all relevant sources, sinks, and storage within the spatial and temporal boundary of the population, system, or activity of interest. Calculated as CO$_2$ equivalent using the relevant 100-year global warming potential (GWP100)." Formerly, the definition is more concise, that the carbon footprint is defined as the amount of greenhouse gas emission set caused by individuals, events, organizations, or products, expressed as carbon dioxide equivalents [14, 15]. The best practice for organizations and individuals seeking carbon neutral status requires reducing and/or avoiding carbon emissions first so that only unavoidable emissions are offset. Carbon neutrality can be achieved in two ways:

i. Balancing carbon dioxide released into the atmosphere from burning fossil fuels. One of the methods is to use renewable energy as an energy source so it can create the same amount of energy with less or no emissions [16].

ii. Carbon offsetting by paying others to remove or sequester 100% of the carbon dioxide emitted from the atmosphere [17]. The example of this method is either planting trees, or "carbon project" funding, which can prevent future GHGs emission or buying carbon credit to remove them. Carbon offsetting is often used alongside energy conservation measures to minimize energy use [8].

Based on Figure 2.1, 64% of global CO$_2$ emissions are energy- or process-related, while 36% are due to deforestation, agriculture, or decay. Among these energy- or process-related emissions, 35% are from industry, 31% from building and 27% from transport [1]. In terms of building and transportation, there are perhaps 75% of design and technology improvements that can be made to help reduce emissions.

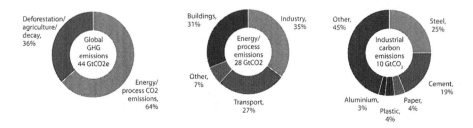

FIGURE 2.1   Sources of global CO$_2$ emission [1].

However, technical solutions for energy efficiency are known, but their implementation depends on political will and public motivation, so it is important to raise public awareness on this matter. However, for industrial emissions, many systems are already highly optimized, and the demand for materials is anticipated to double in the next 40 years [1]. Hence the innovations that relate to energy efficiency and emission mitigation are very important for the time being.

For most of the materials used to provide buildings, infrastructure, equipment, and products, global stocks are still sufficient to meet anticipated demand, but the environmental impacts of materials production and processing, particularly those related to energy, are rapidly becoming critical. In this case, it is not energy efficiency, but rather material efficiency that represents the biggest opportunity. The term "material efficiency" means delivering the same required services with less primary production [1]. It includes the implementation of technologies for material optimization and/or recycling the material to the maximum level.

2.2   CARBON NEUTRAL IN FUEL

Hydrocarbon fuels provide the majority of all transportation energy, and petroleum is the dominant feedstock from which transportation fuels are produced. Hydrocarbons can also be produced from other feedstocks (fossil and biomass), as well as carbon-free energy carriers (such as hydrogen, batteries, and ultracapacitors), which are potentially more sustainable alternatives [18]. The benefits of hydrocarbons over carbon-free energy carriers include higher energy density and the use of existing infrastructure (fuel distribution and vehicles). Despite the increased use of electric vehicles likely to reduce the demand for liquid fuels, hydrocarbons will continue to be needed primarily as fuel in aircraft, ships, and transport vehicles. For most chemical industries, they also provide chemical building blocks, so their widespread use requires a way to produce them sustainably [19].

To the extent that carbon-neutral fuels displace fossil fuels, or if they are produced from waste carbon or seawater carbonic acid, and their combustion is subject to carbon capture at the flue or exhaust pipe, they result in negative carbon dioxide emission and net carbon dioxide removal from the atmosphere, and thus constitute a form of greenhouse gas remediation [20].

Such power to gas carbon-neutral and carbon-negative fuels can be produced by the electrolysis of water to make hydrogen used in the Sabatier reaction to produce methane. This may then be stored to be burned later in power plants as synthetic natural gas, transported by pipeline, truck, or tanker ship, or be used in gas to liquid processes such as the Fischer–Tropsch process to make traditional fuels for transportation or heating [21, 22].

2.2.1   RENEWABLE ENERGY (RE)

Historically, economic development has been strongly correlated with increasing energy use and growth of GHG emissions, and RE can help decouple that correlation, contributing to sustainable development. Though the exact contribution of RE to sustainable development has to be evaluated in a country-specific context, RE offers the opportunity to contribute to social and economic development, energy access, secure energy supply, climate change mitigation, and the reduction of negative environmental and health impacts [23]. However, in a major context, RE technologies can provide important environmental benefits. Maximizing these benefits depends on the specific technology, management, and site characteristics associated with each RE project.

In 2017, the Environment Protection Agency (IEA) reported that the majority of power plant emissions in the United States comes from fossil fuels like coal and natural gas [24]. In contrast, most renewable energy sources produce little to no global warming emissions, and replacing energy with RE will reduce the usage of fossil fuels. Reduced fossil energy consumption leads to reduced direct fossil fuel-based carbon emissions in the electricity sector.

Based on Figure 2.2, on a cumulative basis (2011–2050), combustion-only electric sector fossil fuel-based CO$_2$ emissions under the 80% renewable electricity scenario with incremental technology improvement (RE-ITI) scenario were reduced by approximately 80% on both a direct combustion basis and on a full life cycle basis [25]. Other studies also state that the reduction of CO$_2$ emissions range from 469 to 1,001 g CO$_2$ eq/kWh (fossil fuels) to 4 to 46 g CO$_2$ eq/kWh (all using RE), excluding land use change emissions [26]. To the extent other sectors such as transportation and building could be increasingly electrified, further GHG reductions could be achieved [25].

On the other hand, there are a lot of benefits of RE other than emission reduction which make it favorable to be implemented. These are:

i. RE can contribute to social and economic development. Under favorable conditions, cost savings against the non-use of REs are available, especially in remote areas and poorer rural areas with less centralized access to energy. Costs related to external energy imports can be mitigated through the use of competitive domestic RE technology. In addition, RE can create job opportunities even though the existing studies differ with respect to the magnitude of net employment. Specifically, small-scale renewable energy

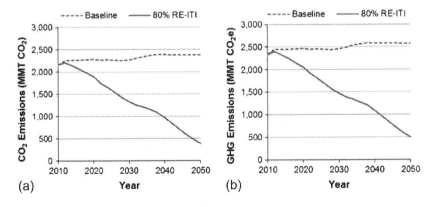

FIGURE 2.2 Greenhouse gas emissions in low-demand baseline and 80% RE-ITI scenarios [27].

projects can help local communities gain energy services and improve their living standards, thus collectively contributing to poverty reduction and increasing their resilience to climate change. For host countries, such initiatives will also contribute to meeting the country's development objectives and strengthen the country's developing capacity for the long term to limit CO_2 emissions into the air [23, 28].

ii. RE can help accelerate access to energy. About 1.4 billion people live without access to electricity and there are an additional 1.3 billion using traditional biomass. However, the basic level of access to modern energy services can provide significant benefits to the community or household. In many developing countries, the decentralized grid based on RE and the inclusion of RE in the central power grid have grown and increased power supply. In addition, non-electric RE technology also offers the opportunity for the modernization of energy services; for example, utilizing solar energy for water heating, biofuel for transport, biogas and modern biomass for heating, cooling, cooking, and lighting, and wind for water pumping. The number of people who have no access to modern energy services is expected to remain unchanged unless there is a relevant domestic policy implemented, which may be supported or complemented by appropriate international assistance. Therefore, the significance of this strategy is to demonstrate that there is a viable energy access solution for the poor in developing countries and to encourage this solution to be replicated and enhanced in various fields [29].

iii. RE options can contribute to a more secure energy supply, although specific challenges for integration must be considered. RE deployment might reduce vulnerability to supply disruption and market volatility if competition is increased and energy sources are diversified. Scenario studies indicate that concerns regarding secure energy supply could continue in the future without technological improvements within the transport sector [30].

iv. The sustainability of bioenergy, particularly in terms of life cycle GHG emissions, is influenced by land and biomass resource management practices. Changes in land and forest use or management that, according to a considerable number of studies, could be brought about directly or indirectly by biomass production for use as fuels, power, or heat, can decrease or increase terrestrial carbon stocks. Moreover, biomass energy also provides an irreversible reduction effect by reducing carbon dioxide at the source, but it can produce more carbon per unit of energy than fossil fuels unless biomass fuel is generated sustainably [31].

2.2.2   CARBON CAPTURE AND SEQUESTRATION (CCS)

Carbon capture and sequestration (CCS) is defined as the removal of CO_2 directly from industrial or utility plants before storing it in a secure stockpiling medium, and it is amongst the most imperative techniques that can be employed to decrease CO_2 emissions [32]. The reason for CCS technology is to empower the generation of biofuels while reducing the emission of CO_2 into the environment and consequently mitigate global warming [7, 33, 34]. Because of the world's substantial supply of fossil fuels, there is a strong necessity to investigate the best capture technique of CO$_2$ from fossil fuel burning as a procedure to keep it out of the environment. Latest estimates suggest that widespread deployment of carbon capture and storage (CCS) could account for up to one-fifth of the needed global reduction in CO$_2$ emissions by 2050 [35]. This will enhance the removal of excess CO$_2$ from the air and, at the same time, convey a low pollution energy source to society as we are now concentrating on the development of energy systems that are efficient, clean, and conservative. However, CCS technology faces numerous issues and challenges before it can be successfully deployed and become a reality. These include the capture technology, storage method, investment challenge for operational cost, and also the public awareness of carbon capture and storage.

Such challenges include how to utilize the concentrated CO$_2$ stream produced from the capture technologies, reducing the effects of impurities on the operating system, scaling technologies up to the level of 500 MW fossil fuel–burning power plants, and retrofitting technologies to existing facilities. However, the innovative design of equipment could solve many of the challenges associated with scaling up and retrofitting technologies [36]. In almost all cases, the presence of impurities (mainly NOx and Sox) reduces the CO$_2$ capture efficiency of the system. Pretreatment stages to remove impurities require additional equipment and lead to increased operating costs. Development of a system that efficiently and simultaneously removes NOx, SOx and CO$_2$ would significantly reduce the costs associated the pretreatment stage of CO$_2$ capture systems. Innovative design of absorbents or adsorbents could possibly address this issue [36].

Yet, corporate decision-makers are finding that CCS investments are challenging. Common investment challenges include operational concerns, such as the high cost of capture technologies, technological uncertainties in integrated CCS systems, and underdeveloped regulatory and liability regimes. Understanding CCS as a corporate technology strategy challenge can help us move beyond the usual list of operational barriers to CCS and make public policy recommendations to help overcome them [35].

To explore public awareness of carbon capture and storage (CCS), a study must be conducted. Attitudes toward the use of CCS and the determinants of CCS acceptance were studied in China in July 2009 based on face-to-face interviews with participants across the country. The result showed that the awareness of CCS was low among the surveyed public in China, compared to other clean energy technologies. The regression model revealed that, in addition to CCS knowledge, respondents' understanding of the characteristics of CCS, such as the maturity of the technology, risks, capability of CO$_2$ emission reductions, and CCS policy were the possibility factors in predicting the acceptance of CCS among public. The findings suggest that integrating public education and communication into CCS development policy would be an effective strategy to overcome the barrier of low public acceptance [37].

2.2.3 RECYCLING CO$_2$ INTO SYNTHETIC FUEL

As a direct replacement for petroleum-based hydrocarbons, biofuels, and fossil carbon-derived synthetic fuels are receiving the most attention. Their sustainability depends largely on the source of the feedstock and, in the case of fossil carbon-based fuels, on the availability of carbon capture and storage technologies and sites. Carbon dioxide captured from large industrial sources could be utilized if there are proper carbon neutral technologies and innovation can be implemented in particular industries [3]. Similar hydrocarbons can also be produced without using fossil fuels or biomass. Synthetic fuel can be produced from sustainable or nuclear energy used to hydrogenate waste CO_2 recycled from power plant flue exhaust gas or derived from carbonic acid in seawater. The synthesizing process can be powered from renewable energy sources such as wind turbines, solar panels, and hydroelectric power stations [19, 38]. Such power sources are potentially carbon neutral because they do not result in a net increase in atmospheric greenhouse gases [18].

The possible pathways from the feedstocks to hydrocarbon fuels is presented in Figure 2.3. Except for direct sunlight-driven processes, the energy collection stage can be considered external to the process; other pathways are not tied to a specific energy source but rather to intermediate heat or electricity. The energy can drive the dissociation of either CO_2 or H_2O, or both, resulting in energy-rich gas mixtures that are readily converted to convenient fuels [18]. Based on the figure, the synthetic fuel production process has several stages: (a) collection of energy, (b) collection of the oxides, H_2O and CO_2, (c) dissociation of the oxides, and (d) fuel synthesis from the products of stage (c). Within each stage, there are many technology options.

In the long term, the capture of CO_2 from the atmosphere would enable a closed-loop hydrocarbon fuel cycle [18, 39]. Based on the figure, through the direct reuse of CO_2 from an industrial plant, it can reduce half of CO_2 emissions; however, closed-loop carbon recycling via air capture of CO_2 will result in near zero net emissions [18]. In consequence, recycling CO_2 into a hydrocarbon fuel would open a new

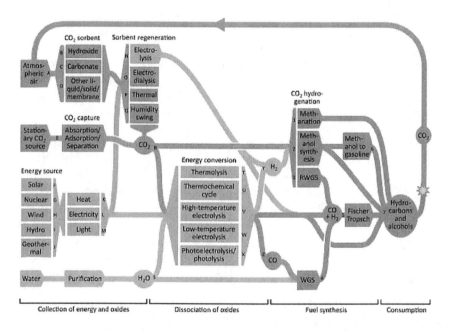

FIGURE 2.3 The possible pathway map of CO_2 and H_2O conversion to hydrocarbon [18].

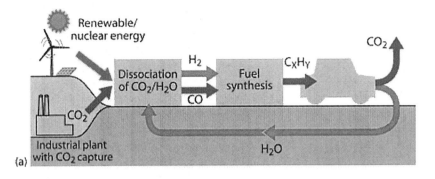

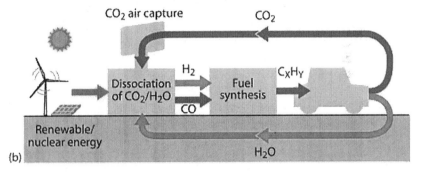

FIGURE 2.4 The synthetic fuel cycle of CO$_2$ and H$_2$O a) through reuse of CO$_2$ and b) continuous close loop carbon recycling via air capture of CO$_2$ [18].

sector, the transportation fuel sector, to renewable energy, which was previously not accessible to renewable energy sources apart from biomass. Synthetic fuels can be made specifically when sufficient power supply is available. This will reduce the demand for energy storage systems designed to receive and return electrical power (for example, hydropower storage, compressed air storage, battery, etc.) [40].

The fuel cycles of carbon-neutral hydrocarbons from various sources, alongside a hydrogen fuel cycle, are illustrated in Figure 2.4. A more thorough analysis of the sustainability of various energy carriers, including CO$_2$-recycled synthetic fuels, is a topic of future work [18].

2.3 CARBON NEUTRAL FOR BUILDING

Energy used to operate buildings is one of the most significant sources of greenhouse gases (GHG) being released into the atmosphere, making it impossible to achieve the necessary reductions in anthropogenic emissions without implementing significant reductions in building sector emissions. Even though this can be done using current technologies, it requires a new set of skills and a transformation in how we think about buildings and how we design and produce them.

Different sources estimate that GHG emissions from the building sector represent ~33% of emissions worldwide, according to the UNEP Sustainable Buildings and

Climate Initiative [13]. However, from another source, greenhouse gas (GHG) emissions from the building sector represent 25% of total emissions and 19% of all global 2010 GHG emissions from electricity and heat production. It has doubled since 1970 to reach 9.18 Gt CO_2eq in 2010, as shown in Figure 2.5 [41].

The IPCC states that measures to reduce GHG emissions from buildings fall into one of three categories: (a) reducing energy consumption and embodied energy in buildings; (b) switching to low-carbon fuels, including a higher share of renewable energy; or (c) controlling the emissions of non-CO_2 GHG. A very large number of technologies that are commercially available and tested in practice can substantially reduce energy use while providing the same services and often substantial co-benefits. This has been reinforced by multiple reports that always state that efficiency is the most cost-effective way to reduce energy consumption and carbon emissions in buildings [42]. Implementing energy efficiency standards in appliances has been among the most effective policies to reduce overall energy consumption [13].

In consequence, for the building-related emissions, it is important to emphasis the development of an envelope that works seamlessly with the heating, ventilation, and air conditioning (HVAC) system, providing carbon-free heating, cooling and daylighting whenever possible. This method is dynamic and can be adapted and further customized by the user, including different strategies as appropriate [13].

Significant reductions in emissions will be possible only if reductions in emissions due to energy consumption in buildings are implemented. However, in addition to energy, buildings also emit GHGs during their construction by the water they use and through the waste they produce. Emissions are produced by the building and its interactions with the environment around it, which can be affected by the building fabric and materials or the building's inputs and outputs (operation, construction, water, waste, and even the site work around the building). These factors can be described by the Equation 2.1 [13]:

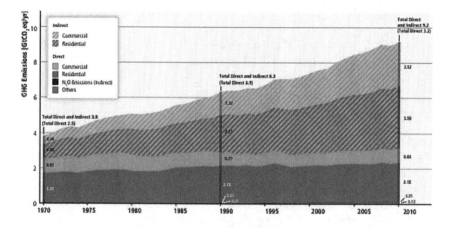

FIGURE 2.5 Direct and indirect emission from building subsector of electricity and heat production [41].

$$T_{be} = O_e + C_e + W_e + W_a \qquad \text{(Equation 2.1)}$$

where:
- T_{be} is the total building emissions
- O_e is the operation emissions (energy)
- C_e is the construction emissions
- W_e is the water emissions
- W_a is the waste emissions

Site work in the landscape can also release carbon sequestered in the ground or plants. Emissions from transportation are also very important and are affected by building location, but they are not produced by the building itself or its inputs and outputs, and so they are not considered in this section [13].

Carbon-neutral design is a movement focusing on creating buildings that reduce or eliminate carbon dioxide emissions throughout a building's life cycle [43]. A typical building generates carbon emissions throughout its life span. Before construction, carbon is emitted when raw materials are processed into building materials and products. During construction, this occurs when building materials are transported to the construction site, when the site is excavated for construction, and when equipment and tools are operated. Once construction is complete, carbon emissions are generated from the building's operation, including mechanical systems, lighting, and appliances. Still more carbon is emitted at the end of the building's life, as the building materials are reused, recycled, or sent to landfill [13].

2.3.1 CARBON NEUTRAL IN THE DESIGN, CONSTRUCTION, AND OPERATION PHASE

Carbon neutrality in the building subsector may be accomplished by implementing innovative sustainable design strategies, generating on-site renewable power, and/or purchasing renewable energy. The first step is to improve efficiency by implementing energy conservation measures; the next step is the implementation of passive heating and cooling strategies; then high-performance HVAC systems; and finally the installation of on-site renewable energy systems. If there are still emissions remaining, they can be offset. This is because renewables are more expensive than energy conservation measures and require additional space that might not be available onsite.

In order of that, some designers use a rule of thumb of an 80% reduction from a reference building, meeting the remaining 20% with renewable energy and/or carbon offsets. If emissions from other sources such as water and waste are to be offset, then the building should become a plus energy building to compensate for the additional emissions. A truly carbon-neutral building must be energy positive; this is because a carbon-neutral building is one in which total emissions are equal to zero. There are differences based on the emissions that are considered, but the definition is the same: emissions produced must be offset by renewable sources. If only emissions from energy are considered, then Equation 2.2 is used. In this case, the emissions from non-renewable energy are offset by the generation of renewable energy (or partially by carbon offsets) [13].

$$0 \geq T_{be} = O_e - R_s \qquad \text{(Equation 2.2)}$$

where:
- T_{be} is the total building emissions
- O_e is the operation emissions (energy)
- R_s is the renewable strategies

In the construction phase, usually water, waste, and construction are also considered in addition to energy. In order of that, the emissions from these must be added to the emissions from energy and Equation 2.3 is used. In this case, the emissions offset from the production of renewable energy must offset all the other emissions, and the building is net positive for energy while still being net zero for carbon.

$$0 \geq T_{be} = O_e + C_e + W_e + W_a - R_s \qquad \text{(Equation 2.3)}$$

where:
- T_{be} is the total building emissions
- O_e is the operation emissions (energy)
- C_e is the construction emissions
- W_e is the water emissions
- W_a is the waste emissions
- R_s is the renewable strategies

The construction industry ranked third in CO_2 production with 131 million metric tons of CO_2 equivalent produced in 2002 in the United States. The same study showed that 76% of the greenhouse gas emissions from the construction industry are produced during the combustion of fossil fuels of stationary and mobile equipment on the construction site [44]. These numbers show that reducing the emissions from construction equipment and vehicles (the largest fuel users) could have an important impact. In 2012, there was a methodology to calculate emissions for each individual construction activity using project data and a schedule from the contractor combined with crew, productivity, and cost data, and the results showed that concrete work was responsible for 45% of the total CO_2 emissions [44]. The results in a different study stated that the majority of all emissions were generated early in the construction process, and that they were primarily due to earthwork and site preparation activities [45].

On top of that, if transportation is also considered for emissions, then it should also be included and Equation 2.4 is used. The building needs to produce even more renewable energy to offset emissions from transportation.

$$0 \geq T_{be} = O_e + C_e + W_e + W_a + T_e - R_s \qquad \text{(Equation 2.4)}$$

where:

T$_{be}$ is the total building emissions
O$_e$ is the operation emissions (energy)
C$_e$ is the construction emissions
W$_e$ is the water emissions
W$_a$ is the waste emissions
T$_e$ is the transportation emissions
R$_s$ is the renewable strategies

For this mitigation measure, renewable strategies (R$_s$) are very important as they will offset the emissions generated by the building. If total building emissions, including renewables, are lower than or equal to zero ($0 \geq T_{be}$), then the building is considered carbon neutral [13].

2.4   CARBON NEUTRAL IN THE PRODUCTION INDUSTRY

The development of sustainable materials is extremely important in order to address and meet the challenges presented by the global energy demand of an increasing population associated with environmental degradation. In consequence, the new production materials should also ideally be low cost, scalable, industrially and economically attractive, and based on renewable and highly abundant resources. Therefore, it is important to develop novel, high performance and sustainable materials to solve the challenge of providing renewable energy for all whilst limiting environmental impacts [46].

2.4.1   STRATEGIES TO IMPROVE MATERIAL EFFICIENCY IN PRODUCTION INDUSTRY

Material efficiency is defined as "providing material services with less material production and processing" [47]. Based on an analysis of strategies to improve material efficiency in these five key materials, Allwood's LCMP Group have created a "Material Manifesto" [1], which includes the following six actions to make the future of material use more sustainable. These are:

   i. Use less metal by design optimization: We could make big savings by optimizing the design of metal components. It is essential that energy considerations are incorporated within the design phase of a product since the majority of a product's environmental impact is determined during this phase [48, 49]. The materials used by industry are often designed in a regular shape to make production easier and more efficient. But this means that they often use more material than they must. The optimization of beam designs, for example, can save up to 30% with a similar reduction in the emissions caused by production. Similar techniques could be applied to the production of components for cars; the "rebar" used to reinforce concrete; steel cans for food storage, and so on.
   ii. Reduce yield losses: At least 25% of liquid steel and 40% of liquid aluminum never makes it into products. Instead, it is cut off as scrap in manufacturing.

One extreme example is the aluminum wing skin used for airplanes: 90% of the metal produced in this process ends up as "swarf," or aluminum scrap. The researchers found that this is often the result of habit rather than necessity. Clothing manufacturers have, for example, actually derived the algorithms needed to make sure that rolls of fabric are used to maximum effect. Manufacturers could do the same thing with the metal they receive. The team calculated that reducing yield losses through this and other techniques would cut CO_2 emissions by about 16% in the steel industry and 7% in the aluminum industry.

iii. Divert manufacturing scrap: Metal scraps are usually sent for recycling, which means melting it (energy-intensive process). But, in fact, it can be reused to produce other items such as iron scraps coming from blanking skeletons. Approximately 60 megatons of steel is scrapped on this basis every year. So the scrap steel can be cut in half if these skeletons go to smaller component manufacturers who can use what's left over.

iv. Reuse old components before recycling at all: Old components are often recycled although they can be reused directly. Car dismantlers are an example of good practice by breaking the damaged or old vehicles and using the components again. However, steel in construction remains the largest potential asset for reuse. Although the beams from dismantled buildings are usually recycled they can be used again; for example, when destroying the builds the steel girder can be reused completely. It just needs to be unbolted and cleaned before reuse. This is because the steel does not degrade with use. Reusing means we can avoid all the melting, casting, and re-rolling of old steel.

v. Extend the lives of products: Consumers can extend the lifetimes of their products by using them for their full useful lives by repairing, maintaining, upgrading, or exchanging defective parts of goods [50, 51]. Most demand for products in developed economies isn't to expand the overall stock, but to replace existing items. The researchers advocate modifying products rather than replacing them wholesale and urging manufacturers to develop adaptable designs that would help this process. This requires a change in thinking and an end to planned obsolescence.

vi. Reduce final demand: The option of a fall that will not be seen by policy makers, except during wartime, is to reduce the final demand. People may not want to make these changes to their simple lifestyles, but that's not to say they could not do it if they had to.

2.4.2 Carbon-Based Products to Enhance Carbon Neutrality in Material Manufacturing

Carbon-based structures are the most versatile materials used in the modern field of renewable energy (in both generation and storage) and environmental science (for purification/remediation). In the context of sustainable materials development, after oxygen, carbon is the most abundant element in the biosphere. Nature utilizes these elements coupled with hydrogen to provide the basis for renewable energy storage (as carbohydrates, for example). However, there is a need to develop increasingly more sustainable variants of classical carbon materials (e.g., activated carbons, carbon nanotubes, carbon aerogels, etc.) [46].

In this regard, perhaps the simulation of some respects of the natural carbon cycles/production, utilization of natural, coupled with simpler, lower-energy synthetic processes can contribute in part to the reduction of greenhouse gas emissions. Moreover, it also can be considered as crucial parameters in the development of sustainable materials manufacturing. Carbon-based systems are increasingly performing a major role in emerging renewable energy conversion technologies; for example, electrodes in energy storage devices, electro- catalysis, photocatalysis, heterogeneous catalysis, biofuels, etc. [52] Carbon materials are also extensively used in water purification, gas separation (e.g., CO$_2$ capture)/storage, and as a soil additive [53]. Furthermore, the importance and potential of carbon-based materials have been recognized such as carbon nanotubes, which can be made from CO$_2$ by molten carbonate electrolysis [54] and graphene, which can be made from carbon atoms [55]. As a consequence, interest in carbon-based materials and their applications is encountering its most rapid development and represents a very important topic in modern materials science [46].

2.5 CONCLUDING REMARKS

Due to the current condition of climate change, mitigation strategies are very important in order to reduce its impact to society. In this case, the authors suggested carbon neutral or carbon neutrality as mitigation strategies. There are three strategies proposed, which are carbon neutral in fuel (including renewable energy, carbon capture and sequestration, and synthetic fuel), carbon-neutral building (during the design and construction phases), and carbon neutral in the production industry (material efficiency and carbon-based product). It is expected that this strategy will somehow reduce CO$_2$ emissions into the air and thereby reduce the impact of global warming on society. Moreover, this strategy is also important to prevent more GHGs being released into the air in the future.

REFERENCES

1. J. Allwood, *Sustainable Materials – With Both Eyes Open*. Cambrige, UK: UIT Cambridge, 2012.
2. B. S. Fisher *et al.*, "Issues related to mitigation in the long term context", In: B. Metz, O.R. Davidson, P.R. Bosch, R. Dave, and L. A. Meyer, eds. *Climate Change 2007: Mitigation, Contribution of Working Group, III to the Fourth Assessment Report of the Inter-governmental Panel on Climate Change*. Cambridge, UK and New York: Cambridge University Press, 2007.
3. F. A. Rahman *et al.*, "Pollution to solution : Capture and sequestration of carbon dioxide (CO$_2$) and its utilization as a renewable energy source for a sustainable future", *Renewable and Sustainable Energy Reviews*, vol. 71, pp. 112–126, 2017.
4. H. Y. Levine *et al.*, "Residential and commercial buildings", In: *Climate Change 2007: Mitigation. Contribution of Working Group III to the Fourth Assessment Report of the Intergovernmental Panel on Climate Change*. Cambridge, UK, and New York: Cambridge University Press, 2007.

5. IPCC, "Summary for policymakers", In: *Climate Change 2007: Mitigation. Contribution of Working Group III to the Fourth Assessment Report of the Intergovernmental Panel on Climate Change*. Cambridge University Press, 2007.

6. The Royal Society, *Geoengineering the Climate: Science, Governance and Uncertainty*. London, UK: The Royal Society, 2009.

7. C. Fu, and T. Gundersen, "Carbon capture and storage in the power industry: Challenges and opportunities", *Energy Procedia*, vol. 16, pp. 1806–1812, 2012.

8. K. Smith, *The Carbon Neutral Myth: Offset Indulgences for Your Climate Sins*. Amsterdam: Transnational Institute, 2007.

9. S. Gössling, "Carbon neutral destinations: A conceptual analysis", *Journal of Sustainable Tourism*, vol. 17, pp. 17–37, 2009.

10. D. W. Parker, *Service Operations Management: The Total Experience*, 2nd edn. Edward Elgar Publishing, 2018.

11. F. Yamin, *What Is 'Carbon Neutrality' – And How Can We Achieve It by 2050?* The Elder, 2014.

12. DECC, *Guidance on Carbon Neutrality*. London, UK: Department of Energy and Climate Change, 2009.

13. P. La Roche, *Architectural Design*. Boca Raton, FL: CRC Press, 2011.

14. L. A. Wright, S. Kemp, and I. Williams, "'Carbon footprinting': Towards a universally accepted definition", *Carbon Management*, vol. 2, pp. 61–72, 2011.

15. S. Zubelzu, and R. Á. Fernández, *Carbon Footprint and Urban Planning: Incorporating Methodologies to Assess the Influence of the Urban Master Plan on the Carbon Footprint of the City*. Springer, 2016.

16. L. J. Martin, "Carbon neutral – What does it mean?", www.eejitsguides.com, 2006.

17. A. C. Revkin, "Carbon-neutral is hip, but is it green?", www.nytimes.com, 2007.

18. C. Graves, S. D. Ebbesen, M. Mogensen, and K. S. Lackner, "Sustainable hydrocarbon fuels by recycling CO_2 and H_2O with renewable or nuclear energy", *Renewable and Sustainable Energy Reviews*, vol. 15, pp. 1–23, 2011.

19. F. S. Zeman, and D. W. Keith, "Carbon neutral hydrocarbons", *Philosophical Transaction of the Royal Society Series A*, vol. 48, pp. 3901–3918, 2008.

20. A. Goeppert, M. Czaun, G. K. S. Surya Prakash, and G. A. Olah, "Air as the renewable carbon source of the future: An overview of CO_2 capture from the atmosphere", *Energy and Environmental Science*, vol. 5, pp. 7833–7853, 2012.

21. H. W. Pennline, E. J. Granite, D. R. Luebke, J. R. Kitchin, J. Landon, and L. M. Weiland, "Separation of CO_2 from flue gas using electrochemical cells", *Fuel*, vol. 89, pp. 1307–1314, 2010.

22. R. J. Pearson *et al.*, "Energy storage via carbon-neutral fuels made from CO_2, water, and renewable energy", *Proceedings of the IEEE*, vol. 100, pp. 440–460, 2012.

23. O. Edenhofer *et al.*, *Renewable Energy Sources and Climate Change Mitigation – Summary for Policy Makers and Technical Summary*, 2012.

24. EIA, "How much of U.S. carbon dioxide emissions are associated with electricity generation?", www.eia.gov, 2018.

25. D. Arent *et al.*, "Implications of high renewable electricity penetration in the U.S. for water use, greenhouse gas emissions, land-use, and materials supply", *Applied Energy*, vol. 123, pp. 368–377, 2014.

26. N. Y. Amponsah, M. Troldborg, B. Kington, I. Aalders, and R. L. Hough, "Greenhouse gas emissions from renewable energy sources: A review of lifecycle considerations", *Renewable and Sustainable Energy Reviews*, vol. 39, pp. 461–475, 2014.

27. National Renewable Energy Laboratory (NREL), "Renewable Electricity Futures Study", *U.S. Department of Energy*, vol. 1, p. 280, 2012.

28. B. Lloyd, and S. Subbarao, "Development challenges under the Clean Development Mechanism (CDM)—Can renewable energy initiatives be put in place before peak oil?", *Energy Policy*, vol. 37, pp. 237–245, 2009.

29. WWF, "Renewable energy: The best option for people and the planet", http://climate-energy.blogs.panda.org, 2015.
30. IEA, *Contribution of Renewables to Energy Security*, 2007.
31. S. Zafar, "Biomass energy and sustainability", www.bioenergyconsult.com, 2016.
32. J. Mack, and B. Endemann, "Making carbon dioxide sequestration feasible: Toward federal regulation of CO$_2$ sequestration pipelines", *Energy Policy*, vol. 38, pp. 735–743, 2010.
33. C. Azar, K. Lindgren, E. Larson, and K. Möllersten, "Carbon capture and storage from fossil fuels and biomass – Cost and potential role in stabilizing the atmosphere", *Climatic Change*, vol. 74, pp. 47–79, 2006.
34. H. Herzog, D. N. Golomb, "Carbon capture and storage from fossil fuel use", *Encyclopaedia of Energy*, vol. 1, pp. 1–11, 2004.
35. F. Bowen, "Carbon capture and storage as a corporate technology strategy challenge", *Energy Policy*, vol. 39, pp. 2256–2264, 2011.
36. B. P. Spigarelli, and S. K. Kawatra, "Opportunities and challenges in carbon dioxide capture", *Journal of CO$_2$ Utilization*, vol. 1, pp. 69–87, 2013.
37. H. Duan, "The public perspective of carbon capture and storage for CO$_2$ emission reductions in China", *Energy Policy*, vol. 38, pp. 5281–5289, 2010.
38. N. MacDowell *et al.*, "An overview of CO$_2$ capture technologies", *Energy and Environmental Science*, vol. 3, pp. 1645–1669, 2010.
39. K. S. Lackner, "Capture of carbon dioxide from ambient air", *The European Physical Journal Special Topics*, vol. 176, pp. 93–106, 2009.
40. M. Fasihi, D. Bogdanov, and C. Breyer, "Long-term hydrocarbon trade options for the Maghreb region and Europe-renewable energy based synthetic fuels for a net zero emissions world", *Sustainability*, vol. 9, p. 306, 2017.
41. O. Lucon *et al.*, "Buildings", In: *Climate Change 2014: Mitigation of Climate Change. Contribution of Working Group III to the Fifth Assessment Report of the Intergovernmental Panel on Climate Change.* Cambridge University Press, pp. 671–738, 2014.
42. M. Jennings, N. Hirst, and A. Gambhir, *Reduction of Carbon Dioxide Emissions in the Global Building Sector to 2050.* London, UK: Grantham Institute – Climate Change and Environment, 2011.
43. P. M. La Roche, *Carbon-Neutral Architectural Design*, 2nd edn. Boca Raton, FL: CRC Press, 2017.
44. I. Arocho, W. Rasdorf, and J. Hummer, "Methodology to forecast the emissions from construction equipment for a transportation construction project", *Construction Research Congress 2014*, May 13, 2014, pp. 140–149, 2014.
45. B. Kim, H. Lee, H. Park, and H. Kim, "Greenhouse gas emissions from onsite equipment usage in road construction", *Journal of Construction Engineering and Management*, vol. 138, pp. 982–990, 2012.
46. M.-M. Titirici *et al.*, "Sustainable carbon materials", *Chemical Society Reviews*, vol. 44, pp. 250–290, 2015.
47. P. Gilbert, P. Wilson, C. Walsh, and P. Hodgson, "The role of material efficiency to reduce CO$_2$ emissions during ship manufacture: A life cycle approach", *Marine Policy*, vol. 75, pp. 227–237, 2017.
48. J. R. Duflou, and W. Dewulf, "Eco-impact anticipation by parametric screening of machine system components an introduction to the EcoPaS methodology", In: *Product Engineering: Eco-Design, Technologies and Green Energy*, Dordrecht: Springer Netherlands, pp. 17–30, 2005.
49. Y. Seow, N. Goffin, S. Rahimifard, and E. Woolley, "A 'Design for Energy Minimization' approach to reduce energy consumption during the manufacturing phase", *Energy*, vol. 109, pp. 894–905, 2016.
50. K. Schanes, S. Giljum, and E. Hertwich, "Low carbon lifestyles: A framework to structure consumption strategies and options to reduce carbon footprints", *Journal of Cleaner Production*, vol. 139, pp. 1033–1043, 2016.

51. T. Cooper, *Longer Lasting Products: Alternatives to the Throwaway Society.* Gower Publishing, Ltd, 2010.

52. N. Linares, A. M. Silvestre-Albero, E. Serrano, J. Silvestre-Albero, and J. García-Martínez, "Mesoporous materials for clean energy technologies", *Chemical Society Reviews*, vol. 43, 7681–7717, 2014.

53. R. J. White, *Porous Carbon Materials from Sustainable Precursors.* Royal Society of Chemistry, 2015.

54. M. Johnson *et al.*, "Carbon nanotube wools made directly from CO_2 by molten electrolysis: Value driven pathways to carbon dioxide greenhouse gas mitigation", *Materials Today Energy*, vol. 5, pp. 230–236, 2017.

55. A. Geim, and K. Novoselov, *Single Sheets of Carbon Open Up a New Field of Research.* IOP Institute of Physics, 2011.

3 Greenhouse Gases Solutions in Roads and Transportation

Thanwa Filza Nashruddin and
Md. Maniruzzaman A. Aziz

CONTENTS

3.0 INTRODUCTION

The construction of road has many implications for the environment, especially in its demand for large amounts of material and energy consumption. As the world's roads and highways systems continue to grow in mileage and traffic volume, it is important to sustainably construct roads and highways that have a low environmental impact. The energy consumption and greenhouse gas (GHG) emissions from the construction of highways has caused public concern, making it necessary to assess the related environmental impacts.

Annually, a mass volume of greenhouse gases (GHG) are released through implanting asphalt pavements all over the world. Previous studies generally classified the environmental burdens of constructing asphalt pavements into four categories, including the depletion of fossil fuels, global warming, acidification and photo oxidant formation [1, 2]. Road construction projects generate considerable amounts of GHG emissions such as carbon dioxide (CO_2) due to the large-scale use of heavy-duty diesel (HDD) construction equipment [3, 4], as well as extensive earthworks and earthmoving operations [5]. Santero and Horvath [6] researched the main factors that contribute to pavement GHG emissions by dividing the construction into several components: materials extraction and production, transportation, onsite equipment and traffic delays. A study conducted by Zapata and Gambatese [7] showed that the main consumption of energy (48%) from extraction to asphalt placement occurs during the mixing and drying of aggregate used for the pavement. Moreover, the production of bitumen accounts for about 40% of the total energy consumption.

Therefore, it is important for road and highway industry to maintain, support or endure the long-term maintenance of pavement and also promote sustainability in road and highway construction by implementing three types of criteria that play an important role in the promotion of green, sustainable road construction (shown in Figure 3.1).

Air pollution control: A high amount of emissions associated with road construction and maintenance activities is often associated with the upstream emissions embodied in the materials used [8]. These materials primarily include asphalt, concrete and steel. A hot mix asphalt (HMA) plant is the main manufacturing industry to produce a high quantity of asphalt material. The asphalt cement, or bitumen, has a high carbon content. The average carbon content of asphalt cement is about 82%, and asphalt cement makes up about 5% of an asphalt pavement, with the rest being aggregates (i.e., stone, sand and gravel) [9]. Facilities in hot mix asphalt (HMA) plants like plants management for stockpile and baghouse of filter only priority to trap the particulate matter. Thus, catalytic methanation conversion method is a promising technique to treat GHG, especially CO_2 gas that releases from the HMA stack, and convert it to the natural gas methane. This method can promote green sustainability in the roads industry and concurrently reduce global warming problems.

Asphalt technology in material: The use of secondary (recycled) aggregates instead of primary (virgin) materials has helped in easing landfill pressures, reducing the need for extraction, protecting the environment and minimizing the consumption of original resources. Resource recycling techniques can be used to fully convert "waste" to another "new resource" and offering environmental protection, conserving the natural resources by using recycled waste products. This can upgrade the quality of conventional road material. Most importantly, these green roads and highways can eliminate waste and reduce the energy required during the construction of pavements.

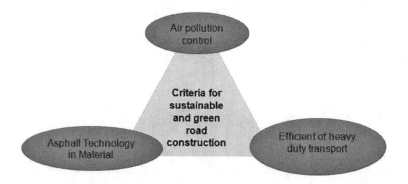

FIGURE 3.1   Criteria for sustainable and green road construction.

Efficient heavy-duty transport: Environment protection agency (EPA) reports show that 74% of GHG emissions are due to the combustion of fossil fuels, which includes the burning of gasoline, diesels and any other hydrocarbon fuel used by construction equipment. Whereas only 24% (i.e., 31 MMTCO$_2$e) of emissions were produced because of electrical power generation by the U.S. construction industry in 2002 [10]. An emission of GHG varies with respect to the type of equipment and its usage. This is due to the difference of fuel that is used in heavy construction equipment, capacity and consumption and largely depends on the type of activity and working hours. The introduction of biofuel, such as biodiesel, in construction equipment can save energy and lower emissions.

Developing novel materials and technologies to integrate greener material, waste and recycled materials into the production cycle of asphalt mixtures is a solution that improves both the sustainability and cost-efficiency of the asphalt pavement industry. Also, the encouragement of sustainable practices and good technology equipment needs to be widely implemented in the construction of roads and highways.

3.1   CURRENT PROBLEM OF ROADS AND HIGHWAYS CONSTRUCTION

Roads are built up in several layers, consisting of subgrade, sub base, base and surface layer. These layers together constitute the pavement. The pavement can be constructed from a wide variety of materials and mixtures of materials consisting of gravel, stone, asphalt, concrete or improved soils. Thus, the roads construction industry plays an important role in ensuring that road pavements are safe, smooth, quiet, durable, economical and made of sustainable materials.

Nearly 90% of paved roads in the world are made of asphalt mixtures; the remaining 10% is made up of either Portland cement concrete or composites of hot mix asphalt over Portland cement concrete or vice versa [7, 11]. Construction of new roads frequently implies interference with the environment, such as the alignment of the road and the necessity of huge amounts of natural aggregates needed for construction. The road sector is a concern because roads consume substantial amounts of materials [12].

The construction of a road needs thousands of tons of material, such as aggregate rock, concrete, asphalt and steel as the main resource. Also required is the consumption of fuel to power the construction equipment. One mile of two-lane asphalt road with an aggregate base can require up to 25,000 tons of aggregate rock (aggregates are the most mined resource in the world and are almost entirely non-renewable). In terms of GHG emissions, between the pavement and sub-base, all the mining, transporting, heating, earthwork and paving work to construct the single-lane mile of freeway will emit enough pollution to equal up to 1,200 tons of CO$_2$. That is about the same as the total annual emissions of 210 passenger cars. According to Stotko [13], the drying aggregate process is responsible for approximately 60%, the largest share, of the total energy consumed at an asphalt plant, whereas the evaporation of the water from these aggregates requires only 25% of the total energy requirements. This difference in energy requirements varies with the moisture percentage content. An asphalt plant, which produces approximately 100 t/h and consumes approximately 300 MJ/t, was evaluated in the previous study [13, 14].

According to Chehovits and Galehouse [15] aggregate production includes the quarrying, hauling, crushing, and screening and lead to the fuel energy consumption for aggregate production ranges from 30 to 40 MJ/t, and GHG emissions ranges from 2.5 to 10 kg CO_2/t. Besides, asphalt binder production includes crude oil extraction, transport and refining. Energy consumption for asphalt binders has been determined to be 4900 MJ/t, and GHG emissions are 285 kg CO_2/t [15]. Other than that, manufacturing asphalt includes all steps involved with handling, storing, drying, mixing and preparation of materials for installation [16]. The energy consumed varies depending on the specific material or product type. Working with the mix at such high temperatures produces GHG emissions, as well as other chemical pollutants that affect air quality [17, 18]. Table 3.1 has represented energy use and carbon dioxide emissions during the manufacture of asphalt pavement by Chappat and Bilal [19].

Figure 3.2 illustrates the energy use and GHG emissions produced from the road project. The environmental burdens come from aggregate acquisition and processing. Energy consumption for aggregate production includes rock blasting, quarrying, hauling, crushing and screening. The GHG emissions, such as CO_2, CH_4, NO and SF_6, are mainly from the energy consumption of machinery and explosives. In an aggregate quarry, the main energy consumption corresponds to electricity and diesel

TABLE 3.1

Energy use and carbon dioxides emissions during manufacture of asphalt pavement [19]

Product	Energy use (MJ/t)	CO_2 emissions (kg/t)
Bitumen	4900	285
Cement	4976	980
Emulsion	3490	221
Crushed aggregate	40	10
Pit – run aggregate	30	2.5
Hydraulic road binder	1244	245
Water	10	0.3
Quicklime	9240	2500
Fuel	35	4.0
Production of Hot Mix Asphalt (HMA)	289	23
Production of Warm Mix Asphalt (WMA)	234	20
Production of Cold Mix Asphalt (CMA)	14	1.0
Surface milling of asphalt for Reclaimed Asphalt Pavement (RAP)	12	0.8
In-situ cold recycling stabilization	15	1.13
In-situ soil cement stabilization	12	0.8
Laying of Hot Mix Asphalt	9	0.6
Laying of Cold Mix Asphalt	6	0.4
Cement concrete road paving	2.2	0.2
Lorry transport (km/t)	0.9	0.06

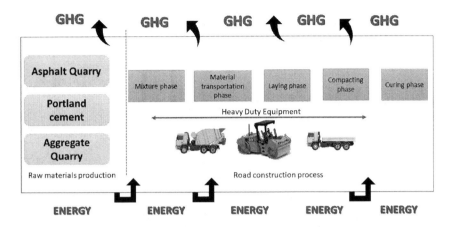

FIGURE 3.2   Energy and GHG emissions boundary for road construction project [20].

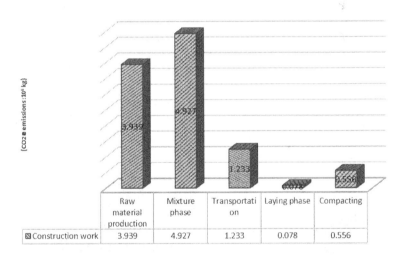

FIGURE 3.3 CO_2e emission of road construction [20].

used by the plants and equipment. The electricity is produced from other energy sources such as coal and gas.

For the 20 km-long asphalt pavement case study, the total GHG emissions of asphalt construction are shown in Figure 3.3, 97.19% of the total GHG emissions are due to the mixture mixing phase and raw materials production phase, whereas 54.01% are from the mixture mixing phase, and 43.18% are from the raw materials production phase. About 1.35% of the total GHG emissions are due to the raw material and mixture transportation phase. Only 0.86% and 0.61% of the total GHG emissions are due to the laying phase and the compacting phase [20].

Growing attention to world issues regarding GHGs emission has introduced the concept of sustainable development as the solution toward this issue. The road construction sector plays an essential role in improving the environment by continuing to improve the environmental performance of the country's pavements and infra-structure. Because of its products' longevity, the road construction industry is in an important position to support environmental benefits, both through everyday jobsite practices and through lasting structural pavement improvements. Three crucial cri-teria that play an important role in promoting sustainable green road construction will be discussed in next section.

3.2 AIR POLLUTION CONTROL

A high amount of emissions associated with road construction and maintenance activities is often associated with the upstream emissions embodied in the materials used [8]. These materials primarily include asphalt, concrete and steel. An HMA plant is the main part of the manufacturing industry to produce a high quantity of asphalt material. The asphalt cement, or bitumen, has high carbon content. The aver-age carbon content of asphalt cement is about 82%, and asphalt cement makes up about 5 % of an asphalt pavement, with the rest being aggregates (i.e., stone, sand and gravel) [9]. The main sources of emissions from asphalt plants are from the dry-ers, mixers and hot bins, which emit large quantities of particulate matter and other gases (Figure 3.4). Other sources include storage silos, loading and unloading opera-tions, hot oil heaters (used to heat the asphalt storage tanks) and yard emissions (the fugitive gases from the mixtures in truck beds). Other types of pollutants include the noise of the equipment and machines, the released odors and the contamination of groundwater caused by leakages [21].

FIGURE 3.4 Emission release from plants.

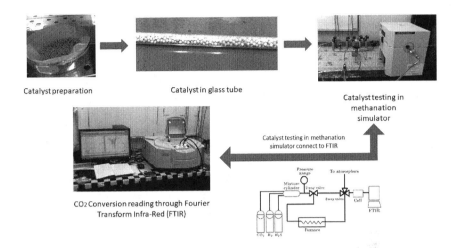

FIGURE 3.5   Catalytic activity measurement in laboratory.

Many researchers have come up with an idea for how to reduce vast amounts of GHG emissions from road production and construction. Basically, to save energy and natural resources for asphalt manufacturing, the use of reclaimed asphalt pavement (RAP) as a combination with new materials along with a recycling agent has been shown to have better (or similar) results compared to mixtures made with all new materials. In the meantime, the installation of bag-house filters as dust removal equipment can reduce the amount of larger particles released into the atmosphere. However, those methods can only control particle dust and save natural resources and energy.

The best way to control GHG emissions from road material production (i.e., quarry plants) is by using the chemical conversion method, which is the catalytic methanation conversion system. While this method is still new in the road industry, it is already successfully testing at Connaught Bridge, Selangor, Malaysia. This power plant is fueled by natural gas. The methanation reaction is a promising method for the purification of natural gas, in which the acid gas of CO_2 is eliminated by catalytic conversion [22]. The advantage of catalytic technology is the utilization of CO_2 present to produce methane gas. Based on this idea, the catalytic methanation conversion system is being studied and tested in the lab before its application to the real asphalt plant (Figure 3.5).

The catalytic methanation conversion system can utilize CO_2 emissions from asphalt production to the renewable gas of methane (CH_4) gas. This green technology promotes saving, sustainable development, and it can be recycled [23]. Methane gas is a colorless, odorless gas with a wide distribution in nature. It is the principal component of natural gas and is claimed to be more environmentally friendly [24].

3.3   ASPHALT TECHNOLOGY IN MATERIAL

Extensive use of natural aggregates in construction projects has been gradually depleting this resource near areas where aggregates are in high demand. The need for resource conservation has increased the demand to introduce substitute materials for natural aggregates.

The energy is required to produce the heat used to manufacture hot mix asphalt. Thus, reducing that amount of heat is one of the main targets to reduce energy consumption, which in turn reduces greenhouse gas emissions and environmental impact. According to the reduction of production temperature in asphalt mixtures, the following advantages are obtained: (1) reduced fuel consumption; (2) less carbon dioxide emission; (3) longer paving season; (4) longer hauling distance; (5) reduced oxidation of asphalt; (6) early opening to traffic; and (7) a better working environment.

Green road initiatives to reduce GHG emissions toward asphalt material must come up with the low-temperature mixtures representing substantial energy savings and the mitigation of emissions such as warm mix asphalt (WMA). Also, recycled asphalt pavement such as reclaimed asphalt pavement (RAP) and recycling cold mix asphalt are promising at reducing natural resources extraction and energy consumption through a low temperature process. The following subsection will discuss the latest asphalt technology to bring about a good impact on the environment and control GHG emissions.

3.3.1 Warm Mixed Asphalt

Warm mix asphalt (WMA) is a variation of traditional HMA. WMA technologies are processes or additives for HMA that allow mixture production and placement to occur at temperatures lower than conventional hot mix asphalt. The energy consumption for different processes of warm mix asphalt technologies compared to hot mix asphalt have been studied by many researchers. In general, 20–70% of energy is saved in the production process because less heating is necessary [25], and up to 30% energy savings can be obtained in the pavement compaction of warm mix asphalt due to lower asphalt viscosity [26]. The mixing temperature of WMA production is usually about 30–60°C lower than HMA production (Figure 3.6). Thus, WMA production can reduce around 30% energy consumption, resulting in lower emissions of harmful gases and dust [14, 27].

Moreover, less energy is involved and, during the paving operations, the temperature of the mixture is lower, resulting in improved working conditions for the workers and an earlier opening date for the road. One concern of the WMA is to keep good workability, which is the workability that completely depends on temperature and additives, which can create proper asphalt even at lower mixing temperatures. The production of additives needed for WMA have significant impacts [29]. The current WMA additives are classified as (1) organic, (2) foaming and (3) chemical. With organic additives, the viscosity of asphalt is reduced at the temperature above the melting point to produce asphalt mixtures at lower temperatures. Generally, (1) Sasobit, (2) Asphaltan-B, (3) Licomont BS-100 and (4) Cecabase RT are the organic WMA additives to have been commercialized in road design [30]. Table 3.2 presents the selected study results of WMA incorporated with the potential additives which can save energy and reduce the emissions.

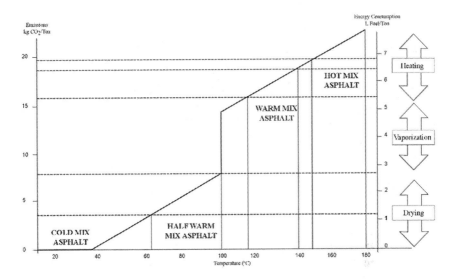

FIGURE 3.6 Temperature classification for different type of asphalt [28].

TABLE 3.2
WMA with different additives

Additives	Temperature (°C)		Energy saving (%)	CO_2 emission reduced (%)	References
	*M	*C			
Zeolites (Dose: 2–6%)	120	–	–	7.2–34	[31]
	140	–	–	2.5–35	
	180	–	–	4.9–33	
CaZ (Dose: 2–6%)	120	–	–	9.7–53	[31]
	140	–	–	7.1–68	
	180	–	–	11.8–7	
Fischer–Tropsch wax addition	–	–	12	–	[11, 32]
chemical tension-active additive	–	–	12	–	[11, 32]
WMA foam (2-part process: soft asphalt added first then hard, foamed asphalt)	43–63	–	–	30–98	[25]
Aspha-Min (Zeolite (21% water) 0.3% by mixture weight)	30	–	–	75–90	
Sasobit (Paraffin wax 0.8 to 0.3% by weight of asphalt)	18–54	–	–	–	
Evoterm (Emulsion (70% asphalt with additives))	50–75	–	–	40–60 °	

Note: *M refers to Mixing, *C refers to Compaction, *L refers to Laying, – refers to not available

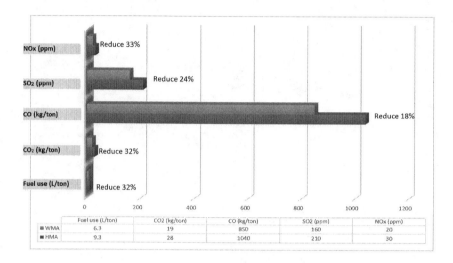

FIGURE 3.7 Emissions and fuel consumption for production of WMA and HMA mixtures in the plant [33].

A case study by Kim et al. [33] has been summarized in Figure 3.7, showing that the decreased production temperatures by using WMA with the selected additives has led to energy savings of 32%, which results in a 32% reduction of CO_2, an 18% reduction of CO, a 24% reduction of SO_2 and a 33% reduction of NO_X compared to HMA.

3.3.2 RECLAIMED ASPHALT PAVEMENT

To decrease energy consumption is to use waste materials in infrastructure construction and rehabilitation. When an asphalt pavement reaches the end of its design life, the road surfacing is milled, creating a milled waste material known as reclaimed asphalt pavement (RAP). The RAP materials contain aggregate and asphalt that are transported to an asphalt plant for recycling. Since aggregate materials are non-renewable natural resources, the primary benefit of using RAP is to reduce the demand for the extraction of new aggregate and disposal of waste materials in landfill.

Moreover, the use of RAP in asphalt pavements has economic savings and environmental benefits [34, 35]. RAP is a useful alternative to virgin materials because it reduces the use of virgin aggregate and the amount of virgin asphalt binder required in the production of hot mix asphalt (HMA) [36]. For a higher RAP content (above 25%), design and production takes extra effort, but the savings in using the higher RAP contents significantly outweigh the added costs. Blending to meet gradation and the appropriate binder grade in the final product are keys to successful mix design, production and performance [34]. The most crucial characteristic of RAP material that affects the properties and performance of recycled mixtures is the aging of its binder.

TABLE 3.3
Marshall parameter for 20% RAP and virgin mixes [39]

Mixes	Binder content (%)	Bulk density (g/cc)	Marshall stability (kN)	Flow (mm)	Air Voids (%)	VFB (%)
20 % RAP	5.00	2.36	14.40	3.49	4.48	73.7
Virgin	5.70	2.35	13.50	2.30	3.57	75.0

Sebaaly et al. [37] indicated that the utilization of a certain percentage of RAP increases the performance properties of mixes. This investigation had already been supported by Paul [38], who indicated that incorporating certain percentages of RAP results in no significant change to the performance of mixes. Based on a study by Pradyumna et al. [39], the results for the optimum binder content (OBC) for both virgin and 20% RAP mixes was found to be 5.7 %. In the case of the 20% RAP mix, the fresh binder to be added comes out to be 5.0% compared to the 5.7% in the virgin mix. The Marshall parameters obtained are presented in Table 3.3 [39].

Thus, the introduction of RAP as the modified binder can improve the performance of HMA. In fact, flexible pavements require significant amounts of energy for the production of bituminous binders, drying aggregates and, subsequently, the production of bituminous mix at a Hot Mix Plant (HMP). The heating of bituminous binder, aggregates and production of huge quantities of HMA releases a significant amount of GHGs and harmful pollutants. Alternative technologies using recycled roads in road construction and maintenance can help to reduce the consumption of fuel and aggregates.

3.3.3 COLD MIX ASPHALT

Cold mix asphalt (CMA) has several advantages over the common HMA, particularly the fact that it requires no heat to manufacture or lay over the pavement. This technology is a promising alternative material for a wide range of paving applications, from preventive maintenance and repair to new pavement construction. In southern China, many pavements are located in moist, highland mountainous areas where ice freezing usually occurs, which can easily lead to a pitted, loose surface with many potholes [40]. To solve these problems, and recover the traffic capacity of a road in a short time, the maintenance department usually applies CMA as an emergency repair.

Moreover, the constant need to repair localized pavement failures, and the nature of the repairs normally required in terms of material quantities and timing at which the repairs need to be executed, make necessary the use of materials that can be stored and used when needed. Although the behavior of HMA is somewhat well understood and there is extensive knowledge about its performance, the inability to produce and store HMA in small quantities to be used as required makes them unsuitable for pothole and other localized repairs. To fill this need, CMA is normally used, as its characteristics allow users to have immediate availability in stock, and use it when and where needed [41].

CMA mixtures produced with virgin aggregate mixed with emulsified asphalt instead of hot asphalt binder have possible environmental benefits, including reduced energy consumption and emissions due to storing the emulsified asphalt at ambient temperatures, and the ability to use aggregates from the stockpile without drying [42]. Also, CMA can be stockpiled and has a longer working life, meaning it can be transported longer distances and placed in locations generally inaccessible or impractical for more traditional methods [43].

Selected addictive emulsion in cold mix asphalt has improved the performance of cold mix asphalt in terms of adhesion properties and moisture damage resistance. Polyvinyl acetate emulsion (PVAC-E) was mixed with a local aggregate in CMA and found to have a compressive strength improved by 31% compared to the values obtained with the unmodified cold mix asphalt [44]. Moreover, most researchers have focused on measuring workability characteristics. Estakhri and Button [45] report the use of unconfined compression tests and suggest limiting criteria from this test for assessing workability. However, some research efforts have studied rutting performance in CMA products. Rosales-Herrera et al. [46] evaluated locally produced and proprietary CMA patching materials using slump tests for workability assessment and stability assessment in the laboratory using Hamburg Wheel Tracking tests and local Texas Stability Tests.

Several researchers have done testing toward CMA in other aspects of stability and properties. Benedito et al. [47] selected fiber to reinforce CMA, and they found the addition of fiber was responsible for a small variation in mixture strength parameters, as well as for substantial drops in the mixture resilient moduli when compared to plain mixtures. Al-Busaltan et al. [48] used waste materials to mix a new CMA, and found it had superior mechanical properties compared to traditional HMA.

Although there are different tests to characterize CMA, most characterization methods for these materials focus on mix workability rather than performance, mostly because these materials are often used as a temporary repair. However, research has been done to develop CMA that can be used as a permanent solution [3], although the ability to match the performance of its hot mixed counterpart has been elusive [40].

3.4   EFFICIENCY OF HEAVY-DUTY TRANSPORT

A U. S. Environmental Protection Agency (EPA) report shows that 74% (i.e., 100 MMTCO$_2$e) of GHG emission is due to the fossil fuel combustion, which includes the burning of gasoline, diesel and any other hydrocarbon fuel by construction equipment. Whereas 24% (i.e., 31 MMTCO$_2$e) of emissions are produced because of electrical power generation for the construction sector by the U.S. construction industry in 2002 [10]. An emission of GHG varies with respect to the type of equipment and its usage. This is due to the difference of fuel that is used in heavy construction equipment, capacity and consumption, and largely depends on the type of activity and working hours.

Fuel characteristics of the vehicle/equipment become the main contributor to the emission of GHGs. GHGs are directly caused by road vehicles that need to burn lots of carbonaceous fuels to run [54]. Transportation of material is a subcategory of road construction activity that brings about a high impact in terms of the environment (i.e., CO_2 emissions) and the economy (i.e., energy consumption and fuels prices). All materials and mixtures were generally to be hauled by heavy-duty vehicles (HDVs). Usually, the U.S. EPA's motor vehicle emissions simulator (MOVES) was used to determine the average fuel consumption and airborne emissions factors for operating diesel-powered, single-unit short-haul trucks and long-haul combination trucks [49].

A study by Abou-Senna et al. [50] integrated MOVES and found that the CO_2 produced by heavy-duty diesel trucks (HDDV) was 25866kg, 19478kg and 23912kg due to the parameter of average speeds, link drive schedules and operating mode distributions (OPMODEs). Graham et al. [51] evaluated the amount of GHGs generated through HDVs by using diesel fuel and liquid natural gas (LNG) fuel. The experimental results showed that diesel fuels from HDVs produced 1631gkm^{-1} of CO_2, while HDVS-LNG fuels produced 1355gkm^{-1} of CO_2. Thus, conventional fuels used by material transportation of HDVs releases a significant amount of GHGs into the environment.

Therefore, renewable energy from biofuels is an alternative fuel source used to control emissions and energy consumption in a different mode to that of a heavy/medium/small truck, transportation material quarrying and hauling in road construction activities. Biofuels that are predominantly produced from biomass are explored and deliberated as friendly, renewable energy alternatives to save the non-renewable conventional fuels (coal, petroleum, natural gas, gasoline and diesel). Countries such as Brazil, the United States, Germany, Australia, Italy and Austria are already implementing the concept of the bio-refinery to produce renewable energy, which can control CO_2 emissions and use cleaner energy sources (i.e., bio-ethanol or blended diesel and biodiesel). Other countries such as Malaysia use blended fuel, such as a combination of 5% palm oil and 95% regular diesel [52]. It is expected that this trend will grow in the coming years and that more countries will use biofuels as part of their long-term strategy to achieve sustainability in heavy-duty transport and road construction [53].

Biofuel production offers a great opportunity for material transport in multi-functional ways to upgrade fuel efficiency [54] and control of air emissions. For example, CO_2 released from the running of fossil fuels in ethanol plants can be captured by using microalgae. Generally, this microscopic plant is cultivated in large open ponds and removed with flue gas or pure CO_2. The CO_2 emissions are captured from power plants in the form of small bubbles, as shown in Figure 3.8. Then, the alga biomass generated is separated from the liquid phase by using mechanical, chemical, gravitational and/or combination processes. The algal oil is extracted by disruption of the algae cells via chemical or physical methods. The oil extracted from the algal biomass can be processed by using transesterification to produce biodiesel for energy and transportation [55].

With the increasing concern over limiting GHG emissions, much attention has been paid to the material transportation (i.e., heavy-duty trucks) and improving their fuel economy as well as reducing their CO_2 emissions [56]. The best way to solve

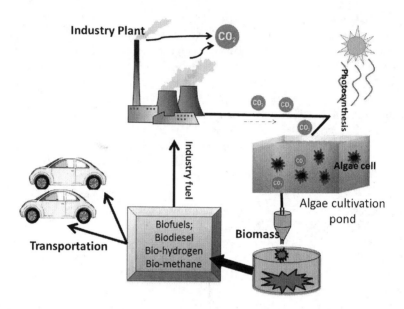

FIGURE 3.8 Carbon dioxide (CO_2) conversion to fuels by microalgae [55].

the problem regarding the environmental impact in materials transportation is by adopting new alternatives that can reduce the consumption of conventional fuels. Therefore, biofuels are a good alternative as a substitute for conventional fuels in material transportation because of the ability to produce low CO_2 emissions in combustion, giving a positive socioeconomic impact [57, 58].

3.5 CONCLUDING REMARKS

With the pollution of the environment, lack of raw materials and annual escalation of energy consumption, green road and highway technology has attracted greater attention. The terms "Green," "Sustainable Development," "Environmental Impact," "Energy Efficiency," "Global Warming," "Greenhouse Gases" and "Eco-Efficiency" are becoming more widely recognized and used. The Green roads system has been developed as a method to assess roadway sustainability. Green roads enable owners, consultants and contractors to make informed decisions by providing a sustainability performance metric for roadway design and construction. The system defines roadway sustainability attributes, provides a system for the evaluation of roadway sustainability and includes a collection of sustainable design and construction practices. To obtain low energy and GHG demand from the road construction industry in future, it is imperative to think about the importance of carbon emissions caused by road construction processes, and three crucial criteria play an important role in promoting green, sustainable road construction; air pollution control, asphalt technology in material and the efficient use of heavy-duty transport must all become best practice in road industry.

REFERENCES

1. K. F. Saberi, M. Fakhri, and A. Azami, "Evaluation of warm mix asphalt mixtures containing reclaimed asphalt pavement and crumb rubber", *Journal of Cleaner Production*, vol. 165, pp. 1125–1132, 2017.
2. M. Mazumder, V. Sriraman, H. H. Kim, and S.-J. Lee, "Quantifying the environmental burdens of the hot mix asphalt (HMA) pavements and the production of warm mix asphalt (WMA)", *International Journal of Pavement Research and Technology*, vol. 9, pp. 190–201, 2016.
3. J. Krantz, W. Lu, T. Johansson, and T. Olofsson, "Analysis of alternative road construction staging approaches to reduce carbon dioxide emissions", *Journal of Cleaner Production*, vol. 143, pp. 980–988, 2017.
4. A. M. Hajji, and P. Lewis, "Development of productivity-based estimating tool for energy and air emissions from earthwork construction activities", *Smart and Sustainable Built Environment*, vol. 2, pp. 84–100, 2013.
5. R. Kenley, and T. Harfield, "Greening procurement of infrastructure construction: Optimizing mass haul operation to reduce greenhouse gas emissions", In: *Proceedings of the CIB W78–W102 International Conference, Sophia Antipolis*, France, pp. 26–28, 2011.
6. N. Santero, E. Masanet, and A. Horvath, *Life-Cycle Assessment of Pavements: A Critical Review of Existing Research and Literature*, 2010.
7. P. Zapata, and J. A. Gambatese, "Energy consumption of asphalt and reinforced concrete pavement materials and construction", *Journal of Infrastructure Systems*, vol. 11, pp. 9–20, 2005.
8. R. B. Noland, and C. S. Hanson, "Life-cycle greenhouse gas emissions associated with a highway reconstruction: A New Jersey case study", *Journal of Cleaner Production*, vol. 107, pp. 731–740, 2015.
9. APA, *Carbon Footprint: How Does Asphalt Stack Up?*, Asphalt Pavement Alliance, 2010.
10. M. S. L. M. Waris, M. F. Khamidi, and A. Idrus, "Reducing the greenhouse gas emissions from onsite mechanized construction", *Australian Journal of Basic and Applied Sciences*, vol. 8, pp. 391–398, 2014.
11. L. P. Thives, and E. Ghisi, "Asphalt mixtures emission and energy consumption: A review", *Renewable and Sustainable Energy Reviews*, vol. 72, pp. 473–484, 2017.
12. H. Birgisdóttir, K. A. Pihl, G. Bhander, M. Z. Hauschild, and T. H. Christensen, "Environmental assessment of roads constructed with and without bottom ash from municipal solid waste incineration", *Transportation Research Part D: Transport and Environment*, vol. 11, pp. 358–368, 2006.
13. O. Stotko, "Energy and related carbon emission reduction technologies for hot mix asphalt plants", In: *Tenth Conference on Asphalt Pavements for Southern Africa—CAPSA*, Winterton, South Africa, 2011.
14. A. Almeida-Costa, and A. Benta, "Economic and environmental impact study of warm mix asphalt compared to hot mix asphalt", *Journal of Cleaner Production*, vol. 112, pp. 2308–2317, 2016.
15. J. Chehovits, and L. Galehouse, "Energy usage and greenhouse gas emissions of pavement preservation processes for asphalt concrete pavements", In: *Proceedings on the 1st International Conference of Pavement Preservation*, Newport Beach, California, pp. 27–42, 2010.
16. J. E. Johnston, *Site Control of Materials: Handling, Storage and Protection*. Elsevier, 2016.
17. M. del Carmen RubioF. Moreno, M. J. Martínez-Echevarría, G. Martínez, and J. M. Vázquez, "Comparative analysis of emissions from the manufacture and use of hot and half-warm mix asphalt", *Journal of Cleaner Production*, vol. 41, pp. 1–6, 2013.

18. T. Blankendaal, P. Schuur, and H. Voordijk, "Reducing the environmental impact of concrete and asphalt: A scenario approach", *Journal of Cleaner Production*, vol. 66, pp. 27–36, 2014.

19. M. Chappat, and J. Bilal, *The Environmental Road of the Future: Life Cycle Analysis*, vol. 9. Paris: Colas SA, pp. 1–34, 2003.

20. F. Ma, A. Sha, R. Lin, Y. Huang, and C. Wang, "Greenhouse gas emissions from asphalt pavement construction: A case study in China", *International Journal of Environmental Research and Public Health*, vol. 13, p. 351, 2016.

21. R. Myers, B. Shrager, G. Brooks, and K. Weant, *Hot Mix Asphalt Plants Emission Assessment Report, Report No. EPA-454/R-00e019*, North Carolina, 2000.

22. W. A. W. A. Wan Abu Bakar, R. Ali, A. A. A. Kadir, S. J. M. Rosid, and N. S. Mohammad, "Catalytic methanation reaction over alumina supported cobalt oxide doped noble metal oxides for the purification of simulated natural gas", *Journal of Fuel Chemistry and Technology*, vol. 40, pp. 822–830, 2012.

23. S. Toemen, W. A. W. A. Bakar, and R. Ali, "Investigation of Ru/Mn/Ce/Al$_2$O$_3$ catalyst for carbon dioxide methanation: Catalytic optimization, physicochemical studies and RSM", *Journal of the Taiwan Institute of Chemical Engineers*, vol. 45, pp. 2370–2378, 2014.

24. J. L. Adgate, B. D. Goldstein, and L. M. McKenzie, "Potential public health hazards, exposures and health effects from unconventional natural gas development", *Environmental Science and Technology*, vol. 48, pp. 8307–8320, 2014.

25. Ó. Kristjánsdóttir, S. T. Muench, L. Michael, and G. Burke, "Assessing potential for warm-mix asphalt technology adoption", *Transportation Research Record: Journal of the Transportation Research Board*, vol. 2040(1), pp. 91–99, 2007.

26. L. Drüschner, "Experience with warm mix asphalt in Germany", In: *Hovedemne På Forbundsudvalgsmøde i Sønderborg, Guest report in conference*. Sønderborg: NVF-rapporter, 2009.

27. A. Vaitkus, D. Čygas, A. Laurinavičius, V. Vorobjovas, and Z. Perveneckas, "Influence of warm mix asphalt technology on asphalt physical and mechanical properties", *Construction and Building Materials*, vol. 112, pp. 800–806, 2016.

28. J. A. D'Angelo, E. E. Harm, J. C. Bartoszek, G. L. Baumgardner, M. R. Corrigan, J. E. Cowsert, *et al.*, *Warm-Mix Asphalt: European Practice*. Alexandria, VA: American Trade Initiatives, 2008.

29. R. Vidal, E. Moliner, G. Martínez, and M. C. Rubio, "Life cycle assessment of hot mix asphalt and zeolite-based warm mix asphalt with reclaimed asphalt pavement", *Resources, Conservation and Recycling*, vol. 74, pp. 101–114, 2013.

30. R. M. Anderson, G. Baumgardner, R. May, and G. Reinke, "Engineering properties, emissions, and field performance of Warm Mix Asphalt Technologies", *Interim Report*, pp. 9–47, 2008.

31. A. Sharma, and B.-K. Lee, "Energy savings and reduction of CO$_2$ emission using Ca(OH)$_2$ incorporated zeolite as an additive for warm and hot mix asphalt production", *Energy*, vol. 136, pp. 142–150, 2017.

32. N. Bueche, and A.-G. Dumont, "Energy in warm mix asphalt", In: *Eurasphalt & Eurobitume Congress*, Istanbul, 2012.

33. Y. Kim, S. Hwang, S. Kwon, K. Jeong, S. Yang, and Y. Kim, "Laboratory and field experiences of low energy and low carbon-dioxide asphalt pavement in Korea", In: *Pavements and Materials: Recent Advances in Design, Testing and Construction*, pp. 131–138, 2011.

34. D. E. Newcomb, E. Arambula, F. Yin, J. Zhang, A. Bhasin, W. Li, *et al.*, *Properties of Foamed Asphalt for Warm Mix Asphalt Applications*, 2015.

35. R. Izaks, V. Haritonovs, I. Klasa, and M. Zaumanis, "Hot mix asphalt with high RAP content", *Procedia Engineering*, vol. 114, pp. 676–684, 2015.

36. A. Copeland, *Reclaimed Asphalt Pavement in Asphalt Mixtures: State of the Practice*, 2011.
37. P. Sebaaly, and R. Shrestha, "A literature review on the use of recycled asphalt pavements in HMA mixtures", In: *Final Report, Western Regional Superpave Center*, University of Nevada, Reno, 2004.
38. H. R. Paul, "Evaluation of recycled projects for performance", *Asphalt Paving Technology*, vol. 65, pp. 231–254, 1996.
39. T. A. Pradyumna, A. Mittal, and P. K. Jain, "Characterization of reclaimed asphalt pavement (RAP) for use in bituminous road construction", *Procedia – Social and Behavioral Sciences*, vol. 104, pp. 1149–1157, 2013.
40. M. Guo, Y. Tan, and S. Zhou, "Multiscale test research on interfacial adhesion property of cold mix asphalt", *Construction and Building Materials*, vol. 68, pp. 769–776, 2014.
41. L. G. Diaz, "Creep performance evaluation of Cold Mix Asphalt patching mixes", *International Journal of Pavement Research and Technology*, vol. 9, pp. 149–158, 2016.
42. P. Jarrett, A. Beaty, and A. Wojcik, "Cold-mix asphalt technology at temperatures below 10 C (with discussion)", In: *Association of Asphalt Paving Technologists Proceedings*, 1984.
43. C. Ling, A. Hanz, and H. Bahia, "Measuring moisture susceptibility of Cold Mix Asphalt with a modified boiling test based on digital imaging", *Construction and Building Materials*, vol. 105, pp. 391–399, 2016.
44. L. E. Chávez-Valencia, E. Alonso, A. Manzano, J. Pérez, M. E. Contreras, and C. Signoret, "Improving the compressive strengths of cold-mix asphalt using asphalt emulsion modified by polyvinyl acetate", *Construction and Building Materials*, vol. 21, pp. 583–589, 2007.
45. C. K. Estakhri, and J. W. Button, "Test methods for evaluation of cold-applied bituminous patching mixtures", *Transportation Research Record: Journal of the Transportation Research Board*, vol. 1590(1), pp. 10–16, 1997.
46. V. I. Rosales-Herrera, M. D. Seifert, and J. A. Prozzi, "Field trial assessment and performance prediction of containerized cold patching mixtures", In: *Proceedings of the 86th Annual Meeting of the Transportation Research Board*, Washington, DC, 2007.
47. B. de S. Bueno, W. R. Da Silva, D. C. de Lima, and E. Minete, "Engineering properties of fiber reinforced cold asphalt mixes", *Journal of Environmental Engineering*, vol. 129, pp. 952–955, 2003.
48. S. Al-Busaltan, H. Al. Al Nageim, W. Atherton, and G. Sharples, "Mechanical properties of an upgrading cold-mix asphalt using waste materials", *Journal of Materials in Civil Engineering*, vol. 24, pp. 1484–1491, 2012.
49. H. Balat, and E. Kırtay, "Hydrogen from biomass–present scenario and future prospects", *International Journal of Hydrogen Energy*, vol. 35, pp. 7416–7426, 2010.
50. United States Environmental Protection Agency, *"Inventory of U. S. Greenhouse Gas Emissions and Sinks: 1990–2009"*. Washington, DC: US EPA, 2011.
51. H. Abou-Senna, E. Radwan, K. Westerlund, and C. D. Cooper, "Using a traffic simulation model (VISSIM) with an emissions model (MOVES) to predict emissions from vehicles on a limited-access highway", *Journal of the Air and Waste Management Association*, vol. 63, pp. 819–831, 2013.
52. L. A. Graham, G. Rideout, D. Rosenblatt, and J. Hendren, "Greenhouse gas emissions from heavy-duty vehicles", *Atmospheric Environment*, vol. 42, pp. 4665–4681, 2008.
53. Z. Helwani, M. R. Othman, N. Aziz, W. J. N. Fernando, and J. Kim, "Technologies for production of biodiesel focusing on green catalytic techniques: A review", *Fuel Processing Technology*, vol. 90, pp. 1502–1514, 2009.
54. M. J. Haas, A. J. McAloon, W. C. Yee, and T. A. Foglia, "A process model to estimate biodiesel production costs", *Bioresource Technology*, vol. 97, pp. 671–678, 2006.

55. R. Hickman, O. Ashiru, and D. Banister, "Transport and climate change: Simulating the options for carbon reduction in London", *Transport Policy*, vol. 17, pp. 110–125, 2010.
56. G. Scora, K. Boriboonsomsin, and M. J. Barth, "Effects of operational variability on heavy-duty truck greenhouse gas emissions", In: *89th Annual Meeting of the Transportation Research Board*, Washington, DC, 2010.
57. F. A. Rahman, M. M. A. Aziz, R. Saidur, W. A. W. A. Bakar, M. R. Hainin, R. Putrajaya, and N. A. Hassan, "Pollution to solution: Capture and sequestration of carbon dioxide (CO_2) and its utilization as a renewable energy source for a sustainable future", *Renewable and Sustainable Energy Reviews*, vol. 71, pp. 112–126, 2017.
58. IEA, *Energy Technology Perspectives 2012*. OECD/IEA International Energy Agency, 2012.

4 Recent Clean Hydrogen Production Technologies

Bawadi Abdullah and Ali Awad

CONTENTS

4.0 INTRODUCTION

The world's population has grown manyfold in recent years. It is expected that this increasing population will reach 8 billion by 2030 and, with a continuous rise, will reach 11 billion in 2100 according to the UN DESA report [1]. Similarly, global energy requirements are also rising drastically with this increase in population and the modernization of the world. As shown in the International Energy Outlook Report of 2017, total energy consumption has risen from 575 quadrillion to 736 quadrillion between 2015 and 2017, an increase of around 28% [2]. These growing energy demands are being fulfilled by various energies, including renewables and non-renewables. The renewable energies are the world's fastest-growing sources, with consumption increasing around 2% per year from 2015 to 2040. Though the use of renewable resources is still an innovative approach, non-renewable fuels including fossil fuels are currently in use and will account for approximately 77% of energy usage until 2040. It is reported that natural gas, being the most abundant fossil fuel, will be facing an increased consumption rate of 1.4% per year from 2015 to 2040. However, the consumption of liquid fuels will fall from 33% in 2015 to 31% in 2040 [2]. Subsequently, the primary resources of world energy demand are coal, natural gas and crude oil. These resources are dominating the world energy market, which is worth around 1.5 trillion dollars. The usage of fossil fuels is increasing periodically every year, thus creating a dilemma when the limited non-renewable resources of energy will diminish. Shahriar et al. [3] predicted that fossil fuels including coal, crude oil and natural gas would be depleted after 100 years. The uncertain future of fossil fuels is the primary motivation of current research for alternative energy resources. However, fossil fuels will still play an active role in the global energy market for at least a couple of decades.

The continuous combustion of fossil fuels in industries to meet their energy demands leads to the emission of pollutants and greenhouse gases, such as oxides of carbon, nitrogen and sulfur (CO, CO_x, NO_x and SO_x). Around 82% of global warming is due to the untreated post-combustion gases of fossil fuels during the generation of electricity. These are mainly the oxides of carbon, estimated to rise to 40 GT by the year 2030. The United States, Japan and Germany are some of the top countries emitting the highest amounts of carbon dioxide (CO_2), as seen in Table 4.1 [4].

Numerous solutions have been cited by scholars to overcome the increasing concentrations of greenhouse gases by designing effective CO_2 capture methods, such as adsorption, membrane separation, cryogenic separation and chemical absorption. Another efficient technique is the conversion of CO_2 into value-added products (methanol, methane, etc.) via photocatalysis under visible light conditions; however, to synthesize a photocatalyst that is visible-light-driven, having a narrow band gap is still challenging [6–8]. Therefore, it is concluded that the future of fossil fuels is still uncertain due to their limited resources and dangerous greenhouse effects caused by post-combustion reaction gases. Time is needed to produce some highly energy-efficient green fuels that not only fulfill the growing energy demands but also serve mitigators of the greenhouse effect. The influence of H_2 on the current energy infrastructure has been the center of research around the globe, and it is believed that H_2 will soon replace all existing fuels [9].

H_2 is predicted to be the most promising, energy-efficient and clean fuel. It can be converted into electricity efficiently via fuel cells with zero emissions of greenhouse gases and hazardous species, such as volatile organic compounds. Nowadays, H_2 has been used in many industrial applications like NH_3 production, CH_3OH

TABLE 4.1
Comparison between pollutant emissions in industrialized countries

State	Pollutants caused by combustion (10^3 Ton)				
	SO_2	N_xO_y	Suspended Particles	CO	CO_2
United States	23,200	20,300	8300	77,400	4,166,000
Japan	1314	1435	–	--	831,000
Germany	3200	3100	725	8650	666,000
UK	4670	1812	442	8891	517,000
France	3460	1847	278	5200	404,000
Italy	3205	1506	433	5487	322,000
Spain	3756	792	1521	3780	198,000
Holland	450	525	150	1450	130,000
Belgium	856	317	267	839	107,000

[5]

production plants and oil refineries [10]. The advantages of H_2 include its clean nature, abundance in the universe, the fact that it is easily storable and that it only produces H_2O during combustion. H_2 has established itself as a vital source of energy, and scholars are working for its commercialization due to its environmentally friendly nature. [9].

H_2 is a fundamental element comprising one proton and electron while being an abundant element in the universe. However, it is always found in a combined form with other elements. The high bonding rate of H_2 with O_2, C and N_2 explains its highly active nature [11]. Consequently, it must be extracted from other primary energy sources like coal, natural gas, water or other heavy hydrocarbons. Global statistics illustrate that a significant amount of H_2 is produced by the reforming of natural gas, i.e., 48% electrolysis of water gives 30%, 18% from burning petroleum products and 4% by coal. The significant contribution to the production of H_2 is from natural gas since there are vast reservoirs of CH_4 in deep-sea beds, especially in industrialized countries like the United States [12]. Natural gas has been named as the primary energy contributor since the early 1920s in Malaysia. In 2008 alone, approximately over 2.5 trillion m^3 of natural gas reserves were found in Malaysia, specifically in Sarawak (East Malaysia). Moreover, the natural gas reserves in Malaysia are the largest in Southeast Asia and the 12th largest in the world [13].

There are various methods for the production of H_2 such as steam reforming (SRM), partial oxidation (POM), dry reforming (DRM) and thermocatalytic decomposition of CH_4 (TCD). Among these methods, POM, SRM and DRM are cited as indirect methods of H_2 production. In these methods, CH_4 is treated with O_2, H_2O and CO_2 under a given set of reaction conditions to produce the synthesis gas, which is a mixture of H_2 and CO [14]. However, there is a common shortcoming in all these three processes in the form of emissions of the greenhouse gases CO_2 and CO. These gases not only play a significant role in global warming, but they are also very harmful when their mixed feedstock with H_2 is used in low-temperature fuel cells like proton exchange membrane fuel cells (PEMFC) [15]. Keeping greenhouse gas emissions and their economic issues in view, the interest of researchers has moved toward exploring a more optimal, greener process for H_2 production. Thermocatalytic decomposition (TCD) is a practical approach to thermally decomposed CH_4 into H_2 and solid carbon. This method is considered novel and eco-friendly as there is no emission of greenhouse gases during the reaction [16]. Another advantage of this method is that the TCD has produced high-quality carbons in the form of CNF, MWCNF and CNTs as by-products [17].

4.1 METHODS FOR HYDROGEN PRODUCTION

The recent methods of H_2 production are mainly based on steam reforming (SRM), dry reforming (DRM), and partial oxidation (POM) and thermocatalytic decomposition (TCD) of CH_4 and carbon-containing heavy feedstocks. As compared to other feedstocks, CH_4 is an ideal feedstock owing to its abundance in nature, not requiring any further purification process from sulfur and other impurities, as well as having the most substantial heat of combustion [18] and the highest carbon to H_2 ratio [19].

4.1.1 STEAM REFORMING OF CH$_4$ (SRM)

Steam reforming of CH$_4$ is one of the most ancient methods used for the synthesis gas production, which was first described by Motay and Marechal in 1868 [20]. The process involves the catalytic reaction of CH$_4$ and steam into H$_2$ and CO known as synthesis gas. This process includes two types of reaction termed as:

(1) Synthesis gas generation
(2) Reverse water-gas shift reaction [21]

The feedstock is mixed with steam over the metallic catalyst at temperatures of around 973–1273 K as it is highly endothermic. The overall reaction is shown in Equation 4.1 [22].

$$CH_4 + H_2O \rightarrow CO + 3H_2 \qquad \Delta H = 2062 \text{ kJ mol}^{-1} \tag{4.1}$$

Simultaneously, the second reaction taking place is a reverse water gas shift as in Equation 4.2, CO reacts again with steam to generate additional H$_2$ along with the CO$_2$.

$$CO + H_2O \rightarrow CO_2 + H_2 \qquad \Delta H = -41.2 \text{ kJ mol}^{-1} \tag{4.2}$$

Generally, the Ni-based catalyst is used for SRM to lower down the reaction temperature and increase the H$_2$ yield. Many different types of supports and promoters are also used to optimize the reaction. Table 4.2 shows the catalytic activity and stability

TABLE 4.2

Catalytic activity and stability of different catalysts having different supporters and reaction conditions in FBR for SRM

Metal	Support	W_C (g)	Preparation	T_R (K)	R_T (h)	Conversion CH$_4$ (%)	Ref
Ni	Mg AL/ CrFe$_3$O$_4$	–	I.M	773–973	11	85	[35]
Ru	Ni-Mg-Al	0.20	H.M	673–1073	–	93	[36]
Ni	SiO$_2$	0.05	I.M	973–1073	7	45	[23]
Ni	Fe/ Al$_2$O$_3$	0.05	S.G	823	15	82	[24]
Ni	Mo	0.10	Co.P	973	9	95	[37]
Ni	Al$_2$O$_3$/ SiO$_2$ Mg-Al	0.10–0.20	I.M	873	2	82	[21]
Ni	SiO$_2$	1	Co.P	1023	4	85	[38]
Ni	SiO$_2$	0.05	I.M	973	–	50	[38]
Ni	Pd-YSZ	–	I.M	923	–	96	[39]
Au	NiLa	0.10	I.M	973–1123	96	45	[40]
Ni	Al$_2$O$_3$/ Cao-ZrO$_2$	0.10	I.M	873	30	80	[41]
Ni	Al, Zr, Fe, Ce	0.10	I.M	723–1123	–	98	[42]
Cu	Co Al	0.02	Co.P	873–1073	–	98	[43]

of various Ni-based catalysts with different supports, as well as the effect of reaction conditions stated against each catalyst used [23, 24].

4.1.2 PARTIAL OXIDATION OF CH₄

Partial oxidation of CH_4 (POM) is a heterogeneous catalytic process, producing synthesis gas with an H_2/CO ratio of 2:1. Unlike SRM, which is highly endothermic and requires extensive heat, POM is mildly exothermic. Thus, POM can be performed with or without a catalyst, therefore it is named as partial catalytic oxidation and no partial catalytic oxidation, as in Equation 4.3 [25].

$$CH_4 + \%_O O_2 \rightarrow CO + 2H_2 \qquad \Delta H = -32 \text{ kJ mol}^{-1} \qquad (4.3)$$

The thermodynamic aspect of POM shows that there are several side reactions taking place in the process of CO, CO_X, H_2 and CNF production. Hence, H_2 purification units must be installed [26] which increases the operating cost of the whole process. Therefore, the industrial application of POM is still under consideration [27]. The catalytic activity and stability of various catalysts employed in the partial catalytic oxidation have been illustrated in Table 4.3.

TABLE 4.3
Catalytic activity and stability of Ni-based catalysts having different supporters and reaction conditions in FBR for POM

Metal	Support	W_C (g)	Preparation	T_R (K)	R_T (h)	Conversion CH_4 (%)	Ref
Ni	CeO_2-ZrO_2-Al_2O_3	0.15 g	Co.P	1073	25	90	[44]
Ni-Co	$CaAl_2O_4/Al_2O_3$	0.05 g	I.M	873–1073	6	80	[45]
Pd	$Al_2O_3/$ CeO_2-ZrO_2	0.50 g	I.M	1073–1173	24	95.3	[46]
Ni	ZrO_2 @ SiO_2	–	S.M	1023	50	98	[47]
Ni	CeO_2	0.30 g	Co.P	973	–	95	[48]
Ni	$MgO/Al_2O_3/$ Sorbacid	0.04 g	I.M	1073	–	95	[49]
Ni	$MgAl_2O_4$	0.02 g	Co.P	1073	50	47	[50]
Ni	CeO_2/ ZrO_2	0.30 g	Co.P	973	20	90	[51]
Pd	SiO_2	0.015 g	I.M	973	7	95	[27]
Ni	CeO_2	0.10 g	I.M	773–973	48	82	[52]
Ni	Ru	0.200 g	I.M	973	–	90	[53]
Ni	SiO_2	0.50 g	I.M	973	9	94	[25]
Ni	Al_2O_3	0.50 g	I.M	1073	50	98	[54]
Ni	Al_2O_3	0.15 g	I.M	973	–	50	[55]

4.1.3   DRY REFORMING OF CH_4 (DRM)

The DRM was first studied by Fischer and Tropsch in 1928 over Ni and Co-based catalysts, and since then, DRM has been continually investigated by numerous researchers. However, a breakthrough that will industrialize this method has not yet been achieved. DRM is an endothermic reaction that requires the operating temperature of 873–1273 K to attain a high equilibrium conversion of the reactants (CH_4 and CO_2) to products (H_2 and CO), as in Equation 4.4 [28–30].

$$CH_4 + CO_2 \rightarrow 2CO + 2H_2 \qquad \Delta H = 247.0 \text{ kJ mol}^{-1} \tag{4.4}$$

Dry reforming is also known as CO_2 reforming, and the primary product of this process is the synthesis gas produced by the reaction of hydrocarbons like CH_4 with CO_2. In ancient times, steam reforming was used to produce the synthesis gas. However, the increasing alarm over the greenhouse gas effect has replaced this process with dry reforming. Thus, greenhouse gases are consumed, and H_2 and CO are produced [31, 32]. Along with the syngas production, reverse water gas shift reaction also occurs during the reaction as given in Equation 4.5.

$$CO_2 + H_2 \rightarrow CO + 2H_2 \qquad \Delta H + -39.5 \text{ kJ mol}^{-1} \tag{4.5}$$

This reaction is highly endothermic and operates at a relatively higher temperature. Thus, to reduce the temperature, metallic catalysts are utilized. Extensive work has been done on the optimization of the process by synthesizing the best catalyst so that the reaction temperature can be lowered and the catalyst deactivation time can be increased as the carbon in the form of by-product covers the active sites of the catalyst. The carbon has a high mechanical strength that destroys the active sites of the catalyst, which results in the decreased surface area and, ultimately, the deactivation of the catalyst. Consequently, to satisfy the research gap, different promoters like ZrO_2, CeO_2 and MgO impregnated on the metallic catalyst are examined in order to increase the yield of syngas, while on the other hand the catalyst deactivation time can be improved [33, 34]. The detailed study of the DRM on different supports and promotes has been furnished in Table 4.4.

$$CH_4 \rightarrow C + 2H_2 \qquad \Delta H = 75.0 \text{ kJ mol}^{-1} \tag{4.6}$$

$$2CO \rightarrow C + CO_2 \qquad \Delta H = -171.0 \text{ kJ mol}^{-1} \tag{4.7}$$

The H_2 production processes mentioned above have been thoroughly explored, and a variety of results are reported. Apart from H_2 as the primary product, the other gaseous products include the oxides of carbon. Hence, the purification of H_2 becomes compulsory for its further application as feedstock. To date, exploring a novel green approach to synthesize H_2 by the reformation of natural gas is receiving a significant boost in the research field.

TABLE 4.4

Catalytic activity and stability of Ni-based catalysts having different supporters and reaction conditions in FBR for DRM

S.No	Metal	Support	W_C (g)	Preparation	T_R (K)	R_T (h)	Conversion CH_4 (%)	Conversion CO_2 (%)	Ref
1.	Ni	Al_2O_3, CeO_2, La_2O_3, ZrO_2	0.1	I.M	823	7	10.4	30.6	[31]
2.	Ni	Al_2O_3	0.20	S.G	1123	48	95.2	96.1	[56]
3.	Ni	ZrO_2, Ce, Ca	0.05	I.M	973	30	80	90	[57]
4.	Ni	CeO_2	0.20	I.M	773–973	25	65	–	[58]
5.	Ni	La, Mg AL Hydrotalcite	–	Co.P	823–1123	50	90	94	[59]
6.	Ni	$Zr-SiO_2$	0.25	I.M	673	–	2	2	[60]
7.	Ni	$MgO-ZrO_2$	0.1	I.M	1123	15	95	97	[61]
8.	Ni	MgO/ZrO_2	0.02	Co.P	823–1023	16	92	94	[62]
9.	Ni	$BaTiO_3$	0.150	S.G	1123	50	93	94	[63]
10.	Ni	MCM42	0.12	H.S	823–1123	30	70	–	[64]
11.	Ni	Ce/ SBA-15	0.020–0.1	I.M	773	12	80	83	[65]
12.	Ni	Mesoporous Silica	0.10	S.G	973	30	90	95	[66]
12.	Ni	ZrO_2/ Al_2O_3	0.6	I.M	973	–	63	57.3	[67]
14.	Ni	MgO	0.05	I.M	973	4	46	58	[68]
15.	Ni	MgO	0.30	Co.P	823–973	5	74	84	[69]

4.1.4 THERMOCATALYTIC DECOMPOSITION OF CH$_4$

Thermocatalytic decomposition (TCD) of CH$_4$ is a single-step process which produces pure H$_2$ and carbon as a by-product [70]. This method is known as an eco-friendly process, and it has been extensively studied by various researchers [71]. The main advantages of TCD include the greenhouse-free H$_2$ production that can be directly used in fuel cells and the production of CNF as a by-product. The reaction proceeds as shown in Equation 4.8 [10].

$$CH_{4\,(gas)} \rightarrow C_{(S)} + 2H_{2\,(gas)} \qquad \Delta H = 74.9 \text{ kJmol}^{-1} \qquad (4.8)$$

TCD is an endothermic reaction and occurs at high temperatures, i.e., 1473 K due to the highly stable tetrahedral structure of the CH$_4$ molecule supported by extremely stable C–H bonds with a bond energy of 434 kJmol^{-1}. Therefore, the applications of the catalyst are mandatory to provide a robust pathway to decrease the activation energy [72]. To improve the reaction kinetics of TCD, metal-based catalysts and carbonaceous catalysts were introduced by numerous researchers. These include transition metal having partially filled d-orbital, activated carbon and carbon black [15]. Over the last few years, countless efforts have been made in the development of the preparation method of a suitable catalyst to optimize TCD of CH$_4$ for CO$_X$-free H$_2$. Various monometallic catalysts with different supports have been reported [73]. Apart from the production of pure H$_2$, the invention of highly ordered carbon in the form of CNF, CNT, BWCNT and MWCNT has been reported by researchers using these catalysts [74]. The morphology of the deposited carbon is mainly dependent upon the metal loadings and reaction parameters. Although the general mechanism of TCD has been proposed to be similar for nearly all metallic materials, the catalytic stability and activity is influenced by catalysis synthesis techniques, pre-treatment of catalyst and, most importantly, the TCD parameters [71, 75]. Ni, Fe and Co are the commonly used transition metals impregnated on Al$_2$O$_3$, SiO$_2$, MgO and La$_2$O$_3$ support [76] as shown in Table 4.5. The comparison of all H$_2$ production methods as shown in Figure 4.1 elaborates that H$_2$ production techniques are classified as indirect and direct methods. SRM, DRM and POM come under the category of indirect methods, as CH$_4$ is treated with H$_2$O, CO$_2$ and O$_2$ under specified reaction parameters resulting in synthesis gas a mixture of H$_2$ and CO. However, there is a common shortcoming in all these processes in the form of greenhouse emissions (CO$_2$ and CO). These gases not only play a major role in global warming, but they also are very harmful when their mixed feedstock with H$_2$ is used in low-temperature fuel cells like proton exchange membrane fuel cells (PEMFC). Hence, TCD is a widely studied technique as it is a single-step process that converts CH$_4$ into CO$_X$-free H$_2$ and elemental carbon, which also has extensive industrial applications. Moreover, TCD occurs at relatively mild reaction conditions as compared to other methods. The only challenge of this method is to lower the operating temperature and coke formation during reactions.

TABLE 4.5

Catalytic activity and stability of metallic catalyst having different supporters and reaction conditions in FBR for TCD

Metal	Support	Wc (g)	Preparation	S.A m² g⁻¹	T_R (K)	R_T (h)	Conversion (%)	Carbon	Ref
10%Ni	Al₂O₃	0.05	I.M(c)	141	853	1	(g)21	–	[77]
25%Ni	2.0%MgO–Al₂O₃	0.025	I.M(c)	149	898	5	(g)05	–	[78]
55%Ni–10%Ce	MgO–Al₂O₃	0.025	I.M(c)	55.06	948	5	(g)12	(j)150	[79]
15%Fe–6%Ni	MgO	0.3	I.M(c)	100	973	3	(g)66	(h)16.26	[80]
30%Fe–15%Ce	Al₂O₃	–	I.M(c)	52.1	973	3	(i)62	–	[81]
50%Ni-5%Pd	Al₂O₃	0.05	I.M(c)	72.75	1023	10	(g)65	(i)360	[82]
11.32%Ni–12.11%Co	La₂O₃	0.3	I.M(c)	63.32	973	5	(g)80	(i)90	[83]
10%Ni–9.5%Mo	Al₂O₃	0.5	I.M(c)	101.2	973	8	(i)80	(i)612	[84]
30%Mn-30%Fe	Al₂O₃	0.3	I.M(c)	123.8	973	3	(g)13	–	[85]
50%Ni–10%Fe–10%Cu	Al₂O₃	0.05	I.M(c)	59.7	1023	10	(g)81	–	[86]
Fe–Mo–Cu	MgO	0.5	I.M(c)	67	1173	–	–	(i)14.2	[87]
50%Ni-5%Cu–5%Zn	MCM–22	1	I.M(c)	19	1023	–	(g)84	(j)900	[88]

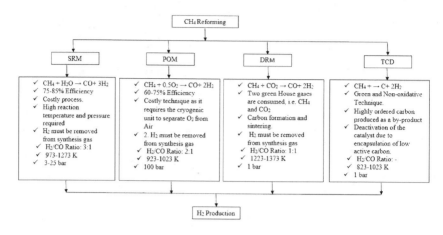

FIGURE 4.1   Comparison of H_2 production techniques.

4.2   CONCLUSION

The industrialization of this world and the alarming global warming situation has opened up new horizons of research in exploring some novel, green and efficient fuels. H_2 has been termed as the fuel of future due to its clean and abundant nature, but it must be separated from various feedstock (mainly natural gas) as it is not found independently in the earth's crust. DRM, SRM and POM are the conventional techniques for H_2 production, but the production of carbon oxides as a by-product has limited their practical applications. Interestingly, TCD is a green technique in which methane is decomposed into elemental carbon and H_2 in a single step. Moreover, the highly ordered carbon in the form of MWCNT, CNF and CNT have plenty of industrial applications. Among all current methods, TCD is the greenest and most practical approach of producing pure hydrogen through hydrocarbon reforming. Nonetheless, the grand challenge of TCD technology is to lower the reaction temperature and coke formation during the reaction.

ACKNOWLEDGMENT

The authors would like to thank Universiti Teknologi PETRONAS for providing financial support for this study.

NOMENCLATURE

T_R: Reaction Temperature
R_T: Reaction Time
$X_{Methane}$:   Conversion of Methane
X_{CO_2}:   Conversion of CO_2
FBR: Fixed Bed Reactor
W_c: Weight of Catalyst

REFERENCES

1. M. Roser, H. Ritchie, and E. Ortiz-Ospina, *"World population growth"*, Our World in Data, 2013: revised in May 2019, accessed in 21/08/2019, URL:< https://ourworldinda ta.org/world-population-growth?utm_content=bufferd03f5&utm_medium=social&ut m_source=twitter.com&utm_campaign=buffer >.
2. U.S. Energy Information Administration, *"International Energy Outlook 2017"*, eia. gov, September 2017, accessed in 21/08/2019, URL:< https://www.eia.gov/outlooks/a rchive/ieo17/ >.
3. S. Shafiee, and E. Topal, "When will fossil fuel reserves be diminished?", *Energy Policy*, vol. 37, pp. 181–189, 2009.
4. Z. Liu, *China's Carbon Emissions Report 2016*. Cambridge, MA: Report for Harvard Belfer Center for Science and International Affairs, 2016.
5. G. Supran, and N. Oreskes, "Assessing ExxonMobil's climate change communications (1977–2014)", *Environmental Research Letters*, vol. 12, p.084019, 2017.
6. M. Eddaoudi, M. J. Zaworotko, P. Nugent, S. Burd, Y. Belmabkhout, and O. Shekhah, "Metal-organic materials (MOMs) for CO_2 adsorption and methods of using MOMs", *Google Patents*, 2017.
7. J. Kothandaraman, A. Goeppert, M. Czaun, G. A. Olah, and G. K. S. Prakash, "Conversion of CO_2 from air into methanol using a polyamine and a homogeneous ruthenium catalyst", *Journal of the American Chemical Society*, vol. 138, pp. 778–781, 2016.
8. M. Canghai, and W. J. Koros, "Composite hollow fiber membranes useful for CO_2 removal from natural gas", *Google Patents*, 2017.
9. I. P. Jain, "Hydrogen the fuel for 21st century", *International Journal of Hydrogen Energy*, vol. 34, pp. 7368–7378, 2009.
10. H. F. Abbas, and W. M. A. Wan Daud, "Hydrogen production by methane decomposition: A review", *International Journal of Hydrogen Energy*, vol. 35, pp. 1160–1190, 2010.
11. M. Momirlan, and T. N. Veziroglu, "The properties of hydrogen as fuel tomorrow in sustainable energy system for a cleaner planet", *International Journal of Hydrogen Energy*, vol. 30, pp. 795–802, 2005.
12. M. Balat, and M. Balat, "Political, economic and environmental impacts of biomass-based hydrogen", *International Journal of Hydrogen Energy*, vol. 34, pp. 3589–3603, 2009.
13. H. C. Ong, T. M. I. Mahlia, and H. H. Masjuki, "A review on energy scenario and sustainable energy in Malaysia", *Renewable and Sustainable Energy Reviews*, vol. 15, pp. 639–647, 2011.
14. Z. Khan, S. Yusup, M. M. Ahmad, V. S. Chok, Y. Uemura, and K. M. Sabil, "Review on hydrogen production technologies in Malaysia", *International Journal of Engineering and Technology*, vol. 10, no. 2, 2010.
15. Y. Li, D. Li, and G. Wang, "Methane decomposition to COx-free hydrogen and nano-carbon material on group 8–10 base metal catalysts: A review", *Catalysis Today*, vol. 162, pp. 1–48, 2011.
16. M. A. Ermakova, D. Y. Ermakov, and G. G. Kuvshinov, "Effective catalysts for direct cracking of methane to produce hydrogen and filamentous carbon", *Applied Catalysis A: General*, vol. 201, pp. 61–70, 2000.
17. I. Suelves, M. J. Lázaro, R. Moliner, Y. Echegoyen, and J. M. Palacios, "Characterization of NiAl and NiCuAl catalysts prepared by different methods for hydrogen production by thermo catalytic decomposition of methane", *Catalysis Today*, vol. 116, pp. 271–280, 2006.
18. J. R. Rostrup-Nielsen, "Catalysis and large-scale conversion of natural gas", *Catalysis Today*, vol. 21, pp. 257–267, 1994.

19. J. H. Lunsford, "Catalytic conversion of methane to more useful chemicals and fuels: A challenge for the 21st century", *Catalysis Today*, vol. 63, pp. 165–174, 2000.

20. A. M. Adris, B. B. Pruden, C. J. Lim, and J. R. Grace, "On the reported attempts to radically improve the performance of the steam methane reforming reactor", *The Canadian Journal of Chemical Engineering*, vol. 74, pp. 177–186, 1996.

21. M. A. Nieva, M. M. Villaverde, A. Monzón, T. F. Garetto, and A. J. Marchi, "Steam-methane reforming at low temperature on nickel-based catalysts", *Chemical Engineering Journal*, vol. 235, pp. 158–166, 2014.

22. L. Barelli, G. Bidini, F. Gallorini, and S. Servili, "Hydrogen production through sorption-enhanced steam methane reforming and membrane technology: A review", *Energy*, vol. 33, pp. 554–570, 2008.

23. Y. Zhang, W. Wang, Z. Wang, X. Zhou, Z. Wang, and C.-J. Liu, "Steam reforming of methane over Ni/SiO_2 catalyst with enhanced coke resistance at low steam to methane ratio", *Catalysis Today*, vol. 256, pp. 130–136, 2015.

24. S. Park, Y. Bang, S. J. Han, J. Yoo, J. H. Song, J. C. Song, *et al.*, "Hydrogen production by steam reforming of liquefied natural gas (LNG) over mesoporous nickel–iron–alumina catalyst", *International Journal of Hydrogen Energy*, vol. 40, pp. 5869–5877, 2015.

25. C. Ding, J. Wang, G. Ai, S. Liu, P. Liu, K. Zhang, *et al.*, "Partial oxidation of methane over silica supported Ni nanoparticles with size control by alkanol solvent", *Fuel*, vol. 175, pp. 1–12, 2016.

26. W.-H. Chen, and C.-T. Shen, "Partial oxidation of methanol over a Pt/Al_2O_3 catalyst enhanced by sprays", *Energy*, vol. 106, pp. 1–12, 2016.

27. L. M. T. S. Rodrigues, R. B. Silva, M. G. C. Rocha, P. Bargiela, F. B. Noronha, and S. T. Brandão, "Partial oxidation of methane on Ni and Pd catalysts: Influence of active phase and CeO_2 modification", *Catalysis Today*, vol. 197, pp. 137–143, 2012.

28. M. Usman, W. M. A. Wan Daud, and H. F. Abbas, "Dry reforming of methane: Influence of process parameters—A review", *Renewable and Sustainable Energy Reviews*, vol. 45, pp. 710–744, 2015.

29. M. B. Bahari, N. H. H. Phuc, B. Abdullah, F. Alenazey, and D.-V. N. Vo, "Ethanol dry reforming for syngas production over Ce-promoted Ni/Al_2O_3 catalyst", *Journal of Environmental Chemical Engineering*, vol. 4, 4830–4838, 2016.

30. O. Omoregbe, H. T. Danh, S. Z. Abidin, H. D. Setiabudi, B. Abdullah, K. B. Vu, and D. N. Vo, "Influence of lanthanide promoters on Ni/SBA-15 catalysts for syngas production by methane dry reforming", *Procedia Engineering*, vol. 148, pp. 1388–1395, 2016.

31. J. Guo, H. Lou, H. Zhao, D. Chai, and X. Zheng, "Dry reforming of methane over nickel catalysts supported on magnesium aluminate spinels", *Applied Catalysis A: General*, vol. 273, pp. 75–82, 2004.

32. F. Fayaz, H. T. Danh, C. Nguyen-Huy, K. B. Vu, B. Abdullah, and D.-V. N. Vo, "Promotional effect of Ce-dopant on Al_2O_3-supported Co catalysts for syngas production via CO_2 reforming of ethanol", *Procedia Engineering*, vol. 148, pp. 646–653, 2016.

33. D. Pakhare, and J. Spivey, "A review of dry (CO_2) reforming of methane over noble metal catalysts", *Chemical Society Reviews*, vol. 43, pp. 7813–7837, 2014.

34. K. Selvarajah, N. H. H. Phuc, B. Abdullah, F. Alenazey, and D.-V. N. Vo, "Syngas production from methane dry reforming over Ni/Al_2O_3 catalyst", *Research on Chemical Intermediates*, vol. 42, pp. 269–288, 2016.

35. A. Vita, C. Italiano, C. Fabiano, M. Laganà, and L. Pino, "Influence of Ce-precursor and fuel on structure and catalytic activity of combustion synthesized Ni/CeO_2 catalysts for biogas oxidative steam reforming", *Materials Chemistry and Physics*, vol. 163, pp. 337–347, 2015.

36. M. Nawfal, C. Gennequin, M. Labaki, B. Nsouli, A. Aboukaïs, and E. Abi-Aad, "Hydrogen production by methane steam reforming over Ru supported on Ni–Mg–Al mixed oxides prepared via hydrotalcite route", *International Journal of Hydrogen Energy*, vol. 40, pp. 1269–1277, 2015.
37. S. S. Maluf, and E. M. Assaf, "Ni catalysts with Mo promoter for methane steam reforming", *Fuel*, vol. 88, pp. 1547–1553, 2009.
38. X. Guo, Y. Sun, Y. Yu, X. Zhu, and C.-j. Liu, "Carbon formation and steam reforming of methane on silica supported nickel catalysts", *Catalysis Communications*, vol. 19, pp. 61–65, 2012.
39. S. M. Lee, and S. C. Hong, "Effect of palladium addition on catalytic activity in steam methane reforming over Ni-YSZ porous membrane", *International Journal of Hydrogen Energy*, vol. 39, pp. 21037–21043, 2014.
40. S. Palma, L. F. Bobadilla, A. Corrales, S. Ivanova, F. Romero-Sarria, M. A. Centeno, and J. A. Odriozola, "Effect of gold on a NiLaO$_3$ perovskite catalyst for methane steam reforming", *Applied Catalysis B: Environmental*, vol. 144, pp. 846–854, 2014.
41. B. Bej, N. C. Pradhan, and S. Neogi, "Production of hydrogen by steam reforming of ethanol over alumina supported nano-NiO/SiO$_2$ catalyst", *Catalysis Today*, vol. 237, pp. 80–88, 2014.
42. M.-W. Lim, S.-T. Yong, and S.-P. Chai, "Combustion-synthesized nickel-based catalysts for the production of hydrogen from steam reforming of methane", *Energy Procedia*, vol. 61, pp. 910–913, 2014.
43. D. Homsi, S. Aouad, C. Gennequin, J. E. El Nakat, A. Aboukaïs, and E. Abi-Aad, "The effect of copper content on the reactivity of Cu/Co$_6$Al$_2$ solids in the catalytic steam reforming of methane reaction", *Comptes Rendus Chimie*, vol. 17, pp. 454–458, 2014.
44. M. Dajiang, C. Yaoqiang, Z. Junbo, W. Zhenling, M. Di, and G. Maochu, "Catalytic partial oxidation of methane over Ni/CeO$_2$–ZrO$_2$-Al$_2$O$_3$", *Journal of Rare Earths*, vol. 25, pp. 311–315, 2007.
45. A. C. Koh, L. Chen, W. K. Keeleong, B. F. Johnson, T. Khimyak, and J. Lin, "Hydrogen or synthesis gas production via the partial oxidation of methane over supported nickel–cobalt catalysts", *International Journal of Hydrogen Energy*, vol. 32, pp. 725–730, 2007.
46. S. Fangli, S. Meiqing, F. Yanan, W. Jun, and W. Duan, "Influence of supports on catalytic performance and carbon deposition of palladium catalyst for methane partial oxidation", *Journal of Rare Earths*, vol. 25, pp. 316–320, 2007.
47. C. Ding, G. Ai, K. Zhang, Q. Yuan, Y. Han, X. Ma, *et al.*, "Coking resistant Ni/ZrO$_2$@SiO$_2$ catalyst for the partial oxidation of methane to synthesis gas", *International Journal of Hydrogen Energy*, vol. 40, pp. 6835–6843, 2015.
48. T. Zhu, and M. Flytzani-Stephanopoulos, "Catalytic partial oxidation of methane to synthesis gas over Ni–CeO$_2$", *Applied Catalysis A: General*, vol. 208, pp. 403–417, 2001.
49. H. Özdemir, M. A. Faruk Öksüzömer, and M. Ali Gürkaynak, "Preparation and characterization of Ni based catalysts for the catalytic partial oxidation of methane: Effect of support basicity on H$_2$/CO ratio and carbon deposition", *International Journal of Hydrogen Energy*, vol. 35, pp. 12147–12160, 2010.
50. H. Özdemir, M. A. F. Öksüzömer, and M. A. Gürkaynak, "Effect of the calcination temperature on Ni/MgAl$_2$O$_4$ catalyst structure and catalytic properties for partial oxidation of methane", *Fuel*, vol. 116, pp. 63–70, 2014.
51. S. Xu, and X. Wang, "Highly active and coking resistant Ni/CeO$_2$–ZrO$_2$ catalyst for partial oxidation of methane", *Fuel*, vol. 84, pp. 563–567, 2005.
52. L. Jalowieckiduhamel, H. Zarrou, and A. D'Huysser, "Hydrogen production at low temperature from methane on cerium and nickel based mixed oxides", *International Journal of Hydrogen Energy*, vol. 33, pp. 5527–5534, 2008.

53. J. A. Velasco, C. Fernandez, L. Lopez, S. Cabrera, M. Boutonnet, and S. Järås, "Catalytic partial oxidation of methane over nickel and ruthenium based catalysts under low O_2/CH_4 ratios and with addition of steam", *Fuel*, vol. 153, pp. 192–201, 2015.

54. C. Ding, J. Wang, Y. Jia, G. Ai, S. Liu, P. Liu, *et al.*, "Anti-coking of Yb-promoted Ni/Al_2O_3 catalyst in partial oxidation of methane", *International Journal of Hydrogen Energy*, vol. 41, pp.10707–10718, 2016.

55. T. de Freitas Silva, C. G. M. Reis, A. F. Lucrédio, E. M. Assaf, and J. M. Assaf, "Hydrogen production from oxidative reforming of methane on Ni/γ-Al_2O_3 catalysts: Effect of support promotion with La, La–Ce and La–Zr", *Fuel Processing Technology*, vol. 127, pp. 97–104, 2014.

56. Z. Hao, Q. Zhu, Z. Jiang, B. Hou, and H. Li, "Characterization of aerogel Ni/Al_2O_3 catalysts and investigation on their stability for CH_4-CO_2 reforming in a fluidized bed", *Fuel Processing Technology*, vol. 90, pp. 113–121, 2009.

57. J.-S. Chang, D.-Y. Hong, X. Li, and S.-E. Park, "Thermogravimetric analyses and catalytic behaviors of zirconia-supported nickel catalysts for carbon dioxide reforming of methane", *Catalysis Today*, vol. 115, pp. 186–190, 2006.

58. M. Rezaei, S. M. Alavi, S. Sahebdelfar, and Z.-F. Yan, "A highly stable catalyst in methane reforming with carbon dioxide", *Scripta Materialia*, vol. 61, pp. 173–176, 2009.

59. H. Liu, D. Wierzbicki, R. Debek, M. Motak, T. Grzybek, P. Da Costa, and M. E. Gálvez, "La-promoted Ni-hydrotalcite-derived catalysts for dry reforming of methane at low temperatures", *Fuel*, vol. 182, pp. 8–16, 2016.

60. L. Yao, J. Shi, H. Xu, W. Shen, and C. Hu, "Low-temperature CO_2 reforming of methane on Zr-promoted Ni/SiO_2 catalyst", *Fuel Processing Technology*, vol. 144, pp. 1–7, 2016.

61. J. Titus, T. Roussière, G. Wasserschaff, S. Schunk, A. Milanov, E. Schwab, *et al.*, "Dry reforming of methane with carbon dioxide over NiO–MgO–ZrO_2", *Catalysis Today*, vol. 270, pp.68–75, 2015.

62. B. M. Nagaraja, D. A. Bulushev, S. Beloshapkin, and J. R. H. Ross, "The effect of potassium on the activity and stability of Ni–MgO–ZrO_2 catalysts for the dry reforming of methane to give synthesis gas", *Catalysis Today*, vol. 178, pp. 132–136, 2011.

63. L. Xiancai, L. Min, L. Zhihua, and H. Fei, "Studies on nickel-based catalysts for carbon dioxide reforming of methane", *Applied Catalysis A: General*, vol. 290, pp. 81–86, 2005.

64. D. Liu, R. Lau, A. Borgna, and Y. Yang, "Carbon dioxide reforming of methane to synthesis gas over Ni-MCM-41 catalysts", *Applied Catalysis A: General*, vol. 358, pp. 110–118, 2009.

65. M. N. Kaydouh, N. El Hassan, A. Davidson, S. Casale, H. El Zakhem, and P. Massiani, "Highly active and stable Ni/SBA-15 catalysts prepared by a "two solvents" method for dry reforming of methane", *Microporous and Mesoporous Materials*, vol. 220, pp. 99–109, 2016.

66. F. Huang, R. Wang, C. Yang, H. Driss, W. Chu, and H. Zhang, "Catalytic performances of Ni/mesoporous SiO_2 catalysts for dry reforming of methane to hydrogen", *Journal of Energy Chemistry*, vol. 25(4), pp.709–719, 2016.

67. S. Therdthianwong, C. Siangchin, and A. Therdthianwong, "Improvement of coke resistance of Ni/Al_2O_3 catalyst in CH_4/CO_2 reforming by ZrO_2 addition", *Fuel Processing Technology*, vol. 89, pp. 160–168, 2008.

68. W. Hua, L. Jin, X. He, J. Liu, and H. Hu, "Preparation of Ni/MgO catalyst for CO_2 reforming of methane by dielectric-barrier discharge plasma", *Catalysis Communications*, vol. 11, pp. 968–972, 2010.

69. R. Zanganeh, M. Rezaei, and A. Zamaniyan, "Dry reforming of methane to synthesis gas on NiO–MgO nanocrystalline solid solution catalysts", *International Journal of Hydrogen Energy*, vol. 38, pp. 3012–3018, 2013.

70. A. Awad, N. Masiran, M. A. Salam, D.-V. N. Vo, and B. Abdullah, "Non-oxidative decomposition of methane/methanol mixture over mesoporous Ni-Cu/Al$_2$O$_3$ Co-doped catalysts", *International Journal of Hydrogen Energy*, vol. 44(37), pp.20889–20899, 2018.

71. I. Suelves, M. Lázaro, R. Moliner, B. Corbella, and J. Palacios, "Hydrogen production by thermo catalytic decomposition of methane on Ni-based catalysts: Influence of operating conditions on catalyst deactivation and carbon characteristics", *International Journal of Hydrogen Energy*, vol. 30, pp. 1555–1567, 2005.

72. U. P. M. Ashik, W. M. A. Wan Daud, and J.-i. Hayashi, "A review on methane transformation to hydrogen and nanocarbon: Relevance of catalyst characteristics and experimental parameters on yield", *Renewable and Sustainable Energy Reviews*, vol. 76, pp. 743–767, 2017.

73. L. Zhou, L. R. Enakonda, M. Harb, Y. Saih, A. Aguilar-Tapia, S. Ould-Chikh, *et al.*, "Fe catalysts for methane decomposition to produce hydrogen and carbon nano materials", *Applied Catalysis B: Environmental*, vol. 208, pp. 44–59, 2017.

74. A. A. Ibrahim, A. H. Fakeeha, A. S. Al-Fatesh, A. E. Abasaeed, and W. U. Khan, "Methane decomposition over iron catalyst for hydrogen production", *International Journal of Hydrogen Energy*, vol. 40, pp. 7593–7600, 2015.

75. A. H. Fakeehaa, A. A. Ibrahima, A. S. Al. Fatesha, W. U. Khana, Y. A. Mohammeda, A. E. Abasaeeda, *et al.*, "Fe supported alumina catalyst for methane decomposition: Effect of Co coupling", *International Journal of Sustainable Water & Environmental Systems*, vol. 8, pp.00–00, 2016.

76. U. P. M. Ashik, W. M. A. Wan Daud, and H. F. Abbas, "Production of greenhouse gas free hydrogen by thermocatalytic decomposition of methane–A review", *Renewable and Sustainable Energy Reviews*, vol. 44, pp. 221–256, 2015.

77. J. F. Pola, M. A. Valenzuela, I. A. Córdova, and J. Wang, "Hydrogen production via methane decomposition using Ni and Ni-Cu catalysts supported on MgO, Al$_2$O$_3$ and MgAl$_2$O$_4$", *MRS Online Proceedings Library Archive*, vol. 1279, pp. 81–93, 2010.

78. A. Rastegarpanah, F. Meshkani, and M. Rezaei, "COx-free hydrogen and carbon nanofibers production by thermocatalytic decomposition of methane over mesoporous MgO· Al$_2$O$_3$ nanopowder-supported nickel catalysts", *Fuel Processing Technology*, vol. 167, pp. 250–262, 2017.

79. A. Rastegarpanah, F. Meshkani, and M. Rezaei, "Thermocatalytic decomposition of methane over mesoporous nanocrystalline promoted Ni/MgO Al$_2$O$_3$ catalysts", *International Journal of Hydrogen Energy*, vol. 42, pp. 16476–16488, 2017.

80. A.-S. Al-Fatesh, S. Barama, A.-A. Ibrahim, A. Barama, W.-U. Khan, and A. Fakeeha, "Study of methane decomposition on Fe/MgO-based catalyst modified by Ni, Co, and Mn additives", *Chemical Engineering Communications*, vol. 204, pp. 739–749, 2017.

81. A. S. Al-Fatesh, A. Amin, A. A. Ibrahim, W. U. Khan, M. A. Soliman, R. L. AL-Otaibi, and A. Fakeeha, "Effect of ce and Co addition to Fe/Al$_2$O$_3$ for catalytic methane decomposition", *Catalysts*, vol. 6, p. 40, 2016.

82. N. Bayat, M. Rezaei, and F. Meshkani, "Hydrogen and carbon nanofibers synthesis by methane decomposition over Ni–Pd/Al$_2$O$_3$ catalyst", *International Journal of Hydrogen Energy*, vol. 41, pp. 5494–5503, 2016.

83. W. U. Khan, A. H. Fakeeha, A. S. Al-Fatesh, A. A. Ibrahim, and A. E. Abasaeed, "La$_2$O$_3$ supported bimetallic catalysts for the production of hydrogen and carbon nanomaterials from methane", *International Journal of Hydrogen Energy*, 41(2), pp.976–983, 2015.

84. A. E. Awadallah, A. A. Aboul-Enein, and A. K. Aboul-Gheit, "Various nickel doping in commercial Ni–Mo/Al$_2$O$_3$ as catalysts for natural gas decomposition to COx-free hydrogen production", *Renewable Energy*, vol. 57, pp. 671–678, 2013.

85. A. H. Fakeeha, A. S. Al-Fatesh, B. Chowdhury, A. A. Ibrahim, W. U. Khan, S. Hassan, *et al.*, "Bi-metallic catalysts of mesoporous Al_2O_3 supported on Fe, Ni and Mn for methane decomposition: Effect of activation temperature", *Chinese Journal of Chemical Engineering*, vol. 26, 1904–1911, 2018.

86. N. Bayat, F. Meshkani, and M. Rezaei, "Thermocatalytic decomposition of methane to COx-free hydrogen and carbon over Ni–Fe–Cu/Al_2O_3 catalysts", *International Journal of Hydrogen Energy*, vol. 41, pp. 13039–13049, 2016.

87. A. E. Awadallah, A. A. Aboul-Enein, M. A. Azab, and Y. K. Abdel-Monem, "Influence of Mo or Cu doping in Fe/MgO catalyst for synthesis of single-walled carbon nanotubes by catalytic chemical vapor deposition of methane", *Fullerenes, Nanotubes and Carbon Nanostructures*, vol. 25, pp. 256–264, 2017.

88. S. K. Saraswat, and K. K. Pant, "Ni–Cu–Zn/MCM-22 catalysts for simultaneous production of hydrogen and multiwall carbon nanotubes via thermo-catalytic decomposition of methane", *International Journal of Hydrogen Energy*, vol. 36, pp. 13352–13360, 2011.

5 Eco-Friendly Sustainable Stabilization of Dredged Soft Clay Using Low-Carbone Recycled Additives

Mohammed Ali Mohammed Al-Bared, Aminaton Marto, and Indra Sati Hamonangan Harahap

CONTENTS

5.0   INTRODUCTION

Rapid urbanization and the increase of population in developed cities around the world require more suitable land for new construction. This development has resulted in the shortage of naturally available resources. Additionally, fast development is taking place in coastal areas that are usually covered by soft soils, particularly in Malaysia [1]. This encourages engineers and practitioners to seek solutions for the problems related to soft soils. The movement and settlement of soft soil induce distress to the superstructures built on them [2]. It was reported in several cities around the world that the influence of soft soils has caused damage to occur in many infrastructural projects. Repairing the damages incurred was reported to have cost several millions of dollars [3, 4]. Nowadays, researchers are looking for a stabilization method that can alter the properties of the soft soil to stand the loads imposed by the structures. This method is believed to be more economical, environmentally friendly and sustainable.

Soft soil is distributed all over the globe and always causes several problems to the structures and buildings constructed on top. One of the problems associated with soft soil is its swelling potential. The swelling potential in soft soil is due to the cation adsorption for the equilibrium of the charge [5]. Soft expansive soil is susceptible to changes in moisture content. The soil swells when water content is gained and shrinks when the water is dried. This process causes cracking to appear on the surface of soft soil that affects the stability of the structures [6]. Soft soils are always found in a moisture content that is almost equal or higher than their liquid limit. They usually exist in areas of high water content that results in their weak shear strength [7]. Soft soils can be characterized as soils that have very low permeability, high compressibility and low shear strength of approximately less than 40 kPa [1]. They are easily molded using finger pressure and unable to stand their self-weight.

Structures that are light or moderately light and encounter soft soil as a foundation soil are very common. The cost involved for deep foundation to bypass the weak layers or the entire strata of weak soil is relatively high. In this case, stabilizing the soft soil to withstand the loads imposed by those structures has become a necessity. The properties of the soft soils can be improved and the total construction cost could be reduced [8]. Soft soil stabilization is an alteration to the physical and chemical properties of soft soils to meet the engineering requirements. Many attempts have been made to stabilize soft soils using various methods such as chemical, hydrological, electrical or biological methods. Chemical stabilization using chemical additives (e.g., [9–13]) is the most used in soil stabilization, especially the one using cement [14]. However, soil improvement using cement is relatively expensive, and the emission of CO_2 during the production of cement pollutes the atmosphere and results in several environmental problems. Besides, soil improvement can be achieved by traditional compaction methods (e.g., [15–18]), or by using biological bacteria (e.g., [19–21]).

The high cost involved and the requirement of using advanced instrumentation in the stabilization process using chemical additives are the main reasons for seeking alternatives [22]. Besides, chemical additives are harmful, poisonous and non-environmentally friendly. Recent studies showed that the adoption of chemical-based additives resulted in underground water table contamination. Therefore, utilizing waste materials in the stabilizing of soft soil is the current practice worldwide in order to achieve cleaner production and green technology. Stabilization using waste is the current trend in this area of research [23–25]. They are being used either alone or together with other additives that act as a binder between the waste materials and the soil particles. Soil treatment using waste materials would contribute in reducing the amount of waste accumulated every day. This research is using ceramic tile waste to treat dredged marine soil. The waste of ceramic tiles is increasing every year, and the amount of waste generated at the factory and in construction sites is massive. It is estimated that the waste in ceramic tile factories during production is between 7–30% [26, 27]. This waste is generally produced in a slurry form that eventually settles at a nearby area. However, it affects the fertility of the soil, consumes large spaces and damages the plantations at the accumulation area [28].

This study aims to introduce an eco-friendly product that can stabilize the problematic soft clay without the necessity of using chemical additives. This research evaluates the effect and suitability of recycled, crushed and powdered ceramic tiles (RCT), to improve the properties of dredged marine soil. Four different sizes of RCT are used in order to evaluate the size effect in stabilizing the dredged clay. Laboratory tests included Atterberg limits, compaction, specific gravity and unconfined compressive strength (UCS) tests. All tests were performed on untreated dredged clay and dredged clay-RCT-treated samples with the four sizes of RCT.

5.1   MATERIALS USED

The materials used in this research were dredged marine clay and recycled, powdered and crushed ceramic tiles. Both materials are considered as waste materials due to their impact on the surrounding environment. In this study, both materials were utilized in order to enhance the properties of the soft clay.

5.1.1   DREDGED MARINE CLAY

Dredged clay was excavated from the construction sites and transported to the disposal sites. The clay then becomes a waste material that consumes large spaces at landfills and illegal dumping sites. It is excavated as unsuitable material because it does not meet engineering requirements. It is usually replaced by stiff soil that is able to stand the construction loads. Besides, it is considered problematic due to its weak behavior, high settlement and plasticity and low shear strength. The dredged clay used in this study was obtained from a disposal site in Johor, Malaysia that was replaced at the original site by laterite soil. It was transported to the laboratory, air-dried, and ground into 2 mm particles. All basic and mechanical tests were carried out in order to assess the properties of the dredged clay. Figure 5.1 shows the dredged clay before and after grinding.

FIGURE 5.1   Dredged clay before (left) and after (right) grinding.

5.1.2   RECYCLED POWDERED TILES

The ceramic tile waste, either at the factory or in the construction sites, is produced in high quantities. The waste in its powder and solid forms affects the fertility of the soil and consumes large spaces. It is aimed to utilize this waste by treating the dredged marine clay. The tiles were collected from construction sites and transported to the laboratory after cleaning the unnecessary materials on its surface. The tiles were prepared for testing in two steps: first, they were crushed by using a hammer, and second, by using an electrical crushing machine. The crushing process resulted in producing tiles of 5 mm size. The Los Angeles abrasion machine was used to produce a mixture that has a size ranging from 2 mm to 0.063 mm. The crushed tiles were sieved using a mechanical shaker and four different sizes were obtained. Those tiles were grouped into two sizes: fine and coarse. The fine size contained 0.063 mm, 0.15 mm and 0.3 mm sizes, while the coarse size contained 1.18 mm size. Those two sizes of ceramic tiles were used to test their suitability and effect in the physical and engineering properties of the dredged marine clay. Figure 5.2 shows the four sizes of the RCT utilized in this study.

FIGURE 5.2   Fine size of recycled tiles (a, b and c) and coarse size of recycled tiles (d).

5.2 TESTING PROGRAM

5.2.1 SAMPLE PREPARATION FOR ATTERBERG LIMIT TESTING

The Atterberg limit tests were conducted in accordance with the BS 1377: Part 2: 1990 [29] for both untreated dredged clay and dredged clay-RCT-treated samples. The clay was oven-dried and sieved using a 0.425 mm sieve mesh. Then it was mixed with RCT and a specific amount of distilled water. The mixture of the clay RCT was kept inside air-tight plastic bags for 24 hours in order to ensure proper distribution of moisture content within the mixture. Finally, the mix was used for both liquid limit and plastic limit determination. This procedure was repeated for all the mix designs of dredged clay-RCT using four percentages of RCT (10, 20, 30 and 40%). Besides, three different sizes of RCT (0.063, 0.15 and 0.3 mm) were used for the evaluation of the Atterberg limit tests.

5.2.2 SAMPLE PREPARATION FOR COMPACTION TESTS

The compaction test was performed on untreated dredged clay and clay-RCT mixture of four different percentages (10, 20, 30 and 40%) and sizes (0.063, 0.15, 0.30, 1.18 mm). The soil was first sieved using a 2 mm mesh, oven-dried to ensure zero moisture content and then approximately 180 kg of soil was used for the testing. The soil was mixed with the RCT and water and kept inside air-tight plastic bags for 24 hours, as shown in Figure 5.3. The testing for all samples was in accordance with BS 1377: Part 4: 1990 [30]. For all sizes and percentages, the maximum dry density (MDD) and optimum moisture content (OMC) were evaluated.

5.2.3 SAMPLING OF UNCONFINED COMPRESSIVE STRENGTH (UCS)

The sampling of the UCS was carried out by using the MDD and OMC values obtained from the compaction tests for the various mix designs and sizes of the clay-RCT. Soil was oven-dried, sieved and mixed with RCT and distilled water. The proportions of the different sizes of RCT were determined based on the dry weight of the dredged clay. The clay-RCT mixture was mixed by hand and palette knives until homogeneity was observed. Then the mixture was molded using a

FIGURE 5.3 Mixture of clay-RCT (left), the compaction mold (right).

stainless steel mold designed for sampling UCS samples. The mold had a dimension of 80 mm height and an inner diameter of 38 mm. Prior to placing the mixture inside the mold, it was lubricated in order to prevent the samples from being damaged at the later stage when it was extruded from the mold. The mixture was placed inside the sampling mold in three equal layers that were compacted 27 blows each. The mixture was compacted using a steel tamper that had a diameter of 37.5 mm [16, 31]. Samples were extruded from the mold using a steel plunger and were kept inside air-tight plastic bottles, as shown in Figure 5.4. The UCS samples were cured for 7, 14 and 28 days inside a controlled humidity chamber as shown in Fig. 5.5 (Left). A total of three samples were used for each test in order to ensure the accuracy of the results. The weight of the samples was measured after the curing period and those samples that had more than 0.5% difference were rejected. The rate of the axial strain was set to 2% per minute. Data acquisition unit (DAU) was used to automatically record the axial deformation and load applied to the samples. Finally the mode of failure for each sample was recorded as shown in Figure 5.5 (Right).

FIGURE 5.4   UCS samples are kept for curing inside plastic bottles.

FIGURE 5.5   Humidity chamber (left) and UCS sample after testing (right).

5.3 RESULTS AND DISCUSSION

This section presents the results of specific gravity, Atterberg limits, compaction tests and unconfined compressive strength. The results are discussed in details in the following sub-sections.

5.3.1 SPECIFIC GRAVITY (SG)

According to Head [32], the specific gravity (SG) for most of the soils ranged from 2.60 to 2.80 Mg/m^3. The SG of clayey soil varies from 2.7 to 2.8 Mg/m^3 while the SG for sands is expected to be 2.65 Mg/m^3. The SG of untreated and treated dredged clay were investigated by adding 10, 20, 30 and 40% of 0.063, 0.15, 0.30 and 1.18 mm RCT, respectively. As shown in Figure 5.6, the untreated dredged clay and RCT particles had the SG of 2.52 and 2.57, respectively. The SG of the clay treated with various proportions of 0.063 mm RCT increased as the additive amount was increased. This increment in SG was due to the increasing amount of fine RCT particles with higher SG. Furthermore, the SG of the dredged clay treated with 0.15 mm RCT also increased with further increment of the RCT. The SG had a sudden jump when 30 and 40% of 0.15 mm of RCT were added, which is due to the higher SG of RCT. Similarly, the behavior of the treated dredged clay with 0.30 mm RCT followed the behavior of 0.063 and 0.15 mm of RCT. However, when a higher quantity of RCT was added (0.30 mm), the value of SG dropped sharply to an approximately similar value of untreated clay. On the other hand, when various proportions of 1.18 mm of RCT were added, the SG only improved slightly with 10% increment and further increment resulted in the drop of the SG value. This is due to the big size of RCT and the high quantity of RCT (20, 30 and 40%).

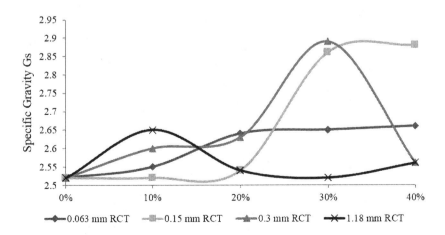

FIGURE 5.6 The specific gravity of the dredged clay and clay-RCT mix designs.

5.3.2 ATTERBERG LIMITS

The Atterberg limit tests included the determination of liquid limit, plastic limit and the plasticity index of untreated dredged clay and clay-RCT mix designs.

5.3.2.1 Liquid Limit

The liquid limit of untreated dredged clay and clay mixed with four percentages of 0.063, 0.15 and 0.30 mm of RCT is shown in Figure 5.7. The addition of all sizes of RCT resulted in a significant reduction in the liquid limit of the dredged clay. The influence of the addition of the different sizes of RCT is observed and investigated. The biggest size resulted in the lowest value of liquid limit followed by 0.15 mm size. Besides, the higher the percentage of RCT, the lower the liquid limit of the mix [33].

5.3.2.2 Plastic Limit

The results of the plastic limit of the dredged clay and the various mix designs of several sizes of RCT are shown in Figure 5.8. Similarly as discussed in the liquid limit sub-section, the higher addition of RCT resulted in the decrease in the value of

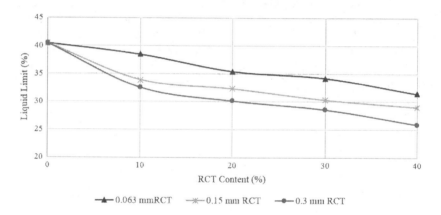

FIGURE 5.7 Liquid limit of untreated dredged clay and clay-RCT mix with several sizes.

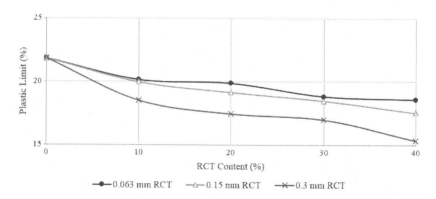

FIGURE 5.8 Plastic limit of untreated dredged clay and clay-RCT mix with several sizes.

the plastic limit of the treated dredged clay [33]. In addition, 0.30 mm size of RCT provided the best results of plastic limit followed by 0.15 mm size.

5.3.2.3 Plasticity Index

The results obtained from the liquid and plastic limits were used to develop the plasticity index curves shown in Figure 5.9. The plasticity index of the clay was steadily reduced with the addition of the different sizes of RCT. It was observed that the grade of 0.30 provided the best plasticity index curve. However, 0.15 mm had an almost similar trend of improvement as 0.30 mm size. For all sizes, a higher percentage of RCT of 40% decreased the plasticity index of the treated clay. Meanwhile, highly plastic soils are considered problematic as they are very difficult to be compacted and their permeability is very low. Hence, reducing the plasticity of soft soil would be able to improve its mechanical properties. It is clearly shown in Figure 5.9 that the plasticity index of the dredged clay was reduced for all sizes of RCT with further increments of RCT as expected [34–36]. The decrement of the plasticity index of the dredged clay by adding the RCT is due to the agglomeration of the clay minerals.

5.3.3 COMPACTION

The compaction tests included the determination of the MDD and OMC for the untreated dredged clay and the clay-RCT mixtures of 0.063, 0.15, 0.30 and 1.18 mm of RCT at 10, 20, 30 and 40%.

5.3.3.1 Maximum Dry Density (MDD)

The maximum dry density (MDD) of the dredged clay and the clay-RCT mixtures with different sizes and percentages is depicted in Figure 5.10. The treatment of the dredged clay with the RCT resulted in an increase of the MDD of the soft clay. The higher the percentage of RCT, the higher the MDD of the treated dredged clay. Additionally, the bigger the size of RCT, the higher the increment of MDD; 1.18 mm size provided significant improvement of MDD [37]. The significant improvement of MDD is due to the replacement of the RCT particles of higher SG with the marine clay particles with lower SG.

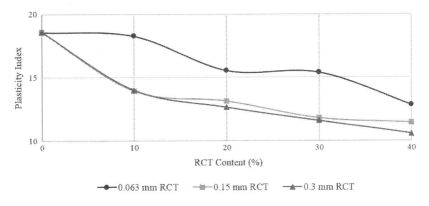

FIGURE 5.9 Plasticity index of untreated dredged clay and clay-RCT mix with several sizes.

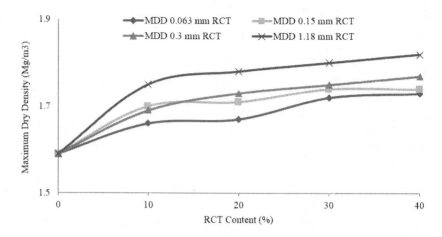

FIGURE 5.10   MDD of untreated dredged clay and clay-RCT mix with several sizes.

5.3.3.2   Optimum Moisture Content (OMC)

The optimum moisture content (OMC) of the dredged clay and the clay-RCT mixtures is shown in Figure 5.11. The OMC of various mix designs (10, 20, 30 and 40%) of four different sizes of RCT (0.063, 0.15, 0.30 and 1.18 mm) are presented and discussed in this section. The increment of RCT led to a continuous decrement in the OMC of the clay-RCT mixtures and the lowest value of OMC which was obtained at 40% RCT of all sizes. The reduction of the OMC of the clay-RCT mixtures is attributed to the decrease in the water holding capacity within the clay particles when the RCT was added [33, 38]. For all RCT sizes, the OMC decreased when the size of RCT increased and the most significant improvement occurred with the addition of 1.18 mm size [37].

5.3.4   Unconfined Compressive Strength

The unconfined compressive strength test (UCS) was used to evaluate the shear strength of the dredged clay and the clay-RCT mixtures through the use of different percentages

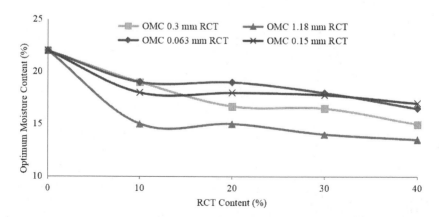

FIGURE 5.11   OMC of untreated dredged clay and clay-RCT mix with several sizes.

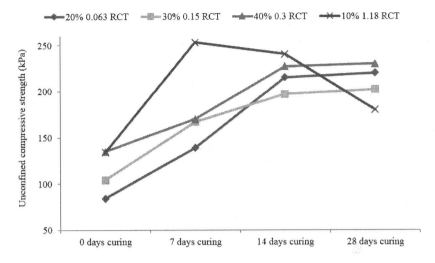

FIGURE 5.12   UCS of untreated dredged clay and clay-RCT mix with several sizes.

and various sizes of RCT as shown in Figure 5.12. This study only presents the optimum percentages that produced good UCS values of each size as a comparison between all the sizes of RCT. The UCS reached the highest value of approximately 250 kPa during 7 days curing time by using 1.18 mm grade. However, the UCS value dropped during 14 and 28 days of curing. This can be attributed to the large size of RCT. Besides, all fine sizes of RCT increase the strength of the clay-RCT and the strength development was also increased within time. Furthermore, for 0.063, 0.15 and 0.30 mm RCT, the optimum percentage of each size was found to be 20, 30 and 40% respectively. A higher percentage of RCT was required when the RCT size turned coarser in order to obtain the highest UCS value before the strength decline. Among the fine sizes of RCT, 0.3 mm was found to be the optimum size at which the strength was the highest.

5.4 CONCLUSION

Recycled ceramic powdered tiles were utilized in this study for sustainable, economical and environmentally friendly stabilization of the dredged soft clay. The laboratory experiments were conducted on different percentages and sizes of dredged clay and clay-recycled ceramic tiles (RCT). Based on the findings of the study, the following conclusions can be drawn:

1. The RCT was able to decrease the plasticity of the dredged clay; the higher the size of RCT, the lower the plasticity.
2. The compaction parameters improved significantly after different proportions and sizes of RCT were added. The maximum dry density increased significantly and the optimum moisture content decreased considerably when 1.18 mm RCT was added.
3. The unconfined compressive strength of the dredged clay increased significantly and permanent strength development was observed with 0.30 mm size of RCT.

ACKNOWLEDGMENT

The authors acknowledge the financial support provided by Universiti Teknologi PETRONAS using the Graduate Assistantship Scheme (GA). The authors would like to express their gratitude to Universiti Teknologi Malaysia for their generous financial support through the Research University Grant Scheme (Q.J. 130000.2522.13H85).

REFERENCES

1. N. O. Mohamad, *et al.*, "Challenges in construction over soft soil – Case studies in Malaysia", *IOP Conference Series: Materials Science and Engineering*, vol. 136, no. 1, pp. 1–8, 2016.
2. A. J. Puppala, and A. Pedarla, "Innovative ground improvement techniques for expansive soils", *Innovative Infrastructure Solutions*, vol. 2, no. 1, pp. 2–24, 2017.
3. D. Basu, A. Misra, and A. J. Puppala, "Sustainability and geotechnical engineering: Perspectives and review", *Canadian Geotechnical Journal*, vol. 52, no. 1, pp. 96–113, 2015.
4. D. Gautam, H. Rodrigues, K. K. Bhetwal, P. Neupane, and Y. Sanada, "Common structural and construction deficiencies of Nepalese buildings", *Innovative Infrastructure Solutions*, vol. 1, no. 1, p. 1, 2016.
5. J. Khazaei, and H. Moayedi, "Soft expansive soil improvement by eco-friendly waste and quick lime", *Arabian Journal for Science and Engineering*, pp. 1–10, 2017.
6. M. A. M. Al-Bared, and A. Marto, "A review on the geotechnical and engineering characteristics of marine clay and the modern methods of improvements", *Malaysian Journal of Fundamental and Applied Sciences*, vol. 13, no. 4, pp. 825–831, 2017.
7. Z. H. Majeed, M. R. Taha, and I. T. Jawad, "Improvement of soft soil using nanomaterials", *Research Journal of Applied Sciences, Engineering and Technology*, vol. 8, no. 4, pp. 503–509, 2014.
8. K. Rangaswamy, "Influence of burnt ash additives on stabilisation of soft clay soils", *Innovative Infrastructure Solutions*, vol. 1, no. 1, p. 25, 2016.
9. N. A. Patel, C. B. Mishra, D. K. Parmar, and S. B. Gautam, "Subgrade soil stabilization using chemical additives", *International Research Journal of Engineering and Technology (IRJET)*, vol. 2, no. 4, pp. 1089–1095, 2015.
10. D. Wu, W. Xu, and R. Tjuar, "Improvements of marine clay slurries using chemical-physical combined method (CPCM)", *Journal of Rock Mechanics and Geotechnical Engineering*, vol. 7, no. 2, pp. 220–225, 2015.
11. M. S. Abid, "Stabilization of soil using chemical additives", *GRD Journals-Global Research and Development Journal for Engineering*, vol. 1, no. 12, pp. 74–80, 2016.
12. G. N. Eujine, S. Chandrakaran, and N. Sankar, "Influence of enzymatic lime on clay mineral behavior", *Arabian Journal of Geosciences*, vol. 10, no. 20, pp. 1–8, 2017.
13. M. U. Sri, and P. M. S. S. Kumar, "An experimental study on laterite soil stabilization using bitumen emulsion", *International Journal of Innovative Research in Science, Engineering and Technology*, vol. 6, no. 2, pp. 2953–2971, 2017.
14. M. A. M. Al-Bared, and A. Marto, "Review on the geotechnical and engineering properties of marine clay and the suitable common stabilization methods", In: *Proceedings of the 2nd International Conference on Separation Technology 2017 REVIEW*, 2017, pp. I1–I3.
15. M. A. M. Al-Bared, A. Marto, I. S. Hamonangan, and F. Kasim, "Compaction and plasticity comparative behaviour of soft clay treated with coarse and fine sizes of ceramic tiles", *E3S Web of Conferences*, vol. 34, pp. 1–9, 2018.
16. Y. Yilmaz, "Compaction and strength characteristics of fly ash and fiber amended clayey soil", *Engineering Geology*, vol. 188, pp. 168–177, 2015.

17. L. S. Wong, S. Mousavi, S. Sobhani, S. Y. Kong, A. H. Birima, and N. I. M. Mohd Pauzi, "Comparative measurement of compaction impact of clay stabilized with cement, peat ash and silica sand", *Measurement*, vol. 94, pp. 498–504, 2016.
18. M. A. M. Al-Bared, and A. Marto, "Evaluating the Compaction Behaviour of Soft Marine Clay Stabilized with Two Sizes of Recycled Crushed Tiles". In: Pradhan B. (ed) *GCEC 2017. GCEC 2017. Lecture Notes in Civil Engineering*, vol. 9, Singapore: Springer, pp. 1273–1284, 2019.
19. D. Kim, and K. Park, "An environmentally friendly soil improvement technology with microorganism", *International Journal of Railway*, vol. 6, no. 3, pp. 90–94, 2013.
20. P. Shahaji, and N. Keshav, and P. D. Nemade, "Assessments of soil properties by using bacterial culture", *International Journal of Innovations in Engineering Research and Technology (IJIERT)*, vol. 2, no. 6, pp. 1–7, 2015.
21. N. Kamaraj, V. Janani, P. T. Ravichandran, D. Nigitha, and K. Priyanka, "Study on improvement of soil behaviour by bio-stabilsation method", *Indian Journal of Science and Technology*, vol. 9, no. 33, pp. 1–5, 2016.
22. S. Pourakbar, A. Asadi, B. B. K. Huat, and M. H. Fasihnikoutalab, "Stabilization of clayey soil using ultrafine palm oil fuel ash (POFA) and cement", *Transportation Geotechnics*, vol. 3, pp. 24–35, 2015.
23. M. Shahbazi, M. Rowshanzamir, S. M. Abtahi, and S. M. Hejazi, "Optimization of carpet waste fibers and steel slag particles to reinforce expansive soil using response surface methodology", *Applied Clay Science*, vol. 142, pp. 185–192, 2017.
24. M. A. M. Al-Bared, A. Marto, N. Latifi, and S. Horpibulsuk, "Sustainable improvement of marine clay using recycled blended tiles", *Geotechnical and Geological Engineering*, vol. 36, no. 5, 3135–3147, 2018.
25. A. Modarres, and Y. M. Nosoudy, "Clay stabilization using coal waste and lime – Technical and environmental impacts", *Applied Clay Science*, vol. 116–117, pp. 281–288, 2015.
26. H. Elçi, "Utilisation of crushed floor and wall tile wastes as aggregate in concrete production", *Journal of Cleaner Production*, vol. 112, pp. 742–752, 2016.
27. A. M. M. Al Bakri, M. N. Norazian, H. Kamarudin, and C. M. Ruzaidi, "The potential of recycled ceramic waste as coarse aggregates for concrete", In: *Malaysian Universities Conferences of Engineering and Technology*, Putra Brasmana, Perlis, Malaysia, pp. 1–3, 2008.
28. N. Zainuddin, N. Z. Mohd Yunus, M. A. M. Al-Bared, A. Marto, I. S. H. Harahap, and A. S. A. Rashid, "Measuring the engineering properties of marine clay treated with disposed granite waste", *Measurement*, vol. 131, pp. 50–60, 2019.
29. BSI 1377: Part 2, "British Standard methods of test for soils for civil engineering purposes: Part 2. Classification tests. London (BS1377)", 1990.
30. BSI 1377: Part 4, "British Standard methods of test for soils for civil engineering purposes: Part 4. Compaction telated tests. London, BS1377, Milton Keynes, U.K.", 1990.
31. A. Ahmed, "Compressive strength and microstructure of soft clay soil stabilized with recycled bassanite", *Applied Clay Science*, vol. 104, pp. 27–35, 2015.
32. K. H. Head, *Manual of Soil Laboratory Testing*, 3rd edn., vol. 1. Boca Raton, FL: CRC Press, 2006.
33. T. G. Rani, C. Shivanarayana, D. Prasad, and G. Prasada Raju, "Strength behaviour of expansive soil treated with tile waste", *International Journal of Engineering Research and Development*, vol. 10, no. 12, pp. 52–57, 2014.
34. A. K. Sabat, "Stabilization of expansive soil using waste ceramic dust", *Electronic Journal of Geotechnical Engineering*, vol. 17, no. Bund. Z, pp. 3915–3926, 4609, 2012.
35. C. Neeladharan, V. Vinitha, B. Priya, S. Saranya, and A. Professor, "Stabilisation of Soil by using tiles waste with sodium hydroxide as binder", *International Journal of Innovative Research in Science*, vol. 6, no. 4, pp. 6762–6768, 2017.

36. J. A. Chen, and F. O. Idusuyi, "Effect of waste ceramic dust (WCD) on index and engineering properties of shrink-swell soils", *International Journal of Engineering and Modern Technology*, vol. 1, no. 8, pp. 2504–8848, 2015.

37. N. K. Ameta, A. S. Wayal, and P. Hiranandani, "Stabilization of dune sand with ceramic tile waste as admixture", *American Journal of Engineering Research*, vol. 02, pp. 133–139, 2013.

38. A. K. Sabat, "Stabilization of expansive soil using waste ceramic dust", *Electronic Journal of Geotechnical Engineering*, vol. 17, pp. 3915–3926, 2012.

6 Acceptable Zone for Green Construction of Compacted Tropical Laterite Soil

Yamusa Bello Yamusa, Kamarudin Ahmad, and George Moses

CONTENTS

6.0 INTRODUCTION

The fast-growing impacts of solid waste are being witnessed globally. Presently, solid waste generated by the world's cities amounts to about 1.3 billion tonnes per year. It is anticipated that this volume will increase to about 2.2 billion tonnes by 2025 [1]. Human activities lead to the generation of waste, with the corresponding depletion of natural resources, loss of energy and the unnecessary emission of carbon dioxide [2]. Likewise, methane, which is a powerful greenhouse gas that is predominantly impactful to the ecosystem in the short-term, is largely sourced from solid waste. One of the main principal sources contributing to the release of leachates in the environment is

85

the municipal solid waste (MSW) in landfill facilities. Worldwide, MSW landfills post a great concern over leachate management [3]. Inappropriate management and disposal of waste can cause a detrimental impact to air, land and water by pollution, which in turn impacts on health, the environment and the economy.

Groundwater contamination caused by leachates in landfill has been a major concern in the tropical regions of the world. Where most of their municipal solid waste management is through landfilling, either in open, control or sanitary sites. To mitigate the negative effects of leachate infiltration through soil barriers, a proper design with the required amount of soil fines content needs to be investigated. The general recommendation is to compact soil meant as a hydraulic barrier on the wet side of optimum because it provides lower hydraulic conductivity. However, care must be taken to ensure that the molding water content is suitably chosen to balance between the volumetric shrinkage strain, unconfined compressive strength and hydraulic conductivity. The reason for this is that a higher molding water content results in lower hydraulic conductivity, whereas it increases volumetric shrinkage and lowers shear strength. Therefore, a natural tropical laterite soil was reconstituted and subjected to laboratory tests to determine its suitability as compacted soil liner used in an engineered sanitary landfill.

In developing countries, MSW is regularly dumped in lowland areas, valleys, waterways, ditches and land next to slums. Environmental threats posed by improper displacement of waste include pollution of surface and groundwater water by leachate, and air pollution from burning of the waste. To overcome the improper disposal of solid waste at a reasonable cost, sanitary landfill is encouraged. Landfills are common final waste repositories which should be engineered and functioned to safeguard the public and environment. Landfills are usually classified as open dumps, controlled dumps, engineered landfills and engineered sanitary landfills or simply sanitary landfills as presented in Table 6.1.

To create a waste disposal site, the use of soils with appropriate geotechnical characteristics must be included to ensure acceptable engineering design with safe conditions in place [5]. Hydraulic barriers (i.e., liners and covers) used in landfill design play an important role in impeding the flow of contaminants. The structural integrity of liners and covers must be guaranteed by ensuring that the constructed amenity has acceptable hydraulic conductivity, tolerable volumetric shrinkage when desiccated and adequate shear strength [6].

According to [7], the commonest soil materials used in the construction of compacted clay liners in sanitary landfills are fine-grained soils. Fine-grained soils are readily available, have good engineering properties when compacted and they are relatively economical. Laterite soils with a high percentage of fines have significant pollutant retention capacity. The criteria in most regulatory agency guidelines and research specified a value of 1×10^{-9} m/s maximum hydraulic conductivity, a maximum of 4% volumetric shrinkage, minimum of 200 kN/m^2 shear strength and a minimum of 30% fines with respect to gradation are required for hydraulic barrier systems [8–21]. Minimum required geotechnical parameters for most soil liner materials are presented in Table 6.2. These guidelines seem to be different from laterite soil with respect to the gradation of 30% fines content. At this percentage requirement, laterite soils are vulnerable to leachate permeation which contaminates the groundwater.

TABLE 6.1
Landfill classification

Landfill Type	Operation and Engineering Measures	Leachate Management	Landfill Gas Management	Reference
Open Dump	Uncontrolled dumping; no engineering measures	Full contaminant release without any restriction	Nil	Daniel [4]
Controlled Dump	Controlled dumping and compaction of waste; surface water monitoring; no daily cover material, no liner in place	Unprotected contaminant release	Nil	Hoornweg and Bhada-Tata [1] and Daniel [4]
Engineered Landfill	Registration and compaction of waste; uses daily cover material; surface and ground water monitoring; infrastructure and liner in place	Containment and leachate recirculation system	Inactive flaring	Hoornweg and Bhada-Tata [1]
Sanitary Landfill	Registration, placement and compaction of waste; uses daily cover; measures for final top cover and closure; proper siting, infrastructure; liner in place and post-closure plan	Containment and leachate treatment system	Flaring with or without energy recovery	Hoornweg and Bhada-Tata [1] and Daniel [4]

TABLE 6.2
Minimum required geotechnical parameters for soil liner materials

Parameter	Author	Recommendation
Grain size	Daniel [4]	Percentage fines: $\geq 20\%$ Percentage gravel: $\leq 30\%$ Maximum particle size: 25–50 mm
	Benson, et al. [22]	Percentage fines: $\geq 30\%$
	Shelley and Daniel [23]	Percentage gravel: $\leq 50\%$
	USEPA [19]	Percentage fines: $\geq 30\%$
	Shakoor and Cook [24]	Percentage gravel: $\leq 50\%$
Atterberg consistency limits	Daniel [4]	Plasticity index: $\geq 7\%$
	Rowe [25]	Plasticity index: $\geq 12\%$
	Jones, et al. [26]	Plasticity index: $< 65\%$ Liquid limit: $< 90\%$
	Benson, et al. [22]	Plasticity index: $\geq 7\%$ Liquid limit: $\geq 20\%$
	USEPA [19]	Plasticity index: $> 10\%$
	USEPA [27]	Plasticity index: $\leq 35\%$

Therefore, due to variations in terms of recommended values of gradations and Atterberg limits, there is a reason to carry out this research in order to establish the recommended values for a green construction of tropical laterite soil liners.

Lateritic soil with a substantial amount of fines is expected to be effective as a barrier material in a sanitary landfill to provide low hydraulic conductivity values of less than 1×10^{-9}m/s [7]. This study shows that 50% fines content is needed to provide a protective barrier that will impede the flow of leachate through the compacted laterite soil which may cause groundwater contamination. Thus, fine-grained soils can achieve the required design state by compaction, either on the wet or dry side of the line of optimum moisture content depending on the type of soil as long as the overall acceptable zone is obtained by superimposing the dry density with molding water content for volumetric shrinkage, unconfined compressive strength and hydraulic conductivity parameters at their permissible limits.

6.1   MATERIAL AND METHODS

The laterite soil used in this study was reconstituted into different gradations. The soil was air-dried and then passed through a BS 4.75mm aperture sieve to remove oversized gravel. The laterite soil was then sieved into three different groups i.e., fines (<0.063 mm), sand (0.063 mm to 2.00 mm) and gravel (>2.00 mm to ≤4.75 mm). Three different gradations of laterite soil samples were investigated as follows:

1. 30%, 40% and 30% of gravel, sand and fines respectively, denoted as L1.
2. 20%, 40% and 40% of gravel, sand and fines respectively, denoted as L2.
3. 10%, 40% and 50% of gravel, sand and fines respectively, denoted as L3.

British Standard light (BSL) compaction in accordance with the BS1377;1990 was used as a baseline. The compaction test is generally carried out to get the relationships between the dry density and water content of soils. This was carried out on all three different gradations i.e., L1, L2 and L3.

Hydraulic conductivity of fine-grained soil samples is measured with a permeameter using a falling head condition [28]. Soil samples were compacted using the BSL at different gradations (L1, L2 and L3) and at different molding water contents of -4%, -2%, 0% (OMC), +2% and +4%. Then the samples were soaked for a minimum of 48 hours until no air bubbles were obviously observed to allow for full saturation inside a water tank. After saturation, each test sample was connected to distilled water through a standpipe. Time and height/distance covered by water in the stand pipe were recorded. This was repeated at various moisture contents for L1, L2 and L3, and the coefficient of permeability or hydraulic conductivity (k) was calculated.

BS 1377 (1990) was adopted in determining the unconfined compressive strength (UCS). The laterite soil samples were compacted and specimens with size 38 mm in diameter and 76 mm high were extruded from the compaction mould using a steel mould of that size. Each sample was then subjected to an unconfined compressive strength test with the rate of axial strain at 2% per minute with respect to the sample length. The test was continued until the maximum value of the axial stress had been passed or the axial strain reached 20% of the soil specimen diameter.

In the volumetric shrinkage, air-dried soil at various gradations was compacted as described in BS1377:1990 at − 4%, − 2%, 0%, +2% and +4% molding water contents with respect to the optimum water content. Then the samples were extruded out of the cylindrical mould and spread on a worktable in the laboratory at a uniformly constant temperature of 26 ± 2°C for 30 days to dry. Vernier caliper accurate to 0.05 mm was used to take daily measurements of the diameters and heights for each sample. Volumetric shrinkage strain (VSS) was computed by using the average diameters and heights [29–32].

6.2   RESULTS AND DISCUSSION

6.2.1   INDEX PROPERTIES

The index properties of the natural soil and particle size distributions of the natural and reconstituted laterite soils in accordance with British Standards [33] are shown in Table 6.3 and Figure 6.1, respectively. The Atterberg limits results revealed a liquid limit of 76%, plastic limit of 42% and plasticity index of 34%. Based on these data, the natural laterite soil is classified as very high plasticity sandy silt with gravel (MV) according to the BS classification. Generally, low hydraulic conductivity is attained from soils with a high liquid limit. The liquid limit of a liner material is recommended to be ≥ 20% [22], although a high liquid limit can cause desiccation cracking in clay soils [7].

The effect of gradation on MDD and OMC of the laterite soil mixtures is shown in Figure 6.2. The MDD generally increased with increasing fines content. This could be due to the fines filling the voids of the soil mixture. As the air voids are squeezed out under the load of the rammer, the finer particles of the soil replaced the air and became denser. Another reason could be that the coarse particles of the soil samples are susceptible to crushing and probably contain air voids, thereby reducing the density. On mixing with 30%, 40% and 50% fines content at OMC, the values of the MDD increased to 1.35, 1.42 and 1.43 Mg/m^3, respectively. This is consistent with Guerrero [34], who demonstrated that an addition of fine content produced an

TABLE 6.3
Index properties of laterite soil

Property	Value
Natural moisture content, %	34
Specific gravity	2.7
% Passing BS 63μm sieve	30
Optimum moisture content, OMC %	30
Maximum dry density, MDD Mg/m^3	1.35
Liquid limit, LL %	76
Plastic limit, PL %	42
Plasticity index, PI %	34
BS classification	MV

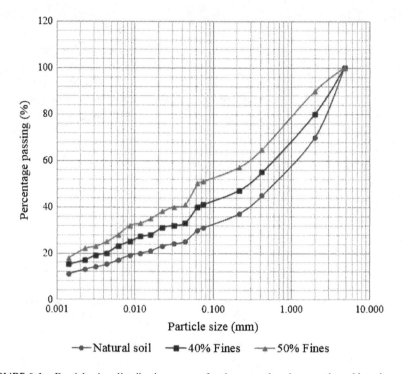

FIGURE 6.1   Particle size distribution curves for the natural and reconstituted laterite soils.

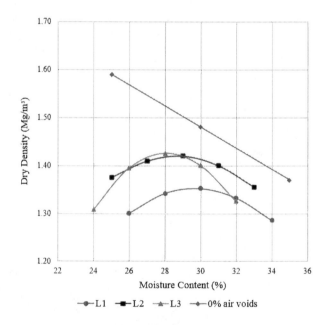

FIGURE 6.2   Dry density versus moisture content.

increment in the maximum dry density in all of the four different types of soils experimented. On the other hand, the OMC decreased with an increasing fines content. The OMC reduced to 30%, 29% and 28% for L1, L2 and L3, respectively, probably due to the sesquioxides of iron and aluminum, which caused a physical cementation of fine particles into coarser aggregations, resulting in a granular structure [13]. By mixing the soil with water, the fines within the soil flocculated, giving the effect of larger particles. In other words, addition of water to the soil caused the aggregation of fines particles forming strongly bonded larger particles. Therefore, L1, L2 and L3 with different amount of fines content will give soil structure that is relatively different from one another.

6.2.2 HYDRAULIC CONDUCTIVITY

6.2.2.1 Effect of Molding Water Content

Figure 6.3 shows a general decrease in hydraulic conductivity, with an increase in molding water content regardless of the soil gradations. At a higher moisture content, the soil structure is more dispersed, attributing a decrease in hydraulic conductivity in agreement with [35]. Likewise, compaction with higher molding water content resulted in soil grading that were devoid of macropores (i.e., pores being filled with water), which conduct flow [11]. The reconstituted laterite soil with 50% fines content compacted on the wet side of optimum gave the maximum hydraulic conductivity value.

6.2.2.2 Effect of Dry Density

The hydraulic conductivity and dry density relationships are presented in Figure 6.4. Regardless of the soil gradations, it can be deduced that k decreases with an increase in dry density, and further decreases after the dry density decreased in values. This could be as a result of the densification of the soil that lead to low permeability.

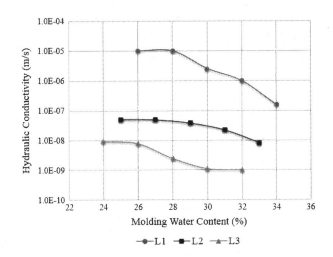

FIGURE 6.3 Hydraulic conductivity versus molding water content.

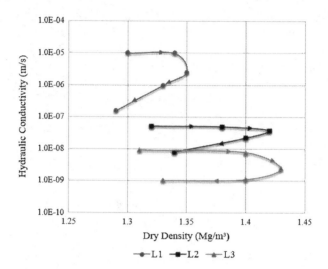

FIGURE 6.4   Hydraulic conductivity versus dry density.

The permissible values of k ≤ 10⁻⁹ m/s were obtained at dry densities of 1.31–1.43 Mg/m³ for L3. The dry densities of the soil increase as the fines content increase for L1, l2 and L3 because the higher the percentage fines the more it fills up the voids present [36]. Low dry density is associated with soils having large, visible voids between pores, while high dry density is associated with soils having small or invisible voids [37].

6.2.3   Unconfined Compressive Strength

6.2.3.1   Effect of Molding Water Content

The relationship of UCS values with molding water content is presented in Figure 6.5. The UCS values decreased with increased molding water content, which agrees with other researchers' findings [6, 38–40]. Such phenomena, regardless of the soil gradation, took place because the soil fabrics were increasingly lubricated, thus leading to a reduction in the shearing resistance due to a loss in cohesion between soil particles [32]. On the dry side of OMC and at OMC, the minimum unconfined compression strength of 200 kN/m² required for soil to be used in compacted soil liners was attained. In addition, this study shows that soils with higher fines content yielded higher UCS values on the dry side of the OMC than soils with lower fines content. Care must be taken to ensure that the molding water content is suitably chosen to balance between the unconfined compressive strength and hydraulic conductivity. This is because higher molding water content results in lower hydraulic conductivity, whereas lower molding water content results in higher unconfined compressive strength [41].

6.2.3.2   Effect of Dry Density

The UCS values vary with dry density as shown in Figure 6.6. As the percentage of fines increased, the dry density likewise increased with the corresponding

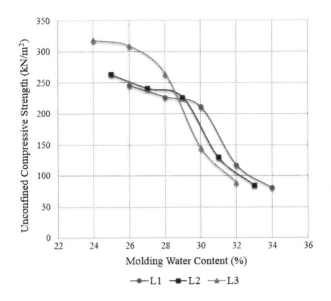

FIGURE 6.5   Unconfined compressive strength versus molding water content.

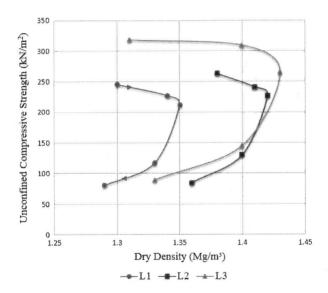

FIGURE 6.6   Unconfined compressive strength versus dry density.

increase in UCS. Soil containing 50% fines recorded the highest dry unit weight and UCS values. The permissible values ≥200 kN/m² shear strength were achieved at dry densities of 1.30–1.35, 1.38–1.42 and 1.31–1.43 Mg/m³ at L1, L2 and L3, respectively.

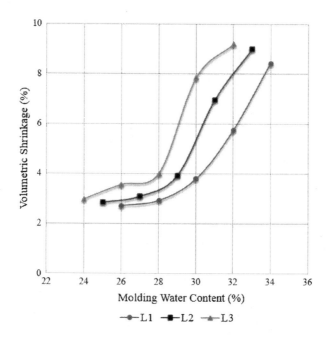

FIGURE 6.7   Volumetric shrinkage versus molding water content.

6.2.4   VOLUMETRIC SHRINKAGE

6.2.4.1   Effect of Molding Water Content

The relationship between VSS and different molding water contents is presented in Figure 6.7. Laterite soil samples prepared at molding water contents on the wet side of optimum recorded unsatisfactory VSS results irrespective of the fines content. However, all the soil specimens prepared on the dry side of the OMCs and OMCs recorded more satisfactory VSS results than those on the wet side of the OMCs. Generally, VSS increased with higher molding water content and this study result is consistent with the results of [31, 42]. Compacted soil specimens at a higher molding water content shrink more when dry, because drying shrinkage in fine-grained soils depends on particle movement due to pore water tension developed by capillary menisci [29, 31, 42]. As higher molding water content is needed to lower the hydraulic conductivity, this can simultaneously affect the volumetric shrinkage when is too high. Thus, care must be taken when selecting the molding water content of laterite soil in order to satisfy the VSS criteria [43].

6.2.4.2   Effect of Dry Density

The relationship of dry density with volumetric shrinkage strain is shown in Figure 6.8. VSS increased with increased dry density and continued increasing after the dry density decreased in values. Thus, VSS increased with an increase in dry density, either on the dry or the wet side of the optimum, which corresponds with the results of Bello [6]. The maximum of 4% volumetric shrinkage was achieved at

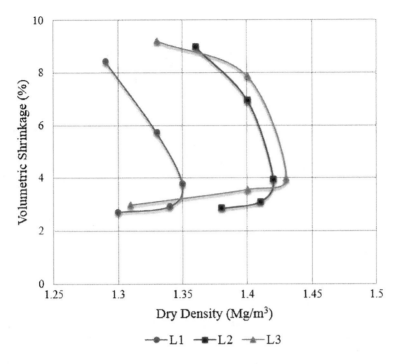

FIGURE 6.8 Volumetric shrinkage versus dry density.

the dry densities between 1.3–1.35, 1.38–1.42 and 1.31–1.43 Mg/m³ at L1, L2 and L3, respectively.

6.2.5 ACCEPTABLE ZONES

A critical step in the design of a compacted soil liner is the determination of the range of acceptable water content and dry unit weight of the soil. In contemporary geo-environmental practice, soil barriers are compacted within these definite ranges of water content and dry unit weight. This requirement is principally needed to attain a minimum dry unit weight for factors controlling the performance of compacted soil liners, especially their hydraulic conductivity [35]. An acceptable zone is obtained by relating the dry density with molding water content for each of the three design parameters at their permissible limits i.e., $k \leq 1 \times 10^{-9}$ m/s, VSS $\leq 4\%$, and UCS ≥ 200 kN/m². The compaction plane was arrived at by using the average density values from k, UCS and VSS test specimens [32, 44]. Table 6.4 presents the acceptable ranges of molding water contents at various fines contents for L1, L2 and L3. Samples L1 and L2 failed to meet the hydraulic conductivity requirement but did meet the shear strength and volumetric shrinkage requirements. Therefore, L1 and L2 are not fit for use as hydraulic barriers. Only samples of L3 satisfy the hydraulic conductivity acceptable ranges of molding water contents. Figures 6.9, 6.10, and 6.11 show acceptable zones of L3 for hydraulic conductivity, shear strength and volumetric shrinkage, respectively. The hatched zones indicate portions on the

TABLE 6.4

Acceptable ranges of molding water contents

Engineering Criteria	Fines Content, %		
	30	40	50
	Molding Water Content Range, %		
k, m/s			24–32
UCS, kN/m²	26–30	25–29	24–28
VSS, %	26–30	25–29	24–28
OAR			24–28

OAR – Overall Acceptable Range

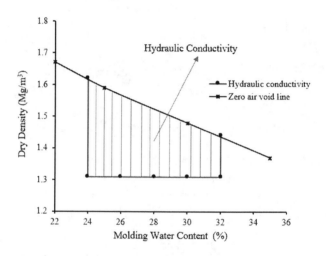

FIGURE 6.9 Acceptable zone for hydraulic conductivity at 50% fines content (L3).

compaction plane showing the limits based on different criteria for the acceptable zones. Furthermore, after an acceptable zone was defined for each of the three design parameters, they are superimposed to obtain the overall acceptable zone as shown in Figure 6.12.

6.3 CONCLUSION

As field investigations have shown, high hydraulic conductivities are usually associated with hydraulic defects in the soil due to construction methodologies. Therefore, compacted soil used as the liner in sanitary landfill needs to be sustainably designed to prevent groundwater pollution, which affects the livelihood of the ecosystem. The basic characteristics of a tropical laterite soil were studied to determine its potential use as liner material in a waste containment facility. Index properties revealed the

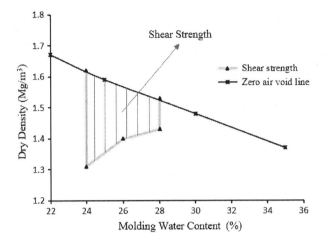

FIGURE 6.10   Acceptable zone for shear strength at 50% fines content (L3).

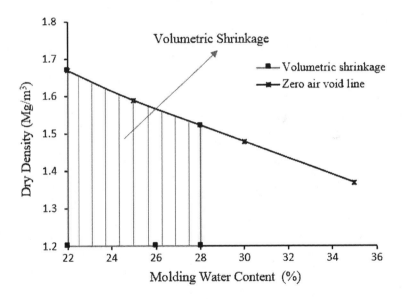

FIGURE 6.11   Acceptable zone for volumetric shrinkage at 50% fines content (L3).

laterite soil as being a very high plasticity sandy silt with gravel. Compaction characteristics showed that MDD generally increased with increasing fines content up to
50% and, on the other hand, the OMC decreased with increasing fines content up
to 50%. Hydraulic conductivity decreases with the increase in molding water content and fines content. However, volumetric shrinkage increases with an increase in
molding water content and fines content. The unconfined compressive strength measured shows an increase in strength with an increase in fines content, but decreases
with an increase in molding water content.

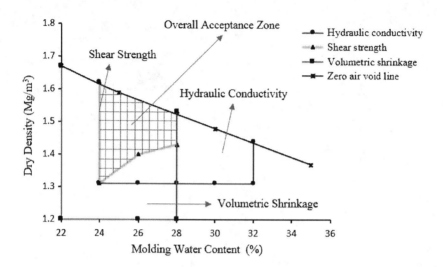

FIGURE 6.12 Overall acceptable zone at 50% fines content (L3).

Hence, care must be taken to ensure that the molding water content is suitably chosen to balance between the k, UCS and VSS. The reason for this is that higher molding water content results in lower k, whereas it lowers UCS and increases VSS. In order to satisfy the k requirement, the VSS and UCS should not be jeopardized as they are both equally important criteria in the design of compacted soil liners and covers.

To find out the optimum fines content of laterite soil, the test samples were compacted at −4%, −2%, 0%, +2% and +4% molding water content relative to optimum using British Standard light compaction. Hydraulic conductivity, volumetric shrinkage and unconfined compressive strength were the three design parameters investigated. In this paper, the effects of fines content required to provide the recommended criteria of ≤ 10^{-9} m/s hydraulic conductivity, ≤ 4% volumetric shrinkage and ≥ 200 kN/m² unconfined compressive strength were established. The three design parameters at their permissible limits were plotted using the dry density against molding water content and obtained the overall acceptable zone. The overall acceptable zone was achieved at molding water contents that ranged between 24–28% and dry densities that ranged between 1.31–1.43 Mg/m³ in which L3 was satisfied. Therefore, the laterite soil used met the regulatory requirement for hydraulic barriers at a minimum of 50% fines. Laterite soils with similar properties should be thoroughly investigated when their fines contents are less than 50%, as they might not impede the migration of leachate through the soil barrier.

ACKNOWLEDGMENT

The authors want to thankfully acknowledge the financial support given by the Universiti Teknologi Malaysia (UTM) under the International Doctoral Fellowship (UTM.J.10.01/13.14/1/128).

REFERENCES

1. D. Hoornweg, and P. Bhada-Tata, *What a Waste: A Global Review of Solid Waste Management. Urban Development & Local Government Unit.* Washington, DC: World Bank, 2012.
2. R. Fragaszy, J. Santamarina, A. Amekudzi, D. Assimaki, R. Bachus, S. Burns, *et al.*, "Sustainable development and energy geotechnology—Potential roles for geotechnical engineering", *KSCE Journal of Civil Engineering*, vol. 15(4), pp. 611–621, 2011.
3. J.-E. Min, M. Kim, J. Y. Kim, I.-S. Park, and J.-W. Park, "Leachate modeling for a municipal solid waste landfill for upper expansion", *KSCE Journal of Civil Engineering*, vol.14(4), pp. 473–480, 2010.
4. D. E. Daniel, *Geotechnical Practice for Waste Disposal.* Springer Science & Business Media, doi:10.1007/978-1-4615-3070-1, 2012.
5. I. Oyediran, and C. Iroegbuchu, "Geotechnical characteristics of some Southwestern Nigerian clays as barrier soils", *IFE Journal of Science*, vol. 15, pp. 17–30, 2013.
6. A. Bello, "Analysis of shear strength of compacted lateritic soils", *The Pacific Journal of Science and Technology*, vol. 12, pp. 425–434, 2011.
7. K. Osinubi, A. Eberemu, A. Bello, and A. Adzegah, "Effect of fines content on the engineering properties of reconstituted lateritic soils in waste containment application", *Nigerian Journal of Technology*, vol. 31, pp. 277–287, 2012.
8. Y. B. Yamusa, K. Ahmad, and N. A. Rahman, "Hydraulic conductivity and volumetric shrinkage properties review on gradation effect of compacted laterite soil liner", *Malaysian Journal of Civil Engineering*, vol. 29, pp. 153–164, 2017.
9. Y. B. Yamusa, K. Ahmad, and N. A. Rahman, "Gradation effect review on hydraulic conductivity and shear strength properties of compacted laterite soil liner", In: *Proceedings of the 11th International Civil Engineering Postgraduate Conference – The 1st International Symposium on Expertise of Engineering Design (SEPKA-ISEED'16), 26–27th September, 2016.* Faculty of Civil Engineering, Universiti Teknologi Malaysia, Johor, Malaysia, 2016.
10. A. Bello, "Acceptable zone for reddish brown tropical soil as liner material", *The Pacific Journal of Science and Technology*, vol. 16, pp. 21–32, 2015.
11. K. Osinubi, G. Moses, and A. Liman, "The influence of compactive effort on compacted lateritic soil treated with cement kiln dust as hydraulic barrier material", *Geotechnical and Geological Engineering*, vol. 33(3), pp. 523–535, 2015.
12. UKEA, "Earthworks in landfill engineering: LFE4", *Design, Construction and Quality Assurance of Earthworks in Landfill Engineering.* UK Environment Agency, 2014.
13. A. Amadi, and A. Eberemu, "Characterization of geotechnical properties of lateritic soil-bentonite mixtures relevant to their use as barrier in engineered waste landfills", *Nigerian Journal of Technology*, vol. 32, pp. 93–100, 2013.
14. K. J. Osinubi, and C. M. Nwaiwu, "Design of compacted lateritic soil liners and covers", *Journal of Geotechnical and Geoenvironmental Engineering*, vol. 132(2), pp. 203–213, 2006.
15. MHLG, Ministry of Housing and Local Government, Criteria for Siting Sanitary Landfills: National Strategic Plan for Solid Waste Management, Vol. 3, Appendix 6B, Kuala Lumpur, Malaysia, 2005.
16. EPA, *Landfill Manuals, Landfill Site Design.* Wexford: Environmental Protection Agency, 2000.
17. DWAF, *Minimum Requirements for Waste Disposal by Landfill, Waste Management Series*, 2nd edn. Department of Water Affairs & Forestry, Republic of South Africa, 1998.
18. EPA, *Environmental Guidelines: Solid Waste Landfills.* Waste Management Branch, Environment Protection Authority, 1996.
19. USEPA, *Solid Waste Disposal Facility Criteria: Technical Manual. EPA 530-R-93-017.* United States Environmental Protection Agency, pp. 1–84, 1993.

20. D. E. Daniel, and Y.-K. Wu, "Compacted clay liners and covers for arid sites", *Journal of Geotechnical Engineering*, vol. 119(2), pp. 223–237, 1993.
21. Y. B. Yamusa, K. Ahmad, and N. A. Rahman, "Effects of gradation on hydraulic conductivity properties of compacted laterite soil liner: A review", *International Journal of Scientific and Engineering Research*, vol. 7, pp. 1088–1091, 2016.
22. C. H. Benson, H. Zhai, and X. Wang, "Estimating hydraulic conductivity of compacted clay liners", *Journal of Geotechnical Engineering*, vol. 120(2), pp. 366–387, 1994.
23. T. L. Shelley, and D. E. Daniel, "Effect of gravel on hydraulic conductivity of compacted soil liners", *Journal of Geotechnical Engineering*, vol. 119(1), pp. 54–68, 1993.
24. A. Shakoor, and B. D. Cook, "The effect of stone content, size, and shape on the engineering properties of a compacted silty clay", *Environmental and Engineering Geoscience*, vol. 27(2), pp. 245–253, 1990.
25. K. R. Rowe, *Clayey Barrier Systems for Waste Disposal Facilities*. London/New York: E & FN Spon, 1995.
26. R. Jones, E. Murray, D. Rix, and R. Humphrey, "Selection of clays for use as landfill liners", *Waste Disposal by Landfill-Green*, vol. 93, pp. 433–438, 1995.
27. USEPA, "Design and construction of RCRA/CERCLA final covers", In: *Soils Used in Cover Systems, EPA/625/4-91/025*. United States Environmental Protection Agency, pp. 1–25, 1991.
28. K. H. Head, and R. J. Epps, *Manual of Soil Laboratory Testing*, Volume 2. Dunbeath: Whittles Publishing, 2011.
29. A. A. Bello, "The use of Standard Proctor for the determination of Shrinkage Properties of Reddish Brown Tropical soil", *Leonardo Journal of Sciences*, vol. 44, pp. 57–68, 2011.
30. A. O. Eberemu, "Desiccation induced shrinkage of compacted tropical clay treated with rice husk ash", *International Journal of Engineering Research in Africa*, vol. 6, pp. 45–64, 2011.
31. K. J. Osinubi, and C. M. O. Nwaiwu, "Desiccation-induced shrinkage in compacted lateritic soils", *Geotechnical and Geological Engineering*, vol. 26(5), pp. 603–611, 2008.
32. K. Osinubi, and G. Moses, "Compacted foundry sand treated with bagasse ash as hydraulic barrier material", In: *Geo-Frontiers 2011: Advances in Geotechnical Engineering*, Edited by Jie Han, PhD, P.E., Daniel E. Alzamora, P.E. American Society of Civil Engineers (ASCE) Book Series, pp. 915–925, 2011.
33. BSI, *Methods of Testing Soil for Civil Engineering Purposes (BS 1377: Part 1–9)*. London: British Standards Institute, 1990.
34. A. M. A. Guerrero, *Effects of the Soil Properties on the Maximum Dry Density Obtained from the Standard Proctor Test*, MSc. Thesis, University of Central Florida, Orlando, Florida, 2004.
35. A. A. Amadi, and A. O. Eberemu, "Delineation of compaction criteria for acceptable hydraulic conductivity of lateritic soil-bentonite mixtures designed as landfill liners", *Environmental Earth Sciences*, vol. 67(4), pp. 999–1006, 2012.
36. Y. B. Yamusa, N. Z. M. Yunus, K. Ahmad, N. A. Rahman, and R. Sa'ari, "Effects of fines content on hydraulic conductivity and morphology of laterite soil as hydraulic barrier", *E3S Web of Conferences, International Conference on Civil & Environmental Engineering (CENVIRON 2017)*, Penang, Malaysia, vol. 34, 2018.
37. A. A. Bello, "Hydraulic conductivity of three compacted reddish brown tropical soils", *KSCE Journal of Civil Engineering*, vol. 17(5), pp. 939–948, 2013.
38. K. Osinubi, A. Bello, C. Quadros, and S. Jacobsz, "Characterization of shear strength of abandoned dumpsite soils, Orita-Aperin, Nigeria", In: *Proceedings of the 15th African Regional Conference on Soil Mechanics and Geotechnical Engineering. Resource and Infrastructure Geotechnics in Africa: Putting the Theory into Practice, Maputo, Mozambique, 18–21 July 2011*, pp. 293–298, 2011.

39. K. J. Osinubi, G. Moses, F. O. P. Oriola, and A. S. Liman, "Influence of molding water content on shear strength characteristic of compacted cement kiln dust treated lateritic soils for liners and covers", *Nigerian Journal of Technology*, vol. 34(2), p. 266, 2015.

40. A. Amadi, A. Eberemu, and K. Osinubi, "Strength consideration in the use of lateritic soil stabilized with fly ash as liners and covers in waste landfills", In: *State-of-the-Art and Practice in Geotechnical Engineering*, vol. 225. American Society of Civil Engineers (ASCE) Geotechnical Special Publication (GSP), pp. 3835–3844, 2012.

41. Y. B. Yamusa, and K. Ahmad, "Evaluating the geotechnical properties of compacted laterite soil at different fines content", *Advanced Science Letters*, vol. 24(6), pp. 4015–4020, 2018.

42. K. J. Osinubi, and A. O. Eberemu, "Desiccation induced shrinkage of compacted lateritic soil treated with blast furnace slag", *Geotechnical and Geological Engineering*, vol. 28(5), pp. 537–547, 2010.

43. B. Y. Yamusa, A. Kamarudin, A. R. Norhan, S. A. Radzuan, Y. Nor Zurairahetty, and R. Ahmad, "Volumetric shrinkage of compacted soil liner for sustainable waste landfill", *Chemical Engineering Transactions*, vol. 63, pp. 613–618, 2018.

44. A. A. Bello, "Design parameters for abandoned dumpsite soil as liner material", *Leonardo Journal of Sciences*, vol. 46, pp. 29–40, 2013.

7 Green Economy and Sustainable Development

*Yusuf Babangida Attahiru, Md. Maniruzzaman
A. Aziz, Khairul Anuar Kassim, and
Wan Azelee Wan Abu Bakar*

CONTENTS

7.0 EXECUTIVE SUMMARY

This proposal highlights the concept of the green economy, which comprises many factors. The following five factors are the major components. These are: (i.) divest from fossil fuels; (ii.) efficient use of energy; (iii.) reduce greenhouse gas (GHG) emissions; (iv.) energy production from waste materials; (v.) use of renewable sources of power. Due to time and budget constraints, this proposal highlights only three from the above. The green economy is often given priority in policies and its emphasis is extremely timely given the recent interest in shifting to the green economy and carbon-neutral materials for the construction of basic facilities. Appropriate planning with new sustainable ideas will help remedy the state of the economy and the environment given the critical issues in the world today. Sustainable development contributes to the fast growth of economic, social, environment, energy and transport sustainability. Carbon neutral materials, hydrogen and bio-fuels can be assessed and produced on the basis of commercial potential. This study will provide a useful reference for carbon-neutral material in sustainable construction developments. It can be concluded that the green economy is all about distributing sufficient resources for the transformation of economic growth while increasing the priority of environmental policies.

7.1   OBJECTIVES OF THE RESEARCH

The objectives of this research are to find or develop policies and measures for global and regional environmental challenges toward the green economy, sustainable development and the shift from consumption to conservation. The specific objectives are shown below:

1. To develop policies and measures toward a successful green economy by creating marketable carbon-neutral products to deploy mineral carbonation through the use of more sustainable developments.
2. To develop natural resources by using carbon-neutral materials and renewable energy for sustainable construction work.
3. To improve functionality because of environmental changes by shifting policy from consumption to a new conservation approach, focused on the consumption of raw materials and renewable energy.
4. To transform greenhouse gasses (GHGs) to hydrogen, clean fuels and renewable energy by fully utilizing greenhouse gasses (i.e., H_2O and CO_2).
5. To develop policy/criteria for the green economy:
   - Develop policy/criteria for carbon-neutral materials
   - Develop policy/criteria for renewable energy
   - Develop policy/criteria for waste to wealth materials
   - Develop policy/criteria for divesting from fossil fuel

Upon completion of this research, a draft protocol is expected to be found or developed as guiding principles in distributing sufficient resources for the better transformation of our economic growth, increasing the priority given to environmental policy.

7.2   METHODOLOGY

The production of hydrogen from natural gas, petroleum or coal coupled with carbon dioxide sequestration is considered a prospective decarburization strategy. The chemical energy of fossil fuel is converted into chemical energy of hydrogen, and carbon dioxide as a byproduct of the reaction before it is captured and sequestered. In the case of charcoal as a feedstock, the hydrogen product is almost entirely originated from water, i.e., with the energy conversion efficiency of 50–60%, whereas in the case of natural gas, half of the hydrogen comes from water and another half from natural gas, i.e., with the energy conversion efficiency of 70–75% [1]. Production of carbon-free energy carriers and transportation fuels, and the assessment of their commercial potential were provided with an overview of art technologies. The transition from the current fossil fuel-based economy to a hydrogen economy was carried out under two elements:

i. Changing the fossil decarburization strategy from one based on CO_2 sequestration to one that involves sequestration and/or the utilization of solid carbon.

ii. The production of carbon-neutral synthetic fuels from bio-carbon and hydrogen generated from water using carbon-neutral sources (nuclear, solar, wind, geothermal).

Therefore, the life cycle analysis model was tested and calibrated as a decision support tool for sustainable construction development in the road industry. The current global market for raw commodities such as cement additives, fillers and iron ore feedstock could be produced by rock and/or industrial waste/byproduct mineralization. It is about 27.5 Gt and can be easily flooded, assuming 10% of the global CO_2 emissions sequestered by carbon dioxide capture and storage by mineral carbonation (CCSM).

Finally, probable waste materials and carbon-neutral materials were identified, and energy oil was converted to develop renewable energy from greenhouse gases (GHGs). Therefore, both energy oil and carbon-neutral materials were also developed from greenhouse gases. Thus, energy oil, renewable energy from greenhouse gases, and carbon-neutral materials were further developed based on policy criteria for the green economy, sustainable development, fossil fuels and carbon-neutral materials for construction by shifting the policy from consumption to conservation.

7.3 GREEN ECONOMY

The main concept of a green economy appears to promise an attractive orientation out of the present crisis of economic development [2]. The green economy is usually given a greater priority in policies, and the environment is viewed as apart from humans. It is not possible to give an accurate definition of the green economy. The green economy approach seeks, in principle, to unite under a single banner the entire suite of economic policies and modes of economic analysis of relevance to sustainable development. In practice, this covers a wide range of literature and analysis that often shares the following starting points. A green economy is based on these five factors: (i.) divest from fossil fuels; (ii.) efficient use of energy; (iii.) reduce greenhouse gas (GHG) emissions; (iv.) energy production from waste materials; (v.) use of renewable sources of power. A green economy creates new opportunities, good, new jobs and stronger communities [3]. A green economy is just like sustainable development, rather an oxymoron which intends to bundle different, partly contradictory, interests and strategies and grant them a certain legitimacy and coherence.

7.4 POLLUTION TO SOLUTION

The major contributor to global warming is human-generated greenhouse gases (GHGs) emissions that pollute the air. GHG emissions are a global issue dominated by the emission of carbon dioxide (CO_2). Notably, CO_2 accounts for an estimated 77% of GHGs and thus has a huge impact on the environment. The capture, sequestration and utilization of CO_2 emissions from flue gas are now becoming familiar worldwide (refer to Figure 7.1). These methods are a promising solution to promote sustainability for the benefit of future generations. Previously, many researchers have focused on capturing and storing CO_2; however, less effort has been spent on finding

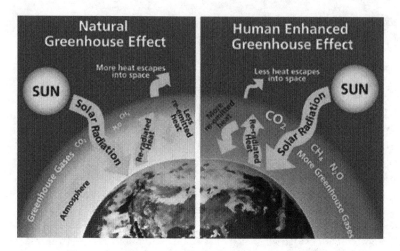

FIGURE 7.1 The effect of greenhouse gases in the atmosphere [4].

ways to utilize flue gas emissions. Moreover, several issues must be overcome in the field of carbon capture and sequestration (CCS) technology, especially regarding the cost, capacity of storage and the durability of the storage reservoir. In addition, this paper addresses new technology in carbon capture and sequestration.

To make CCS technology more feasible, this paper suggests a sustainable method combining CCS and biofuel production using CO_2 as a feedstock. This method offers many advantages, such as CO_2 emission mitigation and energy security through the production of renewable energy. Due to the many advantages of biofuels, the conversion of CO_2 into biofuel is a best practice and may provide a solution to pollution while encouraging sustainability practices.

7.5   CARBON-NEUTRAL MATERIALS

Any available energy source (such as wind, atomic energy, geothermal and solar) can be used for the production of needed hydrogen and chemical conversion of CO_2. Carbon dioxide thus can be chemically transformed from a detrimental greenhouse gas causing global warming, into a valuable, renewable and inexhaustible carbon source of the future allowing for the environmentally neutral use of carbon fuels and derived hydrocarbon products [5]. These products are excellent transportation fuels for internal combustion engines and fuel cells, as well as convenient starting materials for synthetic hydrocarbons and their varied products. For example, liquid methanol is preferable to highly volatile and potentially explosive hydrogen for energy storage and transportation. Carbon dioxide capture and storage by mineral carbonation (CCSM) is a technology that can potentially sequester billions of tons of carbon dioxide (CO_2) per year. Despite this great potential, the costs of CCSM are currently too high for a large deployment of the technology, and new systems are being investigated in an attempt to overcome these limitations [2]. Chemical recycling of carbon dioxide (CO_2) from natural and industrial sources and their varied products can be achieved via its capture and subsequent reductive hydrogenation conversion. Nature's photosynthesis uses the sun's energy with the chlorophyll in plants as a catalyst to recycle carbon dioxide and water into new plant life. Only given sufficient geological time can new fossil fuels form naturally [6].

7.6 SUSTAINABLE DEVELOPMENT

"Sustainable development is a development that the needs of the present without compromising the ability of future generations to their own needs." According to this definition from the WCED (1987), it is clear that the various activities of the construction sector have to be regarded and analyzed when considering sustainable development [7]. Nowadays, principles of sustainability have become mandatory in order to tackle global warming and the associated climate change. The governments of several countries have adequate policies in place with a view to controlling and improving the current state of their construction industries [8]. Sustainable development remains the core principle of international environmental policymaking, and of national environmental planning in many countries. Indeed, the official institutions now promoting green growth insist that it is not a substitute for sustainable development but a way of achieving it [9]. Sustainable development is often presented as being divided into the economy, environment and society [10]. The improving social, economic and environmental indicators of sustainable development are drawing attention to the construction industry, which is a globally emerging sector, and a highly active industry in both developed and developing countries [11]. The three sectors are often presented as three interconnected rings [2]. These three convenient categories make analysis more straightforward. Frequent sustainable development is presented as aiming to bring the three categories together in a balanced way. The model usually shows equal-sized rings in a symmetrical interconnection, even though there is no reason why this should be the case, but if they are seen as separate, as the model implies, different perspectives can often give greater priority to one or the other [12]. The concept of sustainable development is the result of the growing awareness of the global links between mounting environmental problems, socioeconomic issues to do with poverty and inequality and concerns about a healthy future for humanity. It strongly links environmental and socioeconomic issues [6]. The subject of concrete recycling is regarded as being very important in the general attempt for sustainable development in our times [13]. In a parallel manner, it is directly connected with (a) increase of demolition structures past out of performance time, (b) demand for new structures and (c) destruction by drilling, deep excavation of trenches and natural phenomena such as earthquakes.

7.7 CONSUMPTION TO CONSERVATION

The construction industry requires the extraction of vast quantities of materials and this, in turn, results in the consumption of energy resources and the release of detrimental pollutant emissions into the biosphere [14]. Each material has to be fully extracted, processed and finally transported to its place of use [6]. The construction industry has a significant impact on the natural and built environments. Construction activities consume a large amount of energy, natural resources and water while producing a large proportion of waste [15]. There is a large number of factors to consider when comparing energy consumption of different construction processes.

- *Reliable infrastructure:*
  - *optimizing the availability of infrastructure, with its characteristics: available, durable and reliable*
- *Green infrastructure:*
  - *reducing the environmental impact of traffic and infrastructure on the sustainable society, with its characteristics: energy efficient, sustainable and environmental friendly*
- *Safe and smart infrastructure:*
  - *optimizing flows of traffic of all categories of road users and road construction work safety, utilizing the latest technology with its characteristics: easy to access, smart and safe for use*
- *Human infrastructure:*
  - *Harmonizing infrastructure with the human dimensions, with its characteristics: multi-functional, multi-usable and ensure public security*

Finally, the three different structural pavement rehabilitation alternatives were studied and compared using an energy consumption methodology:

- Asphalt overlay
- Reconstruction
- Cold in place recycling with foamed asphalt

The methodology considers different project scenarios by combining expected traffic and soil support values. For each rehabilitation technique and scenario, the construction processes were analyzed and the design layers were transformed to equivalent energy units (MJ/m^2). Different attributes are used for measuring project performance, including economic performance attributes, social performance attributes, and as well as environmental performance attributes.

   i. Flow Chart of Research Activities.
     The flow chart activities of this research work is shown in the Figure 7.2.
   ii. Sustainable Road Development
     Sustainable roads are constructional systems of roads and highways that consist of the planning, design, construction and maintenance phases, which comprise transport and environmental sustainability that is beneficial to transport systems, ecology and urban improvement [11, 16] Carbon neutral materials are used in sustainable roads (green roads) to improve the physical properties of engineering materials and to reduce the carbon emissions released from the vehicle and construction industries [8, 14, 17]. Green economic growth, poverty reduction and social and environmental improvements are the major and overriding priorities in developed and developing countries and are essentially aimed at sustainability targets [10, 12, 18]. Therefore, to achieve the principal purpose of sustainable development

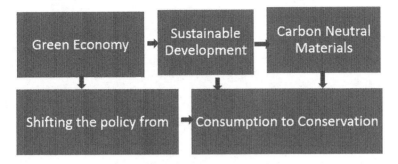

FIGURE 7.2 Flow chart activities of the research work.

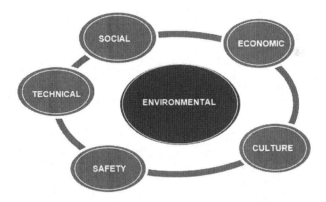

FIGURE 7.3 Principal components of sustainable development in the construction industry [19].

in the construction industry, it is necessary to address the techniques presented in Figure 7.3 [5].

Must consider:
- Economic
- Cultural
- Environmental
- Social
- Technical
- Safety

The use of CO_2 neutralization for fossil and biofuels not only minimizes global warming, but also provides an acceptable, unlimited carbon source for the near future [19–21]. Cyclic transformation should include the conversion of carbon dioxide and methane Figure 7.4 shows a carbon-neutral cycle for different carbon-neutral materials.

Based on current levels of development, different countries have different capacities to introduce and implement policies that cope with transformative change [24, 25].

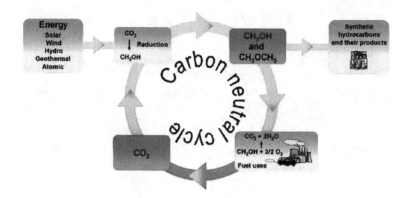

FIGURE 7.4   Cyclic transformation of carbon-neutral materials [22, 23].

7.8   POTENTIAL APPLICATION

This study is all about finding and developing new policies for global and regional environmental challenges toward a successful green economy, sustainable development, and consumption to conservation that are not yet established in the related field. However, this study will cover the possible ways forward in contributing to developing research in the area of the green economy, which is presently limited but highly needed because it appears to promise an attractive orientation out of the present crisis of economic recession. The green economy has been neglected because of the current economic recession. This can be achieved by two solutions: first, to reduce working hours in construction work; second, to interact in structural shifts toward the sustainable development of low productivity. The two green economy strategies are combined together based on their functional capabilities to achieve voluminous high productivity using a simple simulation model of the UK economy.

7.9   RECOMMENDATIONS

- Renewable energy inputs should be kept within the regenerative capacities of the natural resources, should be minimized and should not exceed the minimum strategic levels.
- For a sustainable economy and a cleaner world, some steps should be taken for environmental protection. This can only be achieved with the transformation of low-carbon society and an embracing of a green economy.
- Reduce the consumption of a disproportional amount of natural resources.
- Producers of sustainable energy resources (atomic energy, geothermal and solar) should make these resources more affordable and harmless for human health and the environment.
- Increase the use of wastes or recycled products as raw materials or additives with improved environmental performance
- Materials originated from renewable sources should be used during the sustainable construction process.

- There is a need for shifting the traditional approach of the sustainable construction project from consumption to a new conservation approach that accepts the principles of sustainable development.
- The way of positioning the construction sector into the global approach of sustainable development should be fully clarified and clearly claimed.
- The sustainable construction project should be given more focus on the methods for improving project quality, safety performance and it should be environmentally friendly for the present and future practice of the industry.
- To improve the existing practice of construction implementation toward contributing to sustainable development, all the three dimensions, including economic, social and environmental issues need to be fully concerned in conducting sustainable construction project.
- Use of a sustainability index will greatly simplify the measurement of sustainable development, thereby making a positive contribution to the identification of optimum design solutions and facility operations.
- Sustainability should be integrated into the construction project under the consideration of its life cycle analysis (sustainability would only be possible when construction uses renewable energy resources, and renewable materials or materials recycled from construction wastes).
- To comply with sustainability, the land should not be irreversibly mortgaged.
- When using materials with limited durability, modular construction should be considered.

REFERENCES

1. P. L. Spath, and M. K. Mann, *Life Cycle Assessment of Hydrogen Production via Natural Gas Steam Reforming*. Golden, CO: National Renewable Energy Lab, 2000.
2. P. Moriarty, and D. Honnery, "Intermittent renewable energy: The only future source of hydrogen?", *International Journal of Hydrogen Energy*, vol. 32(12), pp. 1616–1624, 2007.
3. N. D. Oikonomou, "Recycled concrete aggregates", *Cement and Concrete Composites*, vol. 27(2), pp. 315–318, 2005.
4. F. A. Rahman, M. M. A. Aziz, R. Saidur, W. A. W. A. Bakar, M. Hainin, R. Putrajaya, and N. A. Hassan, "Pollution to solution: Capture and sequestration of carbon dioxide (CO_2) and its utilization as a renewable energy source for a sustainable future", *Renewable and Sustainable Energy Reviews*, vol. 71, pp. 112–126, 2017.
5. J. Seppälä, S. Koskela, M. Melanen, and M. Palperi, "The Finnish metals industry and the environment", *Resources, Conservation and Recycling*, vol. 35(1–2), pp. 61–76, 2002.
6. B. K. Gross, I. J. Sutherland, and H. Mooiweer, *Hydrogen Fueling Infrastructure Assessment*. General Motors Corporation, Research & Development Center, 2007.
7. L. Bourdeau, "Sustainable development and the future of construction: A comparison of visions from various countries", *Building Research and Information*, vol. 27(6), pp. 354–366, 1999.
8. J. Kohlmann, and R. Zevenhoven, "The removal of CO_2 from flue gases using magnesium silicates, in Finland", In: *11th International Conference on Coal Science (ICCS 11)*, San Francisco (CA), September, 2001.
9. P. Joseph, and S. Tretsiakova-McNally, "Sustainable non-metallic building materials", *Sustainability*, vol. 2(2), pp. 400–427, 2010.

10. W. Sachs, *Planet Dialectics: Explorations in Environment and Development*. Zed Books Ltd., 2015.

11. V. Malhotra, "Making concrete "greener" with fly ash", *Concrete International*, vol. 21, pp. 61–66, 1999.

12. P. Mehta, "Concrete technology i '-'or sustainable development", In: *Concrete Technology for a Sustainable Development in the 21st Century*, vol. 83. CRC Press, 1999.

13. C. S. Poon, A. T. W. Yu, S. W. Wong, and E. Cheung, "Management of construction waste in public housing projects in Hong Kong", *Construction Management and Economics*, vol. 22(7), pp. 675–689, 2004.

14. W. Cai, Y. Wu, Y. Zhong, and H. Ren, "China building energy consumption: Situation, challenges and corresponding measures", *Energy Policy*, vol. 37(6), pp. 2054–2059, 2009.

15. E. P. Mora, "Life cycle, sustainability and the transcendent quality of building materials", *Building and Environment*, vol. 42(3), pp. 1329–1334, 2007.

16. E. Lauritzen, "The global challenge of recycled concrete", In: *Sustainable Construction: Use of Recycled Concrete Aggregate: Proceedings of the International Symposium Organised by the Concrete Technology Unit, University of Dundee and Held at the Department of Trade and Industry Conference Centre, London, UK, 11–12 November 1998*, pp. 505–519, 1998.

17. A. Brunnengräber, and T. Haas, *Green Economy–Green New Deal–Green Growth: Occupy Rio Plus 20*. W&E Hintergrund, pp. 1–2, 2011.

18. W. H. Scholz, "Processes for industrial production of hydrogen and associated environmental effects", *Gas Separation and Purification*, vol. 7(3), pp. 131–139, 1993.

19. R. Benioff, S. Guill, and J. Lee, *Vulnerability and Adaptation Assessments: An International Handbook*, vol. 7. Springer Science & Business Media, 2012.

20. M. A. Beran, and N. W. Arnell, "Climate change and hydrological disasters", In: *Hydrology of Disasters*. Springer, pp. 41–62, 1996.

21. M. Dresselhaus, G. Crabtree, and M. Buchanan, *Basic Research Needs for the Hydrogen Economy*, Technical Report, Argonne National Laboratory, Basic Energy Sciences, US DOE, 2003.

22. M. Munasinghe, R. Swart, "Social sustainability, indicators and climate change", In: *Climate Change and Its Linkages with Development, Equity, and Sustainability*. Geneva: Jointly Published by LIFE, RIVM and World Bank for IPCC, pp. 95–108, 2000.

23. S. Hug, Z. Karim, M. Asaduzzaman, F. Mahtab, "Adaptation to climate change in Bangladesh: Future outlook", In: *Vulnerability and Adaptation to Climate Change for Bangladesh*. Springer, pp. 125–143, 1999.

24. U. Berardi, "Sustainability assessment in the construction sector: Rating systems and rated buildings", *Sustainable Development*, vol. 20(6), pp. 411–424, 2012.

25. P. R. Berke, "Natural-hazard reduction and sustainable development: A global assessment", *Journal of Planning Literature*, vol. 9(4), pp. 370–382, 1995.

8 Toward Sustainable Landfill Siting: A Case Study for Johor Bahru, Malaysia

Habiba Ibrahim Mohammed,
Zulkepli Majid, and Yamusa Bello Yamusa

CONTENTS

8.0 INTRODUCTION

The increase in economic and population growth has given rise to an increase in the tons of waste generated per day. Sustainable solid waste management is one of the most serious environmental issues faced by many developed and developing countries. This is because of the complexity of the management process as it involves decision makers, government authorities and other related parties [1]. Municipalities are required to maintain a good, sound and healthy environment in order to overcome these challenges. The siting of sanitary landfills is tedious and complicated as it requires various criteria and sub-criteria, such as environmental, socio-economic,

113

physical, etc. Expertise is needed from many different fields including civil engineering, pedology, urban planning and land surveying [2].

Johor Bahru is one of the most populous cities in Malaysia. It is situated along the border with Singapore. It has a population of about 497,097 in 2010 and it is one of the most populous cities in the country. Johor Bahru is also a popular tourist city due to its favorable equatorial climate, and this attracts both local and international tourists. All these factors contribute to waste production and the problems surrounding it, whereby the existing sanitary landfill sites are reaching their capacity. In addition, the well-known social and political opposing factor affecting landfill siting referred to as the "not in my back yard" (NIMBY) behavior of individuals is causing a serious dilemma for policy and decision makers [3]. Therefore, new and sustainable sanitary landfill sites are required for the disposal of municipal solid waste (MSW). This site is expected to protect the well-being of the environment in terms of public health, water resources (i.e., surface and groundwater) and heritage buildings.

Different methods of selecting a site for waste disposal are in existence in the literature. Many researchers from different parts of the world have used various techniques in siting landfills, such as the integration of multiple criteria analysis (MCDA) and GIS [4–7]. In their studies, economic, social and environmental criteria were used in order to select the most suitable sites for landfill. The environmental factor is a very significant factor to consider before constructing a site for MSW disposal because it can affect the ecological and biophysical environment of an area (Alanbari et al., 2014). Furthermore, economic factors need to be considered in the siting process, especially the cost of land acquisition and the monitoring and operation of the site [2, 8].

Presently, GIS is one of the most used advanced technologies in solving environmental issues such as flooding, landslides, earthquakes, urban planning and building information modeling (BIM). GIS plays a vital role in processing and analyzing spatial information due to its competency in dealing with a huge amount of spatial data from various sources [1, 9]. Multi-criteria evaluation (MCE) is used by decision makers to solve the problems and difficulties that arise due to conflicts of interest when siting landfills. The MCE technique helps to break down the decision problems into easier, more understandable components [10]. Currently, the combination of GIS and MCE is an effective method in solving many problems that involve multiple attributes or alternatives [11]. From a literature review, it has been noticed that the use of this technique has tremendously risen in the last few decades [12–14]. Alongside the MCE method, the analytical hierarchical process (AHP) was found to be the most popular and commonly used process in the previous studies that have been carried out [15–18].

There is only one sanitary landfill in Johor Bahru, and it is expected to be closed by 2024 (IRDA). From the thousand tons of waste collected in Johor, a large percentage of it is being disposed in landfill. This waste is being collected from at least three local authorities which are Johor Bahru, Kulai and Johor Bahru Tengah. This objective presents the techniques and methodology used in the assessing and selecting of optimal sites for sanitary landfill for at least 30 years lifespan using MCDA and GIS. The objective of this study is to select suitable sites for landfills within Johor Bahru, using the integration of GIS and MCDA. In addition, future population and waste generation predictions are considered and calculated; the required landfill area to cover the huge volume of MSW produced for at least 30 years was calculated.

8.1 MATERIALS AND METHODS

8.1.1 STUDY AREA

The study area is situated in the southernmost part of Peninsular Malaysia. It lies within latitude 1° 29' 0" N and longitude 103° 44' 0" E as shown in Figure 8.1, with a total land area of about 220km². It covers the administrative boundary of Johor Bahru (JB) which is the capital city of Johor, Malaysia. JB waste generation is about 1.06 kg/person/d and according to estimation is expected to rise to 1.4 kg/person/d by the year 2025 [19]. Solid waste produced in the JB area has risen to nearly 30% from 2005 to 2010 and is estimated to rise to 50% by 2025 [20].

8.1.2 MATERIALS AND METHODS

Landfill siting criteria for both local and international guidelines were identified. Local guidelines, such as the Integrated Solid Waste Management Blueprint for Iskandar Malaysia, and the guideline for waste disposal siting in the National Strategic Plan for Solid Waste Management [21, 22] were used. Other sources include information about landfill siting from a review of the relevant literature, published articles in journals, conference proceedings, books and the Internet.

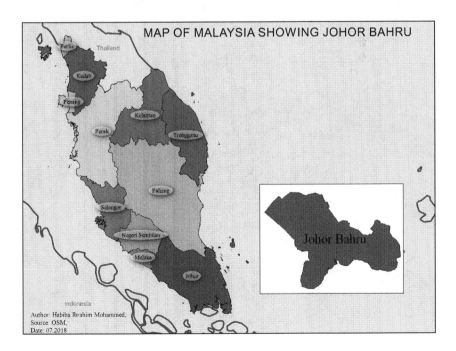

FIGURE 8.1 Map of the study area.

For this study, data was collected from various departments. The Johor Bahru administrative boundary and land use map was obtained from the Iskandar Regional Development Agency (2015). The geological map, with a scale of 1:500,000, was derived from a paper scanned geological map of peninsular Malaysia published by the Director General of geological survey, Malaysia (1985); this was accomplished through digitization. The road and river map were extracted from the digitization of the topographical map series 4551 published in 1996 with a scale of 1:50,000. All data were geo-referenced according to the Kertau RSO projection system. The United States Geological Survey Global Visualization Viewer (USGS GloVis) was used to access other remote sensing data needed for this study from their online archive, such as ASTER GDEM with a spatial resolution of 30 m, which was used to extract a digital elevation model (DEM) and slope information of the study area. ERDAS Imagine software was used in processing and analyzing the satellite images. ArcGIS software was used for digitizing and spatial data analysis. Additionally, several GIS analyses such as buffer zoning, distance, reclassify and overlay analysis were used. In order to evaluate the site selection criterion, MCDA was used to measure the relative importance weight for individual evaluation criteria. MCDA divides the decision problems into smaller understandable parts, analyze each part separately, and then integrate the parts in a logical manner. The flowchart in Figure 8.2 illustrates landfill site selection process in this study.

For the identification process of the sanitary waste disposal, all relevant criteria were recognized from previous landfill site selection studies based on GIS and

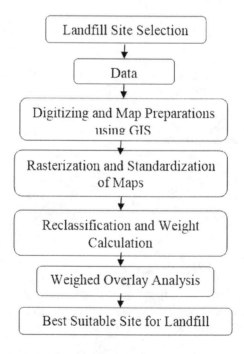

FIGURE 8.2   Flow chart of the methodology.

MCDA [23–27]. In this study, environmental and socio-economic criteria were considered such as water body, road, slope, elevation, residential settlements, forest, vacant and agricultural lands, population, solid waste generation, soil, geology and urban areas. The GIS/AHP method was used to evaluate the best suitable area for siting the landfill. Due to multiple conflicting objectives that normally occur in the decision-making process, AHP was used to help the decision makers to reach the best possible outcome [28]. Criteria weight was calculated by ranking its importance and suitability [11, 29].

8.1.3   POPULATION GROWTH RATE

The growth rate was calculated according to the estimated population of JB from the period of 2000 to 2010, which was extracted from the Department of Statistics Malaysia (web) using the formula below [30].

$$P_e = P_f (1 + r)^n \qquad \text{(Equation 8.1)}$$

where

n = number of years.
r = rate of population growth (in percentage).
P_e = estimated population.
P_f = latest population.

The application of this equation was used to calculate population growth rate for Johor Bahru. The percentage of the growth rate was 3.1%. Furthermore, the daily solid waste generated per person in JB was about 1.06kg/person/d, which is expected to rise to about 1.4kg/person/d by the year 2025 [19].

From Table 8.1, the calculated sum of the projected solid waste from 2010 to 2025 is about 50%. According to Blueprint for Iskandar Malaysia [22], estimated land requirement based on 1,000 tons per day capacity and landfill lifespan of 15 years is 100 hectares excluding the buffer.

TABLE 8.1
Population and solid waste projections for JB [22]

S/No	Year	Population	Solid Waste (tons/year)
1	2010	815,600	315,556
2	2015	952,052	406,574
3	2020	1,104,843	520,215
4	2025	1,493,400	763,128

8.2   RESULTS AND DISCUSSION

According to both local and international guidelines for landfill siting, the site must be located in a safe environment free from polluting soil and water bodies and must be able to prevent the public from the negative long-term effects caused by landfill. It is also not expected to be constructed in a densely populated environment where it can affect public health [25, 31]. For this study, the final suitability map was produced through the AHP pair wise comparison method and assigning weight to each criterion according to its level of significance, after which it was computed in the GIS environment. The criteria, sub-criteria, buffer zones and rankings used are presented in Table 8.2. From the suitability map, it was observed that the environmental criteria have more weight compared to the socio-economic criteria. In addition, population and MSW projected datasets for the study area were also considered to prevent problems related to early closure or inadequate land capacity as mentioned above for waste compaction which can be recognized as an innovative approach in landfill siting study.

8.2.1   Landfill Siting Criteria

8.2.1.1   Surface Water

Sanitary landfill sites cannot be constructed adjacent to water bodies such as rivers, lakes, streams and ponds [24]. This is due to adverse environmental effects which can occur and contaminate the water body as a result of leachate produced from the landfill site in that area [1]. Thus, a 500 m buffer zone was created for each of the water bodies in the study area, in accordance with the Malaysian landfill siting guidelines [21, 32].

8.2.1.2   Distance to Road

The topo-sheet map of the study area was used to extract the road data. The road data includes information on highways, major and minor roads and their connections. This data is necessary because when selecting landfill sites, accessibility of the site must be considered especially for the vehicles used for the collection and disposal of waste. Due to this reason, high scores were given to areas near roads.

8.2.1.3   Slope

The slope is an important criterion when siting a sanitary landfill site. From an economic perspective, the cost of construction in areas with high, steep slopes will be more expensive than in areas with medium slopes. The slope layer for this study was generated from the digital elevation model (DEM). The areas with slopes greater than 15° are considered unsuitable for sanitary landfill [33].

8.2.1.4   Land Use/Land Cover (LU/LC)

The LU/LC map was derived from the Iskandar Regional Development Agency (IRDA). There are different types of land use in the study area. Therefore, values were assigned to each land use type based on its suitability level. Areas like sensitive ecosystems, protected forests and historical sites were assigned with 0 scores, while areas like vacant and agricultural lands were given the scores of 10 and 6, respectively.

TABLE 8.2
Criteria weight and ranking

Main criteria	Weight	Sub-criteria	Buffer zones	Ranking
		Water bodies	<500 m	0
			500–1000	2
			1000–2000	6
			2000–3000	8
			>3000	10
		Soil	Low permeable	8
			Medium permeable	5
			Highly permeable	0
Environmental	0.70	Slope	0–5°	6
			6°–10°	4
			10°–15°	3
			>15°	0
		Elevation	500–1000 m	0
			1000–1500	3
			>1500	5
		Residential	<500 m	0
			500–1000	3
			1000–2000	5
			>2000	9
		Airports	<3 km	0
			3–6 km	6
			6–9 km	7
			9–12 km	8
			>12 km	10
Socio-Economic	0.30	Road	<500	0
			500–1000	2
			1000–2000	6
			2000–3000	5
			>3000	0
		Population	Low density	7
			Medium density	5
			High density	4
		Infrastructures	Power lines	0
			Pipe lines	0
		LU/LC	Agric land	6
			Forest	3
			Vacant land	10

8.2.1.5 Soils

Soil is one of the key criteria to be considered when siting a sanitary landfill. This is to prevent groundwater contamination from landfill leachate. The soils of the study area are divided into three permeability zones which are high, medium and low zones. Therefore, the areas with high permeability rate are considered unsuitable for sanitary landfill sites [34] and were assigned with 0 scores. This is because leachate infiltration is possible to occur there by contaminating both the ground and surface water of the neighboring areas. In addition, areas with low permeability rate were given preference and considered to be suitable for sanitary landfill sites with the score of 10.

8.2.1.6 Geology

There are seven different geological structures present in the study area based on the geological map, three are of intrusive rocks, which are intermediate, acidic and basic (mainly gabbro). Scores were assigned to these geological structures according to their suitability level based on reviewed literature [26, 35], making sure that they were not located in a geological fault area.

8.2.1.7 Residential Areas

This criterion does not permit landfill siting in the area. The presence of any waste disposal site near or within the urban and rural residential area may cause health and environmental problems. According to related literature regarding distance to residential areas [1, 17], the desirable distance from landfill sites to residential areas should be at least 1 km. Therefore, a distance of 500 m and above was taken into consideration as being a suitable landfill location.

8.2.1.8 Population

This is an important factor to consider when siting landfill. The population data was used to identify urban centers with high population density [36]. Therefore, areas with a high density of population were given low scores while areas with a low density of population were given high scores.

Final suitability map was obtained after assigning weight to each criterion through the AHP method and sub-criteria weighting, the ArcGIS was used to reclassify each criteria map, which was converted from vector to raster format. The map algebra tool was further used to execute the analysis by multiplying each criterion with its weight plus another criterion for the suitable landfill sites. The results were classified as being either unsuitable, less suitable, moderately suitable, suitable and most suitable Figure 8.3.

Final suitability map (Figure 8.3) was obtained after assigning weight to each criterion through the AHP method and sub-criteria weighting, the ArcGIS was used to reclassify each criteria map which was converted from vector to raster format. Map algebra tool was further used to execute the analysis by multiplying each criterion with its weight plus another criterion for the suitable landfill sites. The results were classified various classes which are unsuitable, less suitable, moderately suitable, suitable, and most suitable.

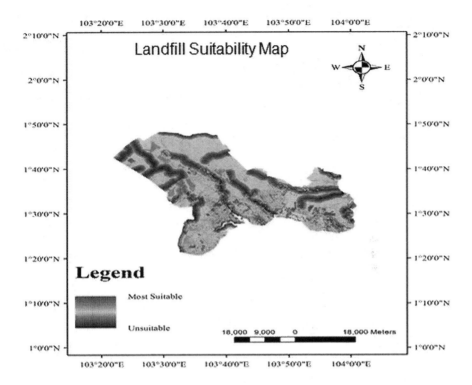

FIGURE 8.3   Landfill suitability map.

8.3   CONCLUSIONS

Rapid population growth and an increase in economic and commercial activities have resulted in an increased amount of regularly produced solid waste. This study proposed an enhanced method in the process of selecting landfill sites in Johor Bahru, which include more siting criteria, such as population and projected solid waste by using GIS and AHP. The use of GIS has increased tremendously over the past few decades in solving environmentally related issues. It can store, retrieve, manipulate and analyze a huge volume of spatial data and display it as valuable information. While the AHP method was implemented to allocate criteria weight to each parameter based on its relative importance in the landfill decision process. Finally, this study was able to compute an optimized site for a sustainable and long-term solution to solid waste management in Johor Bahru, with a total land area calculated to be about 100 hectares to accommodate the generated waste for up to 30 years.

REFERENCES

1. D. Güler, and T. Yomralıoğlu, "Alternative suitable landfill site selection using analytic hierarchy process and geographic information systems: A case study in Istanbul", *Environmental Earth Sciences*, vol. 76, p. 678, 2017.

2. S. Issa, and B. Shehhi, "A GIS-based multi-criteria evaluation system for selection of landfill sites: A case study from Abu Dhabi, United Arab Emirates", International Archives of the Photogrammetry, Remote Sensing and Spatial Information Sciences, Melbourne, Australia, vol. 39, 2012.
3. N. B. Chang, G. Parvathinathan, and J. B. Breeden, "Combining GIS with fuzzy multicriteria decision-making for landfill siting in a fast-growing urban region", *Journal of Environmental Management*, vol. 87, pp. 139–153, 2008.
4. A. Ohri, S. P. Maurya, and S. Mishra, "Sanitary landfill site selection by using geographic information system", vol. 1, pp. 170–180, 2015.
5. Z. G. Rahmat, M. V. Niri, N. Alavi, G. Goudarzi, A. A. Babaei, Z. Baboli *et al.*, "Landfill site selection using GIS and AHP: A case study: Behbahan, Iran", *KSCE Journal of Civil Engineering*, vol. 21, pp. 111–118, 2016.
6. M. Torabi-Kaveh, R. Babazadeh, S. D. Mohammadi, and M. Zaresefat, "Landfill site selection using combination of GIS and fuzzy AHP, a case study: Iranshahr, Iran", *Waste Management & Reseach*, vol. 34, pp. 438–448, 2016.
7. H. I. Mohammed, Z. Majid, N. B. Yusof, and Y. B. Yamusa, "A review of GIS-based and multi-criteria evaluation method for sustainable landfill site selection model", In: Global Civil Engineering Conference. Singapore: Springer, pp. 741–751, 2017.
8. O. B. Delgado, M. Mendoza, E. L. Granados, and D. Geneletti, "Analysis of land suitability for the siting of inter-municipal landfills in the Cuitzeo Lake Basin, Mexico", *Waste Management*, vol. 28, pp. 1137–1146, 2008.
9. T. D. Kontos, D. P. Komilis, and C. P. Halvadakis, "Siting MSW landfills with a spatial multiple criteria analysis methodology", *Waste Management*, vol. 25, pp. 818–832, 2005.
10. J. Malczewski, "GIS-based land-use suitability analysis: A critical overview", *Progress in Planning*, vol. 62, pp. 3–65, 2004.
11. A. Barakat, A. Hilali, M. El Baghdadi, and F. Touhami, "Landfill site selection with GIS-based multi-criteria evaluation technique. A case study in Béni Mellal-Khouribga Region, Morocco", *Environmental Earth Sciences*, vol. 76, p. 413, 2017.
12. A. Afzali, J. Samani, and M. Rashid, "Municipal landfill site selection for Isfahan City by use of fuzzy logic and analytic hierarchy process", *Iranian Journal of Environmental Health Science & Engineering*, vol. 8, p. 273, 2011.
13. B. Nas, T. Cay, F. Iscan, and A. Berktay, "Selection of MSW landfill site for Konya, Turkey using GIS and multi-criteria evaluation", *Environmental Monitoring and Assessment*, vol. 160, pp. 491–500, 2010.
14. N. R. Khalili, and S. Duecker, "Application of multi-criteria decision analysis in design of sustainable environmental management system framework", *Journal of Cleaner Production*, vol. 47, pp. 188–198, 2013.
15. H. Javaheri, T. Nasrabadi, M. Jafarian, G. Rowshan, and H. Khoshnam, "Site selection of municipal solid waste landfills using analytical hierarchy process method in a geographical information technology environment in Giroft", *Journal of Environmental Health Science & Engineering*, vol. 3, pp. 177–184, 2006.
16. G. Wang, L. Qin, G. Li, and L. Chen, "Landfill site selection using spatial information technologies and AHP: A case study in Beijing, China", *Journal of Environmental Management*, vol. 90, pp. 2414–2421, 2009.
17. P. V. Gorsevski, K. R. Donevska, C. D. Mitrovski, and J. P. Frizado, "Integrating multi-criteria evaluation techniques with geographic information systems for landfill site selection: A case study using ordered weighted average", *Waste Management*, vol. 32, pp. 287–296, 2012.
18. A. J. Chabuk, N. Al-Ansari, H. M. Hussain, S. Knutsson, and R. Pusch, "GIS-based assessment of combined AHP and SAW methods for selecting suitable sites for landfill in Al-Musayiab Qadhaa, Babylon, Iraq", *Environmental Earth Sciences*, vol. 76, Article ID: 209, 2017.

19. A. H. Abba, Z. Z. Noor, A. Aliyu, and N. I. Medugu, "Assessing sustainable municipal solid waste management factors for Johor-Bahru by analytical hierarchy process", *Advanced Materials Research*, vol. 689, pp. 540–545, 2013.
20. S. T. Tan, C. T. Lee, H. Hashim, W. S. Ho, and J. S. Lim, "Optimal process network for municipal solid waste management in Iskandar Malaysia", *Journal of Cleaner Production*, vol. 71, pp. 48–58, 2014.
21. Ministry of Housing and Local Government, National Strategic Plan for Solid Waste Management, Putrajaya, Malaysia, 2005.
22. Blueprint for Iskandar Malaysia, Integrated Solid Waste Management Blueprint of Iskandar Malaysia, Johor Bahru, Malaysia, 2010.
23. S. K. Abujayyab, M. Ahamad, A. S. Yahya, and A. H. Saad, "A new framework for geospatial site selection using artificial neural networks as decision rules: A case study on landfill sites", *ISPRS Annals of the Photogrammetry, Remote Sensing and Spatial Information Sciences*, vol. 2, p. 131, 2015.
24. S. Z. Ahmad, M. S. Ahamad, and M. S. Yusoff, "Spatial effect of new municipal solid waste landfill siting using different guidelines", *Waste Management & Research*, vol. 32, pp. 24–33, 2014.
25. M. Eskandari, M. Homaee, S. Mahmoodi, and E. Pazira, "Integrating GIS and AHP for municipal solid waste landfill site selection", *Journal of Basic and Applied Scientific Research*, vol. 3, pp. 588–595, 2013.
26. Ş. Şener, E. Sener, and R. Karagüzel, "Solid waste disposal site selection with GIS and AHP methodology: A case study in Senirkent–Uluborlu (Isparta) Basin, Turkey", *Environmental Monitoring and Assessment*, vol. 173, pp. 533–554, 2011.
27. H. I. Mohammed, Z. Majid, N. B. Yusof, and Y. B. Yamusa, "Analysis of multi-criteria evaluation method of landfill site selection for municipal solid waste management", 2018.
28. T. Saaty, "The analytic hierarchy and analytic network processes for the measurement of intangible criteria and for decision-making", *Multiple Criteria Decision Analysis: State of the Art Surveys*, vol. 1, pp. 345–405, 2005.
29. C. Kara, and N. Doratli, "Application of GIS/AHP in siting sanitary landfill: A case study in Northern Cyprus", *Waste Management & Research*, vol. 30, pp. 966–980, 2012.
30. M. Alanbari, N. Al-Ansari, H. Jasim, and S. Knutsson, "Modeling landfill suitability based on GIS and multicriteria decision analysis: Case study in Al-Mahaweel Qadaa", *Natural Science*, vol. 6, pp. 828–851, 2014.
31. S. Bahrani, T. Ebadi, H. Ehsani, H. Yousefi, and R. Maknoon, "Modeling landfill site selection by multi-criteria decision making and fuzzy functions in GIS, case study: Shabestar, Iran", *Environmental Earth Sciences*, vol. 75, p. 337, 2016.
32. S. Z. Ahmad, M. S. S. Ahamad, and M. S. Yusoff, "A comprehensive review of environmental, physical and socio-economic (EPSE) criteria for spatial site selection of landfills in Malaysia", *Applied Mechanics and Materials*, vol. 802, pp. 412–418, 2015.
33. M. A. Alanbari, N. Al-Ansari, H. K. Jasim, and S. Knutsson, "Al-Mseiab Qadaa landfill site selection using GIS and multicriteria decision analysis", *Engineering*, vol. 06, pp. 526–549, 2014.
34. V. R. Sumathi, U. Natesan, and C. Sarkar, "GIS-based approach for optimized siting of municipal solid waste landfill", *Waste Management*, vol. 28, pp. 2146–2160, 2008.
35. N. Alavi, G. Goudarzi, A. A. Babaei, N. Jaafarzadeh, and M. Hosseinzadeh, "Municipal solid waste landfill site selection with geographic information systems and analytical hierarchy process: A case study in Mahshahr County, Iran", *Waste Management & Research*, vol. 31, pp. 98–105, 2013.
36. E. Aksoy, and B. T. San, "Geographical information systems (GIS) and Multi-Criteria Decision Analysis (MCDA) integration for sustainable landfill site selection considering dynamic data source", *Bulletin of Engineering Geology and the Environment*, vol. 78, pp. 779–791, 2017.

9 Solution of Global Elevated Carbon Dioxide Emission— A Comprehensive Technique and Methodology for Quantification

Md. Maniruzzaman A. Aziz, Khairul Anuar Kassim, Wan Azelee Wan Abu Bakar, Nurul Hidayah Muslim, Azman Mohamed, and M. Ehsan Jorat

CONTENTS

9.0   INTRODUCTION

The quantification of CO_2 since the 1860s has illustrated an increasing concentration at an exponential rate of growth [1]. Similarly, the quantification of CO_2 emissions has been given a lot of attention over the past two decades [2]. In the past, the removal of CO_2 from the atmosphere occurred mainly via photosynthesis, in which crops and other plants naturally consumed CO_2 and sunlight and released oxygen [3, 4],[5].

However, Earth's atmosphere contains carbon dioxide (CO_2), which when present at a lower concentration is considered harmless [6, 7]. Therefore, the rising amount of CO_2 in the atmosphere has changed seasonal variations in most parts of the world [1].

Therefore, this chapter suggests a solution that can be used to reduce the pollution caused by carbon dioxide emissions and also to utilize carbon dioxide to enhance the benefit for future generations [8–10]. In view of this, the following objectives are forwarded:

   i. To develop and improve the quantification of carbon dioxide and determine its effect on the environment through the empirical method, modeling or gas analyzer.
  ii. To improve the utilization of carbon dioxide for future generations by transforming greenhouse gases into clean fuels, hydrogen and renewable energy.
 iii. To improve the environmental challenges toward the consumption of renewable energy.

Consequently, carbon dioxide seems to be the gas compound that is intensifying the most in the atmosphere. It has become the largest contributor (60%) to the atmosphere in recent years, and it is viewed as being a potential agent of global warming [2–4]. Therefore, there is an excessive volume of greenhouse gases (GHGs) in the atmospheric system, and there is a broad consensus that this will have serious consequences in terms of climate change [11, 12]. The industrial flue gas emissions include carbon dioxide (CO_2), nitrogen oxides (Nox), hydrocarbons, carbon monoxide (CO), particulate matter and sulphur dioxide (SO_2), which almost all of these emissions are GHGs [13–15]. Greenhouse gas emissions contain about 77% CO_2 [16–18].

According to recent IPCC reports, the global mean concentration of CO_2 in the atmosphere is now close to 400 ppm; however, the most comprehensive research recommended that the safe level of CO_2 concentration is below 350 ppm [19–21]. Therefore, it is important to focus on the control of CO_2 and to promote sustainable practice in all sectors [22, 23]. A certain amount of GHGs exists in the atmospheric system, helping to absorb thermal radiation from the earth's surface and then re-emitting the radiation back to the earth [24–27].

As such, carbon dioxide can be chemically transformed from a detrimental greenhouse gas into a renewable, valuable carbon source for the future [28–30]. Thus, air pollution largely results from the combustion of fossil fuels through vehicle usage, coal-fired electricity generation and industrial operations [8].

In the year 2013, emissions from fossil-fuel combustion and industrial processes (production of cement clinker, metals and chemicals) totaled 35.7 billion tonnes CO_2. At the same time, the world's economy grew by 3%, showing a partial decoupling between the growth in global CO_2 emissions and the economy [31–33].

Hence, approximately 50% of total CO_2 emissions in the atmosphere are caused by burned fossil fuels, which equates to about 2.5×10^9 metric tonnes of carbon per year, while the residual carbon is absorbed by oceans, plants and land biomass [9].

Therefore, this chapter aims to fill this gap by presenting all the CO_2 quantification techniques in different fields of applications. The chapter also categorizes the CO_2 quantification techniques into four groups: i. atmospheric conditions, ii. materials, iii. machinery and equipment operations and iv. energy consumption [34–36]. Table 9.1 shows the identification of research gap – CO_2 quantification methods.

9.1 METHODS TO QUANTIFY ELEVATED CO_2 IN THE ATMOSPHERE

According to the recent IPCC reports, the global mean concentration of CO_2 in the atmosphere is now close to 400 ppm, however the most comprehensive research recommended that the safe level of CO_2 concentration is below 350 ppm [3, 59, 60]. Therefore, based on our review of the literature, four techniques are used to measure and analyze CO_2 concentration in the atmosphere, as shown in Figure 9.1.

9.1.1 INFRARED (IR) RADIATION: 2μ (MICRON) ABSORPTION BAND

This method is developed to quantify CO_2 concentration in the atmosphere from space. The 2μ absorption band was introduced to replace the usage of high-dispersion methods, such as spectrometers and interferometers, due to its stability compared to the strong absorption band [26]. Additionally, a simple filter photometry material such as glass or quartz is sufficient to assist the measurement of CO_2 [61]. However, due to recent rapid industrial development, plants alone are no longer able to deal with the amount of CO_2 in the atmosphere and remove it naturally [3, 62, 63]. Therefore, Figure 9.2 describes the process of determining CO_2 concentration in the atmosphere. Furthermore, the measurement process uses an optical signal, an electronic signal and a logic system [12].

The process begins with the reflection of solar radiation from the Earth's surface in two narrow wavelengths of 2.1μ. Thus, optical sampling is performed using the 2.06μ absorption band, while in a region with no CO_2, absorption near the 2.14μ absorption band is conducted using a lead sulphide detector [64]. In addition, the bandwidth filter used has a size of nearly 200A. Therefore, the total amount of CO_2

TABLE 9.1
Identification of research gap – CO$_2$ quantification methods

No	Year	Author	Application				Estimation/Quantification Methods				
			Atm	Mat	P/E	Mc/Eq	EM	Man	Empiric	GA	Model
1	1987	Bornemann and [37]	✓				✓				✓
2	2003	Itoh and Kitagawa [38]		✓				✓			
3	2006	Gonzalez et. al [39]		✓					✓		
4	2006	Ross Philips [40]		✓		✓					✓
5	2008	Weber and Matthews [41]		✓							✓
6	2010	Abanda et al. [42]		✓					✓		
7	2010	Ahn, Changburn et al. [5]				✓			✓		✓
8	2011	Ammouri et al. [43]		✓					✓		✓
9	2011	Brockens et al. [44]		✓					✓		
10	2011	Li Hong Xian et al. [45]	✓	✓		✓			✓		✓
11	2012	Mayrwoger et al. [46]				✓	✓				
12	2013	Huang et al.[47]		✓		✓			✓		✓
13	2012	Kim et al.[48]				✓			✓		
14	2012	Ren Z. Et al. [49]			✓	✓			✓		
15	2012	Richardson J. Scott et al. [50]	✓			✓	✓				
16	2013	Evrendilek F. [51]	✓				✓				✓
17	2013	Seo and Kim [52]		✓					✓		
18	2013	Melanta et al. [53]		✓	✓				✓		✓

(Continued)

TABLE 9.1 (CONTINUED)
Identification of research gap – CO_2 quantification methods

No	Year	Author	Application				Estimation/Quantification Methods				
			Atm	Mat	P/E	Mc/Eq	EM	Man	Empiric	GA	Model
19	2013	Lijewski et al. [54]				✓				✓	✓
20	2014	Moussavi and Akhbarnezhad [55]		✓		✓			✓		
21	2014	Wang et al. [56]		✓	✓	✓			✓		
22	2015	Chou et al. [57]		✓	✓	✓			✓		
23	2015	Peng et al. [58]		✓				✓	✓		✓

Note: Atm=atmosphere, Mat=material, Mc=machine, Eq=equipment, EM=electromagnetic, Man=manual, Empiric=empirical method, GA=gas analyzer

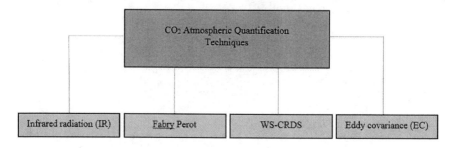

FIGURE 9.1   Material emission quantification techniques.

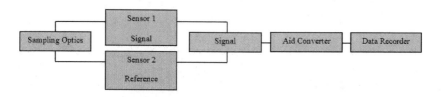

FIGURE 9.2   Flow chart of quantification process using 2μ absorption band [12].

within the understudied atmosphere sample is calculated by the logarithmic ratio of the two existing signals [12].

9.1.2   FABRY–PEROT-BASED BOLOMETER

The Fabry–Perot IR detector is a low-cost, Non-Dispersive Infrared (NDIR) gas sensor that measures dilute CO_2 gas in the atmosphere. This detector has been tested in an NDIR gas sensor system for its ability to measure the CO_2 concentration in a pure nitrogen dioxide (N_2) atmosphere [27]. Therefore, the absorption mechanism is based on the constructive and destructive interference in the spacing layer between two metal mirrors. Hence, the solvent absorbs the CO_2 and is then heated (to around 120°C) before being cooled and recycled continuously for use in the upcoming cycle of the separation process [65, 66]. In addition, CO_2 removed from the solvents is dried, compressed and transported to a safe storage facility [3, 67]. Therefore, the fundamental setup is portrayed in Figure 9.3.

In order to enhance the absorption capacity so that it corresponds to different mediums, the thickness of the spacing layer d is calculated as in Equation 9.1:

$$d = (2m+1)\frac{\lambda}{4} \tag{9.1}$$

where,

    m   represents the order of interference,
    λ represents wavelength.

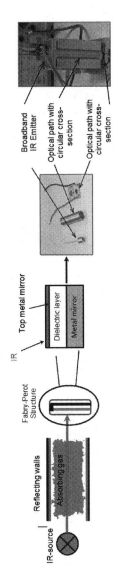

FIGURE 9.3 Non-symmetric Fabry–Perot IR detector with bolometer [27].

Therefore, the interference can be defined as the layer number, i.e., 1 denotes the first layer. The absorbent structure used in the spacing layer provides the values of the wavelength and the index of refraction. Hence, the values depend on the type of material used, e.g., germanium (Ge).

9.1.3 WAVELENGTH SCANNED-CAVITY RING DOWN SPECTROSCOPY

Wavelength Scanned-Cavity Ring Down Spectroscopy (WS-CRDS) is a laser-based technology, which improves stability and concurrent water vapor measurements. Consequently, the WS-CRDS system described consists of Carbon Analyzer Datasheet (CADS) models. The $A_{12}C_{16}O_2$ line of CO_2 is scanned using the WS-CRDS instrument to subtract the total CO_2 amount of the gas sample. However, the instrument cannot detect isotopologue such as $^{13}C^{16}O_2$ in the gas sample; this isotopic variation causes errors in the final value of the total CO_2 content [68]. At a ratio which can change between 0.001 and 0.002, the CO_2 content of the atmosphere can contain up to 1.1% $^{13}C^{16}O_2$. Based on this amount, the error value is calculated to be $2.2 \times 10-5$ at the understudied CO_2 mixing ratio [22].

Recently, most of the high-accuracy results in the field of CO_2 measurements have been made using sensors based on NDIR spectroscopic gas detectors [31,32]. Hence, WS-CRDS has several benefits over the NDIR technique, such as its ability to improve stability, reduce gas calibration and concurrent water vapor measurements and eliminate the need for drying [57].

9.1.4 EDDY COVARIANCE (EC) METHOD

Human activities induce a disturbance on the spatial-temporal dynamics of the atmosphere and the biosphere so that the key to a better perception of this effect is developing new methods to quantify and partition the long-term Net Ecosystem Exchange (NEE) of CO_2 into flux and temporal components [69–71]. Hence, the results may also be useful to improve a scheme to prevent this effect in the unstable nature of the environment [72].

9.2 METHODS TO QUANTIFY ELEVATED CARBON DIOXIDE (CO₂) EMITTED BY MATERIALS

There are numerous ways to quantify CO_2 emissions from materials [73]. Therefore, different parameters have been used to estimate emissions from materials during production and manufacturing, delivery, construction and demolition; hence, there is no available standard for the computation of emissions from materials [73–75]. However, it is important to focus on the control of CO_2 and to promote sustainable practice in all sectors [76]. Thus, a certain amount of GHGs exists in the atmospheric system and helps to absorb thermal radiation from the earth's surface and then re-emits the radiation back to the earth [3, 65, 77].

Therefore, this section will focus on describing two types of quantification techniques, including empirical methods and modelling methods as illustrated in Figure 9.4.

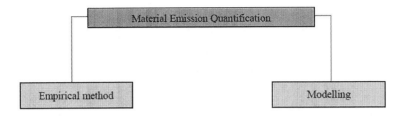

FIGURE 9.4 Material emission quantification techniques.

Hence, the industrial flue gas emissions include carbon dioxide (CO_2), nitrogen oxides (Nox), hydrocarbons, carbon monoxide (CO), particulate matter and sulphur dioxide (SO_2), almost all of which are GHGs [3, 78].

9.2.1 EMPIRICAL METHODS

Numerous empirical methods have been developed to measure material emissions. However, the complexity and uniqueness of each project represent a significant drawback with respect to implementation and standardization [79]. From our review, three empirical methods have been used in several types of construction project:

    i. Quantification of carbon dioxide (CO_2) emissions based on material emission factor;
    ii. Quantification of carbon dioxide (CO_2) based on the carbon footprint estimation tool; and
    iii. Quantification of carbon dioxide (CO_2) based on the construction environmental decision support tool (CEDST).

Therefore, these methods are occasionally integrated with a range of programming tools or carbon calculators [80]. In short, the methods to mitigate CO_2 emissions can be categorized into carbon source-based, carbon emission minimization-based and carbon sink-based methods [3, 57].

9.2.2 MODELLING METHODS

Minx et al. identify three main methods for the calculation of the embodied CO_2 emissions of goods and services:

- Process Life Cycle Assessment (PLCA),
- Input–Output (IO) Life Cycle Assessment (IOLCA), and
- Hybrid Life Cycle Assessment (HLCA) [81].

However, the first one compiles each life cycle inventory of an individual product. Therefore, the data collection can be taken directly from the Life Cycle Inventory (LCI) database.

9.2.2.1 Life Cycle Computational Model

Life cycle assessment has become common practice, especially in carbon and energy database development [82–85]. Therefore, IOLCA is useful to analyze the environmental effects of a product from its inception to disposal, based on the environmental-input analysis [57, 86, 87]. Similarly, Leontief developed input and output analysis to analyze the industrial interdependencies in a national or regional economic system prior to the development of an environmental application [43, 44, 88]. Thus, HLCA is a combined process in which sector inputs and outputs and environmental impact data are used [39][89]. The environmental impact from material extraction and manufacturing is calculated based on Equation 9.2 [90–92]:

$$I_{mi} = \sum_{j=1}^{n} \left(1 + \lambda_j\right) \times m_j \times \mu_{ji}$$ (9.2)

where,

I_{mi} denotes the material extraction and manufacturing impacts, particularly, CO_2, sulphur dioxide (SO_2), CO, NO_x and PM_{10},

j is the material type,

λ_j is the waste material factor during the erection of the building,

m_j is the amount of material in kg and

μ_{ji} is the impact i of the EF from material j during extraction and manufacturing.

9.2.2.2 Integrated Time and Cost Model

Several time cost integration models have been developed to quantify CO_2; however, the correlation between time and cost is not clearly revealed [66, 68, 93]. Therefore, Poh and Tan have integrated a WBS concept into the existing time cost model to overcome the correlation problem [48]. The evaluation is made based on WBS using materials, labor, plant and equipment and subcontract, as shown in Equations 9.3 to 9.5 [22, 94]:

$$T_{L,0} = \frac{Q_w}{P_L \times Q_L}$$ (9.3)

$$C_{L,0} = \sum \left(Q_{L,v} \times UR_{L,v}\right) \times T_{L,0}$$ (9.4)

$$C_0 = C_{P,0} + C_{L,0} + C_{S,0} + C_{M,0}$$ (9.5)

where,

$T_{L,0}$ is the duration required to complete the task,

Q_w is the quantity of work of a task,

P_L is the labor productivity,

Q_L is the quantity of allocated labor and

$C_{P,0}, C_{L,0}, C_{S,0}$ and $C_{M,0}$ are representing CO_2 emissions in plant, labor, subcontract and material, respectively.

9.2.2.3 Input–Output (IO) Model for Expenditure in Carbon Footprint (CF) Calculation

One of the advantages of utilizing the Input–Output (IO) analysis is that, it better represents the overall economics of the supply chain in the environmental assessment for all the resources (goods or services) in a country [95–97].

Furthermore, there are some major errors involved with IO analysis, as a solution to which Multiregional Input–Output (MRIO) is introduced. In theory, MRIO solves the errors through utilizing an explicit model for the various production functions in various production regions (e.g., countries) [93, 98, 99]. Therefore, to reduce the aggregation errors various disaggregated IOTs can be utilized, such as the US benchmark input–output model [100]. According to Leontief, the total output of an economy can be expressed as the sum of its intermediate consumption and final consumption, as shown in Equation 9.6 [101].

$$x = Ax + y \qquad (9.6)$$

where,

A is the economy's production function in matrix form. For total output, Equation 9.7

$$\text{yields:} \, x = (I - A)^{-1} y \qquad (9.7)$$

The production function for the economy can be generalized for an open economy in Equation 9.8:

$$x = \left(A^d + A^m \right) x + y^d + y^m + y^{ex} - m \qquad (9.8)$$

where,

A^d is the domestic portion of the production function (domestic inter-industry demand for domestic goods),

A^m shows domestic use of imports to make domestic products,

y^d, y^m and y^{ex} represent final demands for domestic production, imports and exports and

m represents total imports.

Consequently, common equation models are formed to estimate the income and expenditure elasticity ε of the Household Carbon Footprint (HCF), as defined in Equation 9.9 [102, 103]:

$$\varepsilon = \frac{\bar{y}}{CO_2} \frac{\partial CO_2}{\partial y} \qquad (9.9)$$

where,

$\bar{y}$ represents either yearly expenditure or yearly after-tax income. In the models, the household size is defined by two separate parameters, n_{child} and

n_{adult} where 'n' stands for number and the index 'child' and 'adult' represent the children and the adult group, respectively. Breaking down the parameters in this way helps to better see the corresponding impact on the overall picture since the amount of consumption for each of these groups varies. Thus, the model forms and their associated mean elasticity are shown in the Equations 9.10 to 9.13 [22, 104].

$$CO_2 = a + b + cn_{child} + dn_{adult}, \varepsilon = \frac{\bar{y}}{co_2} b \tag{9.10}$$

$$CO_2 = a + b + cy^2 + dn_{adult} + en_{adult}, \varepsilon = \frac{\bar{y}}{co_2}(b + 2c\bar{y}) \tag{9.11}$$

$$CO_2 = ay^2 \exp\left(bn_{child} + cn_{adult}\right), \quad \varepsilon = \frac{\bar{y}}{CO_2} \varepsilon ay^{e-1} \tag{9.12}$$

$$CO_2 = ay^2 + bn_{child} + cn_{adult}, \varepsilon = \frac{\bar{y}}{CO_2} \varepsilon ay^{e-1} \tag{9.13}$$

9.3 METHODS TO QUANTIFY ELEVATED CARBON DIOXIDE (CO_2) FROM ENERGY CONSUMPTION IN MACHINERY, EQUIPMENT AND OPERATIONS

From our review, we focus on three types of quantification techniques; those that use an empirical method, modelling and gas analyser, as illustrated in Figure 9.5 [105, 106]. Therefore, it is believed that these emissions will continue to increase in the future due to industrial development and economic growth [3, 107, 108].

9.3.1 CARBON EMISSION ESTIMATION MODELS

In empirically based carbon emission estimation models, the calculation of carbon emissions is based on three parameters that are commonly used in the construction industry namely, electricity, fuel and water resource consumptions [109–111].

9.3.1.1 Electricity Consumption

The total CO_2 emissions from electricity consumption throughout the life cycle of buildings can be determined using Equation 9.14 [29]:

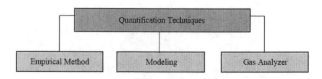

FIGURE 9.5 Construction energy, machinery and equipment emission quantification techniques.

$$P_{CO_2} = \left(P_{manuf} + P_{erect} + P_{occup+renov} + P_{demol} \right) x\delta \qquad (9.14)$$

where,

P_{manuf}	is the CO_2 emissions from electricity consumption in the construction material manufacture and transportation phase,
P_{erect}	is the CO_2 emissions from electricity consumption in the erection phase,
$P_{occup+renov}$	is the CO_2 emissions from electricity consumption in the occupation and renovation phase,
P_{demol}	is CO_2 emissions from electricity consumption in the demolition phase,
$\delta = [C_f/h_f \times C_{CO_2}/C_m]$	is the electricity emission coefficient (kg-CO_2/kWh),
C_f	is the carbon content of a given fuel (kg$_C$/kg$_{fuel}$),
h_f	is the energy content of a given fuel (kWh/kg$_{fuel}$),
C_m	is the carbon mass (kg/mol carbon) and
C_{CO_2}	is the CO_2 mass (kg/mol).

9.3.1.2 Fuel Consumption

The CO_2 emissions caused by fuel consumption can be estimated using Equation 9.15 [112], and this is applicable to the entire life cycle:

$$F_{CO_2} = \sum_{i=1}^{m} \left[\left(F_i x F_i^{ec} \right) - \left(F_i x F_i^{ec} x F_i^{cs} \right) \right] x FCO_i x\varphi \qquad (9.15)$$

where,

FCO_2	is the CO_2 emissions from fuel consumption, m is the number of fuel types,
i	is the types of fuel,
F_i	is amount of fuel types used,
F_i	equals $F_U \times F_{kt}$, F_U is unit of fuel consumed by machines (in L/km or L/h),
F_{kt}	is the transport distance or operational time of the machines,
F_i^{ec}	is the carbon emission coefficients of the fuels,
F_i^{cs}	is the carbon stored ratio of fuel,
FCO_i	is the fraction of carbon oxidized and
φ	is the molecular weight ratio of CO_2 to carbon.

9.3.1.3 Water Resource Consumption

The CO_2 emission from water resource consumption throughout the building's life cycle is shown in Equation 9.16 [113]. Therefore, the conversion coefficient represents the equivalent CO_2 emissions from energy consumption per cubic metre of water managed by water companies [113, 114].

In the short-run, however, biofuels may also increase CO_2 emissions temporarily, through indirect land use change, carbon leakage and crop yield effects [3, 115, 116].

$$W_{CO_2} = \left(W_{manuf} + W_{erect} + W_{occup+renov} + W_{demol} \right) x\delta \qquad (9.16)$$

where,

W_{CO_2} is the CO_2 emissions from water consumption,

W_{manuf} is the CO_2 emissions from water consumed in the construction material manufacture and transportation phase,

W_{erect} is the CO_2 emissions from water consumed in the erection phase,

$W_{occup/renov}$ is the CO_2 emissions from water consumed in the occupation and renovation phase,

W_{demol} is CO_2 emissions from water consumed in the demolition phase and

ρ is the carbon emission coefficient of water resources (kg-CO_2/L) = 0.193, as reported by a water company [58].

Finally, Equation 9.17 shows the overall CO_2 emission calculation for the entire building life cycle:

$$Total_{CO_2} = P_{CO_2} + F_{CO_2} + W_{CO_2} \qquad (9.17)$$

where,

P_{CO_2} is the CO_2 emissions from electricity consumption,

F_{CO_2} is CO_2 emissions from fuel consumption and

W_{CO_2} is CO_2 emissions from water consumption.

9.3.2 DISCRETE EVENT SIMULATION (DES) AND LEWIS MODEL FOR CARBON DIOXIDE (CO_2) ESTIMATION FOR CONSTRUCTION EQUIPMENT

The Discrete Event Simulation (DES) model is an integrated environment for building special-purpose simulation tools for modelling construction systems [59]. Therefore, the DES model requires a duty cycle for a given construction task and the duty cycle is referring to a complete cycle of equipment under a specific task. Meanwhile, to compute the fuel consumption rate as well as the emission discharge rate, the Lewis model is used for different duty cycles with regards to field data [117–120]. Therefore, this model categorizes equipment operations based on a specified engine speed or speed and load for a given period of time [121]. Thus, fuel rates are determined by engine power and engine model year, while emission rates in each duty cycle are calculated based on the fraction of time spent in each mode of operation [111, 122].

9.3.3 EXHAUST ANALYZER

SENSOR is a portable emission analyser which monitors the exhaust emissions concentration while measuring the flow rate of each exhaust gas at the same time [60]. After passing through the analyser probe at 191°C, the exhaust gases are separated and then introduced to the Flame-Ionising Detector (FID) where their Hydro-Carbon (HC) concentration is measured [123]. In the next stage, to measure the concentration of emissions, the temperature is lowered to 4°C [124]. Additionally, the concentration of NO_x emission is then calculated by a non-dispersive ultra violet (NDUV) analyser device, while carbon monoxide (CO) and CO_2 concentrations are measured in the NDIR analyser and O_2 concentration is calculated in the listed order [125]. It is possible to add data sent directly from the vehicle diagnostic system to the central unit of the analyser and use a global positioning system (GPS) signal. For comparative purposes herein, after running the exhaust emission analysis, an on-board diagnostic system signal was utilized to measure the engine speed, vehicle speed, load and the intake air temperature [71, 126, 127]. In fact, it is possible to develop the time-density maps from some of these signals, which will display the actual operating time of the vehicle during field operation [128, 129]. Lastly, Table 9.2 is demonstrating the SENSOR process flow schematically [21].

TABLE 9.2

Flow channel of exhaust gas using gas analyzer (SENSOR) [21]

Model name, application	Air pollutant estimation	Source/ organization	Utility	Limitations
NONROAD 2008, equipment	CO, CO_2, NOx, SOx, HC, PM	U.S EPA	Non road vehicle and diesel equipment	Do not consider alternative fuels and produces country/ national emission level only
URBEMIS, equipment and material	CO, CO_2, NOx, PM, ROG	SQAQMD	Urban project construction	Consider material and equipment only
Palate, equipment and material	CO, CO_2, NOx, PM10, SO_2, leachate	University of California, Berkeley	Lifecycle emission from material and equipment use on pavement construction projects	Accounts for limited variety of construction materials emission specific to pavement construction only and does not capture all major processes within a general construction project
OFFROAD 2007, equipment	CO, CO_2, NOx, SOx, HC, PM	California ARB	Agricultural, construction, garden equipment and recreation vehicles	Only accounts for equipment emission and does not consider alternative fuels
GLOBEIS, biogenic	VOC, CO, soil NOx	California ARB	Emission from biogenic sources	Utilizes a macroscopic approach, i.e., is not suitable for use in project-level emission estimation
COLE, biogenic	Carbon	U.S. EPA	Tool for forest carbon analysis	
CCT, biogenic	Carbon stock	USDA	State-level annualized estimates of carbon stocks on forestland	

TABLE 9.3

Summary of applied CO_2 emission estimation models in the construction sector [20]

Model name, application	Air pollutant estimation	Source/ organization	Utility	Limitations
NONROAD 2008, equipment [49]	CO, CO_2, NOx, SOx, HC, PM	U.S EPA	Non road vehicle and diesel equipment	Do not consider alternative fuels and produces country/ national emission level only
URBEMIS, equipment and material [50]	CO, CO_2, NOx, PM, ROG	SQAQMD	Urban project construction	Consider material and equipment only
Palate, equipment and material [51]	CO, CO_2, NOx, PM10, SO_2, leachate	University of California, Berkeley	Lifecycle emission from material and equipment use on pavement construction projects	Accounts for limited variety of construction materials emission specific to pavement construction only and does not capture all major processes in a general construction project
OFFROAD 2007, equipment [52]	CO, CO_2, NOx, SOx, HC, PM	California ARB	Agricultural, construction, garden equipment and recreation vehicles	Only accounts for equipment emission and does not consider alternative fuels
GLOBEIS, biogenic [24]	VOC, CO, soil NOx	California ARB	Emission from biogenic sources	Utilizes a macroscopic approach, i.e., is not suitable for use in project-level emission estimation
COLE, biogenic [53]	Carbon	U.S. EPA	Tool for forest carbon analysis	
CCT, biogenic [53]	Carbon stock	USDA	State-level annualized estimates of carbon stocks on forestland	

Table 9.3 describes the summary of applied CO_2 emission estimation models in the construction sector [20].

9.4   RECOMMENDATIONS

The selection of material plays an important role in defining CO_2 emissions, which results in variation in emission values. Moreover, the origin, manufacturing process, transportation and quantity of the material all greatly influence the determination of the CO_2 emission factor.

Therefore, energy consumption is another aspect that has a high impact on the environment. Therefore, this chapter has highlighted the methods that can be used to determine energy consumption, mostly on construction sites. Among all the discussed approaches, the exhaust analyser is the most recommended as a way to determine CO_2 concentration as it provides results that are more accurate when compared to those of CO_2 emission modelling and empirical analysis. Finally, the integration of a modelling method, empirical analysis and a gas analyser for quantifying CO_2 emissions is highly recommended as an avenue for future research, particularly in relation to improve the reliability and accuracy of CO_2 emission assessments.

ACKNOWLEDGMENT

The authors would like to thank the Ministry of Higher Education and Universiti Teknologi Malaysia (UTM) in the form of research grant (Vote No. Q.J130000.2522.13H00), otherwise, this study would not have been possible. We shall remain indebted to them for their generosity.

REFERENCES

1. G. Hammond, and C. Jones, "Inventory of Carbon and Energy (ICE). Version 2.0. Sustainable Energy Research Team, Department of Mechanical Engineering, University of Bath", www. bath. ac. uk/mech-eng/sert/embodied, 2011.
2. G. P. Hammond, H. A. Harajli, C. I. Jones, and A. B. Winnett, "Whole systems appraisal of a UK Building Integrated Photovoltaic (BIPV) system: Energy, environmental, and economic evaluations", *Energy Policy*, vol. 40, pp. 219–230, 2012.
3. F. A. Rahman, M. M. A. Aziz, R. Saidur, W. A. W. A. Bakar, M. Hainin, R. Putrajaya, *et al.*, "Pollution to solution: Capture and sequestration of carbon dioxide (CO_2) and its utilization as a renewable energy source for a sustainable future", *Renewable and Sustainable Energy Reviews*, vol. 71, pp. 112–126, 2017.
4. T. Moore, N. Roulet, and J. Waddington, "Uncertainty in predicting the effect of climatic change on the carbon cycling of Canadian peatlands", *Climatic Change*, vol. 40, pp. 229–245, 1998.
5. R. Afroz, M. N. Hassan, and N. A. Ibrahim, "Review of air pollution and health impacts in Malaysia", *Environmental Research*, vol. 92, pp. 71–77, 2003.
6. I. Change, 2006 *IPCC Guidelines for National Greenhouse Gas Inventories*, 2006.
7. S. Kar, A. Behl, A. Shukla, and P. Jain, "Estimation of carbon footprints of bituminous road construction process", *Journal of Civil & Environmental Engineering*, vol. 5, p. 2, 2015.
8. A. Alcorn, *Embodied Energy and CO Coefficients for NZ Building Materials*. The Centre, 2003.
9. R. E. Miller, and P. D. Blair, *Input-Output Analysis: Foundations and Extensions*. Cambridge University Press, 2009.
10. T. Chen, J. Burnett, and C. Chau, "Analysis of embodied energy use in the residential building of Hong Kong", *Energy*, vol. 26, pp. 323–340, 2001.
11. C. Hanson, R. Noland, and K. Cavale, "Life-cycle greenhouse gas emissions of materials used in road construction", *Transportation Research Record: Journal of the Transportation Research Board*, vol. 2287, pp. 174–181, 2012.
12. G. Sinden, "The contribution of PAS 2050 to the evolution of international greenhouse gas emission standards", *The International Journal of Life Cycle Assessment*, vol. 14, pp. 195–203, 2009.

13. H. Le Treut, R. Somerville, U. Cubasch, Y. Ding, C. Mauritzen, A. Mokssit, *et al.*, "Historical overview of climate change", pp. 93–127, 2007.
14. I. Arocho, W. Rasdorf, and J. Hummer, "Methodology to forecast the emissions from construction equipment for a transportation construction project", In: *Construction Research Congress 2014@ sConstruction in a Global Network*, pp. 554–563.
15. S. K. Sharma, "From the Desk of The National President", *Materials Management Review*, 2011.
16. C2ES, *Global Anthropogenic Ghg Emissions by Gas*. Available at: http://www.c2es.org/facts-figures/international-emissions/gas, 2005.
17. P. S. Bakwin, P. P. Tans, D. F. Hurst, and C. Zhao, "Measurements of carbon dioxide on very tall towers: Results of the NOAA/CMDL program", *Tellus B: Chemical and Physical Meteorology,* vol. 50, pp. 401–415, 1998.
18. E. Zusman, A. Srinivasan, and S. Dhakal, *Low Carbon Transport in Asia: Strategies for Optimizing Co-Benefits*. Routledge, 2012.
19. R. Wennersten, Q. Sun, and H. Li, "The future potential for carbon capture and storage in climate change mitigation—An overview from perspectives of technology, economy and risk", *Journal of Cleaner Production*, vol. 103, pp. 724–736, 2015.
20. J. T. Houghton, L. Meira Filho, B. Lim, K. Treanton, and I. Mamaty, *Revised 1996 IPCC Guidelines for National Greenhouse Gas Inventories. v. 1: Greenhouse Gas Inventory Reporting Instructions. v. 2: Greenhouse Gas Inventory Workbook. v. 3: Greenhouse Gas Inventory Reference Manual*, 1997.
21. I. P. O. C. Change, "Climate change 2007: The physical science basis", *Agenda*, vol. 6, p. 333, 2007.
22. F. Evrendilek, S. Berberoglu, N. Karakaya, A. Cilek, G. Aslan, and K. Gungor, "Historical spatiotemporal analysis of land-use/land-cover changes and carbon budget in a temperate peatland (Turkey) using remotely sensed data", *Applied Geography*, vol. 31, pp. 1166–1172, 2011.
23. C. M. Dufournaud, J. J. Harrington, and P. P. Rogers, "Leontief's 'Environmental repercussions and the economic structure...' revisited: A general equilibrium formulation", *Geographical Analysis*, vol. 20, pp. 318–327, 1988.
24. M. Lallanila, *What Is the Greenhouse Effect?* Available at: http://www.livescience.com/37743-greenhouse-effect.html, 2015.
25. S. Kang, and H. Lin, "Wavelet analysis of hydrological and water quality signals in an agricultural watershed", *Journal of Hydrology*, vol. 338, pp. 1–14, 2007.
26. G. P. Hammond, and C. I. Jones, "Embodied energy and carbon in construction materials", *Proceedings of the Institution of Civil Engineers – Energy*, vol. 161, pp. 87–98, 2008.
27. N. R. Council, *Limiting the Magnitude of Future Climate Change*. National Academies Press, 2011.
28. B. Kim, H. Lee, H. Park, and H. Kim, "Greenhouse gas emissions from onsite equipment usage in road construction", *Journal of Construction Engineering and Management*, vol. 138, pp. 982–990, 2011.
29. S.-J. Huang, and C.-T. Hsieh, "Coiflet wavelet transform applied to inspect power system disturbance-generated signals", *IEEE Transactions on Aerospace and Electronic Systems*, vol. 38, pp. 204–210, 2002.
30. C. T. Hendrickson, L. B. Lave, and H. S. Matthews, *Environmental Life Cycle Assessment of Goods and Services: An Input-Output Approach*. Resources for the Future, 2006.
31. S. R. Koirala, R. W. Gentry, P. J. Mulholland, E. Perfect, and J. S. Schwartz, "Time and frequency domain analyses of high-frequency hydrologic and chloride data in an east Tennessee watershed", *Journal of Hydrology*, vol. 387, pp. 256–264, 2010.
32. W. Leontif, *Iinput-Output Econonics*. London, UK: Oxford University Press, 1966.

33. G. Peters, and E. Hertwich, "Production factors embodied in trade: Theoretical development", *NTNU Working Papers* 2004, vol. 5, pp. 1–34, 2004.
34. T. Bruce, and J. Chick, "Energy and carbon costing of breakwaters", In: *Coasts, Marine Structures and Breakwaters: Adapting to Change: Proceedings of the 9th International Conference Organised by the Institution of Civil Engineers and Held in Edinburgh on 16 to 18 September 2009, Thomas Telford Ltd*, pp. 582–590, 2010.
35. G. J. Treloar, P. E. Love, and G. D. Holt, "Using national input/output data for embodied energy analysis of individual residential buildings", *Construction Management and Economics*, vol. 19, pp. 49–61, 2001.
36. G. P. Peters, and E. G. Hertwich, "The importance of imports for household environmental impacts", *Journal of Industrial Ecology*, vol. 10, pp. 89–109, 2006.
37. H.-J. Bornemann, and A. Seidel, "A get-away special experiment to measure the carbon dioxide content of the earth's atmosphere", *Acta Astronautica*, vol. 15, pp. 871–878, 1987.
38. Y. Itoh, and T. Kitagawa, "Using CO_2 emission quantities in bridge lifecycle analysis", *Engineering Structures*, vol. 25, pp. 565–577, 2003.
39. M. J. González, and J. G. Navarro, "Assessment of the decrease of CO_2 emissions in the construction field through the selection of materials: Practical case study of three houses of low environmental impact", *Building and Environment*, vol. 41, pp. 902–909, 2006.
40. R. Phillips, and B. N. Banregionen, "Air pollution associated with the construction of Swedish railways", *Banverket Norra Banregionen, Luleå*, 2006.
41. C. L. Weber, and H. S. Matthews, "Quantifying the global and distributional aspects of American household carbon footprint", *Ecological Economics*, vol. 66, pp. 379–391, 2008.
42. H. Abanda, J. Tah, F. Cheung, and W. Zhou, "Measuring the embodied energy, waste, CO_2 emissions, time and cost for building design and construction", In: *Computing in Civil and Building Engineering, Proceedings of the International Conference*, 2010.
43. A. H. Ammouri, I. Srour, and R. F. Hamade, "Carbon footprint calculator for construction projects", In: ASME 2011 International Mechanical Engineering Congress and Exposition, Denver, Colorado, pp. 813–819, 2011.
44. R. Broekens, M. Escarameia, C. Cantelmo, and G. Woolhouse, "Quantifying the carbon footprint of coastal construction—A new tool HRCAT", In: *ICE Coastal Management* 2011, 15–17 November *2011*, Belfast, 2011.
45. H. X. Li, Z. Lei, Z. Y. Hu, and B. S. Liu, "The empirical research on the estimation of the construction CO_2 emission based on EPA non-road modeling", *Applied Mechanics and Materials*, vols. 71–78, pp. 4943–4948, 2011.
46. J. Mayrwöger, W. Reichl, C. Krutzler, and B. Jakoby, "Measuring CO_2 concentration with a Fabry–Perot based bolometer using a glass plate as simple infrared filter", *Sensors and Actuators B: Chemical*, vol. 170, pp. 143–147, 2012.
47. Y. Huang, B. Hakim, and S. Zammataro, "Measuring the carbon footprint of road construction using CHANGER", *International Journal of Pavement Engineering*, vol. 14, pp. 590–600, 2013.
48. B. Kim, H. Lee, H. Park, and H. Kim, "Framework for estimating greenhouse gas emissions due to asphalt pavement construction", *Journal of Construction Engineering and Management*, vol. 138, pp. 1312–1321, 2012.
49. Z. Ren, V. Chrysostomou, and T. Price, "The measurement of carbon performance of construction activities: A case study of a hotel construction project in South Wales", *Smart and Sustainable Built Environment*, vol. 1, pp. 153–171, 2012.
50. S. J. Richardson, N. L. Miles, K. J. Davis, E. R. Crosson, C. W. Rella, and A. E. Andrews, "Field testing of cavity ring-down spectroscopy analyzers measuring carbon dioxide and water vapor", *Journal of Atmospheric and Oceanic Technology*, vol. 29, pp. 397–406, 2012.

51. F. Evrendilek, "Quantifying biosphere–atmosphere exchange of CO_2 using eddy covariance, wavelet denoising, neural networks, and multiple regression models", *Agricultural and Forest Meteorology*, vol. 171, pp. 1–8, 2013.
52. Y. Seo, and S.-M. Kim, "Estimation of materials-induced CO_2 emission from road construction in Korea", *Renewable and Sustainable Energy Reviews*, vol. 26, pp. 625–631, 2013.
53. S. Melanta, E. Miller-Hooks, and H. G. Avetisyan, "Carbon footprint estimation tool for transportation construction projects", *Journal of Construction Engineering and Management*, vol. 139, pp. 547–555, 2012.
54. P. Lijewski, J. Merkisz, P. Fuc, and P. Daszkiewicz, "The comparison of the exhaust emissions from an agricultural tractor and a truck", *Applied Mechanics and Materials*, pp. 196–201, vol. 391, 2013.
55. Z. M. Nadoushani, and A. Akbarnezha, "A computational framework for estimating the carbon footprint of construction", In: *ISARC. Proceedings of the International Symposium on Automation and Robotics in Construction*, 2014.
56. X. Wang, Z. Duan, L. Wu, and D. Yang, "Estimation of carbon dioxide emission in highway construction: A case study in southwest region of China", *Journal of Cleaner Production*, vol. 103, pp. 705–714, 2015.
57. J.-S. Chou, and K.-C. Yeh, "Life cycle carbon dioxide emissions simulation and environmental cost analysis for building construction", *Journal of Cleaner Production*, vol. 101, pp. 137–147, 2015.
58. B. Peng, C. Cai, G. Yin, W. Li, and Y. Zhan, "Evaluation system for CO_2 emission of hot asphalt mixture", *Journal of Traffic and Transportation Engineering (English Edition)*, vol. 2, pp. 116–124, 2015.
59. M. Reichstein, E. Falge, D. Baldocchi, D. Papale, M. Aubinet, P. Berbigier, *et al.*, "On the separation of net ecosystem exchange into assimilation and ecosystem respiration: Review and improved algorithm", *Global Change Biology*, vol. 11, pp. 1424–1439, 2005.
60. S. S. Brown, H. Stark, and A. Ravishankara, "Applicability of the steady state approximation to the interpretation of atmospheric observations of NO_3 and N_2O_5", *Journal of Geophysical Research: Atmospheres*, vol. 108, pp. ACH6 1–10, 2003.
61. R. J. Lempert, and D. G. Groves, "Identifying and evaluating robust adaptive policy responses to climate change for water management agencies in the American west", *Technological Forecasting and Social Change*, vol. 77, pp. 960–974, 2010.
62. Y. Chen, and Y. Zhu, *Analysis of Environmental Impacts in the Construction Phase of Concrete Frame Buildings*. Department of Construction Management, Tsinghua University, China, 2008.
63. L. Geller, R. Fri, and M. Brown, "Limiting the magnitude of future climate change", National Academies Press, 2011
64. R. A. Herendeen, C. Ford, and B. Hannon, "Energy cost of living, 1972–1973", *Energy*, vol. 6, pp. 1433–1450, 1981.
65. M. Lenzen, "Primary energy and greenhouse gases embodied in Australian final consumption: An input–output analysis", *Energy Policy*, vol. 26, pp. 495–506, 1998.
66. M. K. Wali, F. Evrendilek, T. O. West, S. E. Watts, D. Pant, H. K. Gibbs, *et al.*, "Assessing terrestrial ecosystem sustainability: Usefulness of regional carbon and nitrogen models", Nature & Resources, vol. 35, pp. 21–33, 1999.
67. P. Truitt, *Potential for Reducing Greenhouse Gas Emissions in the Construction Sector*, US Environmental Protection Agency, p. 12, 2009.
68. D. Baldocchi, "'Breathing'of the terrestrial biosphere: Lessons learned from a global network of carbon dioxide flux measurement systems", *Australian Journal of Botany*, vol. 56, pp. 1–26, 2008.
69. R. V. Wendling, and R. B. Lorance, "Integration of schedule and cost risk models", *AACE International Transactions*, p. K91, 1999.

70. L. J. Isidore, and W. E. Back, "Multiple simulation analysis for probabilistic cost and schedule integration", *Journal of Construction Engineering and Management*, vol. 128, pp. 211–219, 2002.

71. E. K. Webb, G. I. Pearman, and R. Leuning, "Correction of flux measurements for density effects due to heat and water vapour transfer", *Quarterly Journal of the Royal Meteorological Society*, vol. 106, pp. 85–100, 1980.

72. S.-C. Wang, and B.-C. Li, "Criterion emission amount of automotive pollution in highway tunnel [J]", *Journal of Chang'an University (Natural Science Edition)*, vol. 1, p. 019, 2005.

73. C. Shang, Z. Zhang, and X. Li, "Research on energy consumption and emission of life cycle of expressway [J]", *Journal of Highway and Transportation Research and Development*, vol. 8, p. 027, 2010.

74. E. Miller-Hooks, S. Melanta, and H. Avetisyan, *Tools to Support GHG Emissions Reduction: A Regional Effort. Part 1-Carbon Footprint Estimation and Decision Support*. Maryland State Highway Administration/US Department of Transportation, 2010.

75. A. Tukker, and B. Jansen, "Environmental impacts of products: A detailed review of studies", *Journal of Industrial Ecology*, vol. 10, pp. 159–182, 2006.

76. S. Deng, and B. Shi, "An experimental study on pollutant emission factors of light duty vehicles in China", *China Environmental Science*, vol. 19, pp. 176–179, 1999.

77. G. Hammond, C. Jones, F. Lowrie, and P. Tse, *Inventory of Carbon & Energy: ICE*. Sustainable Energy Research Team, Department of Mechanical Engineering, University of Bath Bath, 2008.

78. R. Kok, R. M. Benders, and H. C. Moll, "Measuring the environmental load of household consumption using some methods based on input–output energy analysis: A comparison of methods and a discussion of results", *Energy Policy*, vol. 34, pp. 2744–2761, 2006.

79. E. J. Jaselskis, and T. El-Misalami, "Implementing radio frequency identification in the construction process", *Journal of Construction Engineering and Management*, vol. 129, pp. 680–688, 2003.

80. T. J. Fahey, P. B. Woodbury, J. J. Battles, C. L. Goodale, S. P. Hamburg, S. V. Ollinger, *et al.*, "Forest carbon storage: Ecology, management, and policy", *Frontiers in Ecology and the Environment*, vol. 8, pp. 245–252, 2010.

81. L. B. C. Schaefer, *Lehrbuch der Experimentalphysik, Band III Optik*, W. de Gruyter, 1987.

82. Y. S. Touloukian, R. Powell, C. Ho, and M. Nicolaou, "Thermophysical properties of matter-the TPRC data series, Volume 10. Thermal Diffusivity", DTIC Document, 1974.

83. Y. Huang, A. Spray, and T. Parry, "Sensitivity analysis of methodological choices in road pavement LCA", *The International Journal of Life Cycle Assessment*, vol. 18, pp. 93–101, 2013.

84. Y. Poh, and J. Tah, "Integrated duration–cost influence network for modelling risk impacts on construction tasks", *Construction Management and Economics*, vol. 24, pp. 861–868, 2006.

85. A. Horvath, "Pavement life-cycle assessment tool for environmental and economic effects", Personal Communication, vol. 4, p. 2008, 2008.

86. H. Stripple, "Life cycle assessment of road", A Pilot Study *for Inventory Analysis. 2nd Revised Edition*. Report from the IVL Swedish Environmental Research Institute, vol. 96, 2001.

87. D. Senthil Kumaran, S. Ong, R. B. Tan, and A. Nee, "Environmental life cycle cost analysis of products", *Environmental Management and Health*, vol. 12, pp. 260–276, 2001.

88. Y. Huang, and T. Parry, "Pavement life cycle assessment", In Gopalakrishnan, K., W. Steyn, and J. Harvey (Eds),: *Climate Change, Energy, Sustainability and Pavements*. Springer, pp. 1–40, 2014.

89. Y. Huang, T. Parry, M. Wayman, C. Mcnally, Y. Ersson-Sköld, O. Wik, *et al.*, "Risk assessment and life cycle assessment of reclaimed asphalt", In: *Proceedings of the 2012 WASCON Conference*, Gothenburg, Sweden, 2012.

90. H. Chen, J. Winderlich, C. Gerbig, A. Hoefer, C. Rella, E. Crosson, *et al.*, "High-accuracy continuous airborne measurements of greenhouse gases (CO_2 and CH_4) using the cavity ring-down spectroscopy (CRDS) technique", *Atmospheric Measurement Techniques.*, vol. 3, pp. 375–386, 2010.

91. C. Weiland, and S. Muench, "Life-cycle assessment of reconstruction options for inter-state highway pavement in Seattle, Washington", *Transportation Research Record: Journal of the Transportation Research Board*, vol. 2170(1), pp. 18–27, 2010.

92. A. Horvath, "PaLATE: Pavement Life-Cycle Assessment Tool for Environmental and Economic Benefits. Berkeley, CA: University of California, Berkeley", Available at: www.ce.berkeley.edu/~h orvath/palate.html, 2004.

93. R. J. O'Born, H. Brattebø, O. M. K. Iversen, S. Miliutenko, and J. Potting, "Quantifying energy demand and greenhouse gas emissions of road infrastructure projects: An LCA case study of the Oslo fjord crossing in Norway", *European Journal of Transport and Infrastructure Research*, vol. 16(3), pp. 445–466, 2016.

94. T. Hong, C. Ji, M. Jang, and H. Park, "Assessment model for energy consumption and greenhouse gas emissions during building construction", *Journal of Management in Engineering*, vol. 30, pp. 226–235, 2013.

95. N. Bueche, and A.-G. Dumont, "IRF Greenhouse Gas calculator-Analysis and valida-tion", *Infoscience EPFL Scientific Publications* , 2009.

96. J. Minx, T. Wiedmann, and J. Barrett, *Methods Review to Support the PAS Process for the Calculation of the Greenhouse Gas Emissions Embodied in Good and Services*. Department for Environment, Food and Rural Affairs, 2008.

97. S. Abolhasani, H. C. Frey, K. Kim, W. Rasdorf, P. Lewis, and S.-H. Pang, "Real-world in-use activity, fuel use, and emissions for nonroad construction vehicles: A case study for excavators", *Journal of the Air & Waste Management Association*, vol. 58, pp. 1033–1046, 2008.

98. G. J. Treloar, H. Gupta, P. E. Love, and B. Nguyen, "An analysis of factors influencing waste minimisation and use of recycled materials for the construction of residential buildings", *Management of Environmental Quality: An International Journal*, vol. 14, pp. 134–145, 2003.

99. A. R. Board, *EMFAC2007-Calculating Emissions Inventory for Vehicles in California*. California Air and Resource Board. Available at: http://www. arb. ca. gov/msei/onroad/downloads/docs/user_guide_emfac2007.pdf, 2010.

100. S. K. Hoekman, "Biofuels in the US—Challenges and opportunities", *Renewable Energy*, vol. 34, pp. 14–22, 2009.

101. J. I. Steinfeld, T. J. Vickers, and A. Scheeline, *Book Reviews: Cavity Ringdown Spectroscopy: An Ultratrace-Absorption Measurement Technique, Encyclopedia of Spectroscopy and Spectrometry, An Atlas of High Resolution Spectra of Rare Earth Elements for Inductively Coupled Plasma Atomic Emission Spectroscopy*. London, UK: SAGE Publications, 2000.

102. M. Lenzen, "Errors in conventional and input-output—based life—cycle inventories", *Journal of Industrial Ecology*, vol. 4, pp. 127–148, 2000.

103. K. Adalberth, "Energy use during the life cycle of buildings: A method", *Building and Environment*, vol. 32, pp. 317–320, 1997.

104. K. Yahya, and A. Halim Boussabaine, "Eco-costing of construction waste", *Management of Environmental Quality: An International Journal*, vol. 17, pp. 6–19, 2006.

105. P. F. Nelson, A. R. Tibbett, and S. J. Day, "Effects of vehicle type and fuel quality on real world toxic emissions from diesel vehicles", *Atmospheric Environment*, vol. 42, pp. 5291–5303, 2008.

106. M. Lenzen, M. Wier, C. Cohen, H. Hayami, S. Pachauri, and R. Schaeffer, "A comparative multivariate analysis of household energy requirements in Australia, Brazil, Denmark, India and Japan", *Energy*, vol. 31, pp. 181–207, 2006.

107. M. Wier, M. Lenzen, J. Munksgaard, and S. Smed, "Effects of household consumption patterns on CO_2 requirements", *Economic Systems Research*, vol. 13, pp. 259–274, 2001.

108. M. Eisert, K. Lilienbecker, M. Clary, J. Petersen, H. Zhuang, S. Hill, *et al.*, *Environmental Assessment for Conversion of the Existing Aero Club Runway to Emergency Helipad for David Grant Medical Center Travis Air Force Base, Fairfield, California*. U.S. Air Force Center for Engineering and the Environment Travis Air Force Bace, California, 2010.

109. M. Liu, F. Qian, and X. Zhan, "Calculation model for energy carbon emission of building material transportation", In: 2010 *International Conference on E-Product E-Service and E-Entertainment (ICEEE)*, pp. 1–3, 2010.

110. B. Kourmpanis, A. Papadopoulos, K. Moustakas, M. Stylianou, K. Haralambous, and M. Loizidou, "Preliminary study for the management of construction and demolition waste", *Waste Management & Research*, vol. 26, pp. 267–275, 2008.

111. C. W. Bullard, P. S. Penner, and D. A. Pilati, "Net energy analysis: Handbook for combining process and input-output analysis", *Resources and Energy*, vol. 1, pp. 267–313, 1978.

112. C. Ahn, J. C. Martinez, P. V. Rekapalli, and F. A. Peña-Mora, "Sustainability analysis of earthmoving operations", In: Winter Simulation Conference, Austin, TX, pp. 2605–2611, 2009.

113. A. A. Guggemos, and A. Horvath, "Comparison of environmental effects of steel-and concrete-framed buildings", *Journal of Infrastructure Systems*, vol. 11, pp. 93–101, 2005.

114. A. Reinders, K. Vringer, and K. Blok, "The direct and indirect energy requirement of households in the European Union", *Energy Policy*, vol. 31, pp. 139–153, 2003.

115. A. A. Guggemos, and A. Horvath, "Decision-support tool for assessing the environmental effects of constructing commercial buildings", *Journal of Architectural Engineering*, vol. 12, pp. 187–195, 2006.

116. C. W. Bullard, P. S. Penner, and D. A. Pilati, "Net energy analysis: Handbook for combining process and input-output analysis", *CAC* Document No. 214, 1976.

117. A. Matin, P. Collas, D. Blain, C. Ha, C. Liang, L. MacDonald, *et al.*, *Canada's Greenhouse Gas Inventory 1990–2002*. Greenhouse Gas Division, Environment Canada Ottawa, Canada, 2004.

118. D. Hajjar, and S. AbouRizk, "Simphony: An environment for building special purpose construction simulation tools", In: *Proceedings of the 31st Conference on Winter Simulation: Simulation—A Bridge to the Future-Volume 2*, pp. 998–1006, 1999.

119. M. P. Lewis, *Estimating Fuel Use and Emission Rates of Nonroad Diesel Construction Equipment Performing Representative Duty Cycles*. North Carolina State University, 2009.

120. J. Jadhao, and D. Thombare, "Review on exhaust gas heat recovery for IC engine", *International Journal of Engineering and Innovation Technology (IJEIT)*. vol. 2, pp. 93–100, 2013.

121. C. A. C. Coello, and C. S. P. Zacatenco, "List of references on constraint-handling techniques used with evolutionary algorithms", *Information Sciences*, vol. 191, pp. 146–168, 2012.

122. J. Martinez, P. G. Ioannou, and R. I. Carr, "State and resource based construction process simulation", In: *Computing in Civil Engineering*, pp. 177–184, 1994.

123. V. Shahinian, "SENSOR tech-ct update application soft-ware for SEMTECH mobile emission analyzers", In: *Sensors* 4th *Annual* SUN (*SEMTECH User Network*) Conference, p. 2007, 2007.

124. F. S. Chapin III, P. A. Matson, and P. Vitousek, *Principles of Terrestrial Ecosystem Ecology*. Springer Science & Business Media, 2011.

125. S. Zammataro, K. Laych, V. Sheela, M. Rao, Y. Huang, and B. Hakim, "Assessing greenhouse gas emissions in road construction: An example of calculation tool for road projects", In: *PIARC Seminar on Reducing the Carbon Footprint in Road Construction*, pp. 17–19, 2011.

126. H. Bible, and O. Testament, "The University of Solar System Studies" Edwin House Pub Incorporated, 2001.

127. R. Zimmerman, Q. Zhu, and C. Dimitri, "Promoting resilience for food, energy, and water interdependencies", *Journal of Environmental Studies and Sciences*, vol. 6, pp. 50–61, 2016.

128. R. Starkey, "Standardization of environmental management systems: ISO 14001, ISO 14004 and EMAS", *Corporate Environmental Management 1: Systems and Strategies*, pp. 61–89, 1998.

129. G. Labeckas, and S. Slavinskas, "Performance and emission characteristics of a direct injection diesel engine operating on KDV synthetic diesel fuel", *Energy Conversion and Management*, vol. 66, pp. 173–188, 2013.

10 Traffic Pollution: Perspective Overview toward Carbon Dioxide Capture and Separation Method

Md. Maniruzzaman A. Aziz, Khairul Anuar Kassim, Wan Azelee Wan Abu Bakar, Fauzan Mohd Jakarmi, A. B. M. Amimul Ahsan, Salmiah Jamal Mat Rosid, and Susilawati Toemen

CONTENTS

10.0 INTRODUCTION

Traffic pollutants have been building up over time with the increases in traffic volume and population. While the transport sector definitely contributes to global social and economic development, it also affects the environment by emitting pollutants—volatile organic compounds (VOCs), sulfur dioxide (SO_2) nitrogen oxide (NO_x), fine particulate matter (PM2.5), etc.—and greenhouse gases (GHGs)—fluorinated gases (F-gases), nitrous oxide (N_2O), methane (CH_4) and carbon dioxide (CO_2), etc.—into the atmosphere [1–3]. The transport sector heavily relies on fossil fuels and accounts for nearly 60% of the global oil demand. Motor vehicles are a significant source of emissions, and it is estimated that 23% of energy-related emissions come from transportation systems and that 74% of this release is due to regular traffic (i.e., cars, buses and trucks) [4]. Furthermore, traffic-induced pollutants have been reported to have adverse impacts on human health in both developed and developing countries [5]. Stress, frustration and anger develop during traffic jams, while vehicles pollute the environment with emissions that are 29% higher at intersections for slow-moving vehicles during traffic back-ups [6, 7]. Long exposure to traffic pollutants contributes to tiredness, weakness, memory loss, confusion, nausea and loss of appetite [8, 9]. Promoting active travel behaviors (e.g., walking, cycling, using public transport) is the suggested best practice for reducing GHGs for short city trips [10–12].

Currently, more than 1.2 billion motor vehicles are registered throughout the world, and these vehicles account for 70% of carbon monoxide (CO) and 19% of CO_2 emissions globally [13]. China has the largest number of motor vehicles in the world at almost 300 million. China's rapid economic growth has given rise to urbanization, energy consumption and carbon emissions. Now, China consumes more energy and emits more CO_2 than any other country [14, 15]. The transportation sector is one of the main contributors to energy consumption in China. The energy used for transportation accounted for 8.71% of China's energy consumption in 2012 [16]. Private cars generate approximately 88% of the CO_2 emissions from urban transport in Beijing [17]. However, only data for commercial vehicles is recorded as transport-sector energy consumption, while private car energy consumption is recorded under the category of household energy consumption [18]. Thus, the energy consumption

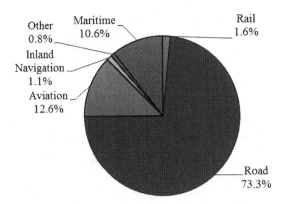

FIGURE 10.1 GHG emissions from the transport sector in Europe by mode in 2014 [20].

and CO_2 emissions of private cars in China are not indicated in transport sector data, which makes it misleading.

There are nearly 291 million vehicles on Europe's roads—approximately one vehicle for every two people [19]. Transport accounts for almost a quarter of Europe's GHG emissions and is the main source of pollution in cities. As shown in Figure 10.1, within this sector, road transport is the main source of emissions, having produced 73.3% of all GHG emissions in 2014 [20]. London, Paris and Moscow are the three European cities which appear on the list of the top ten most traffic-jammed cities in the world [21]. The emission of pollutants and GHGs harm the health and welfare of the earth's inhabitants; to respond to this problem, the European Union (EU), in accordance with the Kyoto Protocol, set regulations for reducing the average emissions of new passenger cars to 140 g CO_2/km in 2008 [22].

According to the US Department of Transportation, there were 269 million motor vehicles—automobiles, buses, trucks and motorcycles—in the United States in 2016 [23]. Even though the United States is recorded as having the third most motor vehicles in the world, it is the world leader in traffic jams, and five of the top ten most traffic-jammed cities in the world are in the United States [21]. Los Angeles has the most congested traffic and the worst driving conditions in the United States. In 2017, drivers spent 102 hours in traffic jams during rush-hour periods [24]. The heavy congestion is mainly due to the very large population, which means that even traveling via public transport takes a long time. The other reason is the number of cars owned in the city; Los Angeles County had an estimated population of 10 million and 7.8 million registered vehicles in 2016 [25]. There are direct and indirect economic costs which are considered for households with regards to driving. Direct costs include the value of the time spent in traffic jams, the cost for additional fuel and the social cost of the released emissions. Indirect costs include increases in the prices of goods and services that are passed along to customers due to traffic jams [21]. Traffic congestion cost the United States nearly $305 billion in 2017 [24]. Figure 10.2 shows the traffic jam data for Los Angeles along Interstate 405 during the Thanksgiving holiday [26].

In fact, the traffic congestion pattern in developing countries such as China is totally different from that in developed countries. Expressways in major regions of

FIGURE 10.2   Los Angeles traffic jam data during Thanksgiving weekend [26].

China have great potential to accommodate potential travel needs [27]. Figure 10.3 shows a traffic jam on the G4 Beijing–Hong Kong–Macau Expressway. At that time, hundreds of thousands of Beijing residents tried to return from their journey on national holidays [28].

In recent years, the largest economy has set targets for the release of new vehicles. The European Union has been a front-runner in the international arena to meet CO_2 emission targets for new vehicles. Compared to the emission targets in other countries—e.g., the United States (99 g CO_2 /km), South Korea (97 g CO_2 /km) and Canada (99 g CO_2 /km)—European Union nations have a lower rate of CO_2 emissions, with a target of 95 g CO_2 /km by the year 2020 [29]. Table 10.1 shows GHGs and pollutant emissions in the road transportation sector in European Union countries with baseline projections for up to the year 2030 [3].

As shown in Figure 10.4, road transport is the main source of CO_2, NO_x, PM2.5 and VOC emissions. From May 2000 to September 2016, four emission limits ("Euro 3," "Euro 4," "Euro 5" and "Euro 6") have been applied, and the NO_x emissions for new diesel cars dropped from 0.5 g/km in 2000 to 0.08 g/km in 2014. All emissions have declined from 2000 to 2015, except for CO_2 emissions, which have continued to increase.

With the global shift toward a low-emission society, countries have launched low-carbon mobility strategies to ensure their sustainable development and growth [30].

FIGURE 10.3   The gridlock that took place on Beijing's busiest and widest highways with as many as 50 lanes queued up near a toll [28].

Furthermore, by the US discovery of an illegal device in Volkswagen vehicles in 2015, several organizations and governments conducted independent emissions testing on cars [31]. Australia is following the European regulatory template to control vehicular emissions. China applied its first vehicle standard—"China 1" (equivalent to "Euro 1" regulations)—in 2000, and the standards are to be tightened gradually over time from "China 1" to "China 5" [32]. Canada has adapted US EPA regulations by reference and will make adjustments automatically as adjustments are made to the US program [33].

Countries have regulated sustainable low-carbon mobility strategies for their development and growth by applying controlling actions. Hence, numerous academic and industrial communities have directed their efforts for reducing CO_2 emissions in the atmosphere. Carbon capture and sequestration (CCS) is considered the most promising technology worldwide for CO_2 removal. The CCS process consists of CO_2 capture and separation, transportation and underground storage [34, 35].

TABLE 10.1

Ghgs and pollutant emissions in the road transportation sector in European Union countries with baseline projections up to 2030*

GHGs and pollutants	Unit	2000	2005	2010	2015	2020	2025	2030
CO_2	10^9kg	835	887	881	894	863	845	807
CH_4	10^6kg	201	133	86	44	24	16	13
N_2O	10^6kg	43	34	29	31	32	33	32
F-gases	10^9kg CO_2eq	0	0	0	0	0	0	0
SO_2	10^6kg	151	33	6	6	6	6	6
NO_x	10^6kg	5901	5010	3918	2700	1645	1113	863
PM2.5	10^6kg	269	238	174	113	78	63	59
VOC	10^6kg	3778	2191	1344	788	535	425	388
NH_3	10^6kg	149	125	97	79	76	74	73

*CORINAIR emission inventory methodology is used for assessment of the quantities.

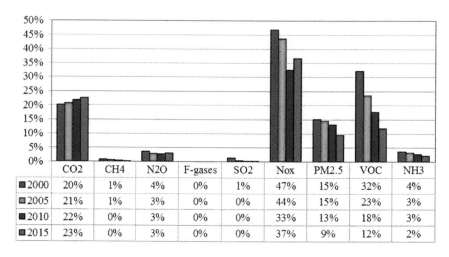

FIGURE 10.4 Comparing GHG and pollutant emissions from the on-road transportation sector in EU countries.

Based on the study of the International Energy Agency, this method is expected to reduce CO_2 emissions by 19% by 2050 [36–38]. However, the development of large CCS demonstration projects, especially in developing countries, has been hindered by high levels of construction. [7]. CCS technologies are focused mainly on power generation, during which CO_2 gas is produced in large quantities. Therefore, this review is intended to give some ideas of how carbon capture might be used directly from the exhaust of automobiles.

10.1   REVIEW OF SEPARATION METHODS

Massive separation and storage methods have been considered among the most reliable options to capture CO_2 on a large scale. Different CO_2 separation methods are currently being researched, and their development mainly includes physical/chemical absorption, adsorption, membrane separation and biotechnology. For instance, Guo et al. [39] studied CO_2 capture from industrial flue gas using microalgae [39, 40], Li et al. [41] focused on absorption and regeneration of aqueous ammonia for CO_2 capture. Subha et al. [42] studied the absorption of lithium silicate by CO_2 at low temperatures. Though there are several reviews focused on CO_2 capture and storage, to the authors' knowledge, there are no reviews for different separation methods with detailed analyses of each. Hence, the objective of this chapter is to provide a holistic study of separation methods by comparing their operations and costs. Table 10.2 shows some recent research on separation methods in the literature.

10.2   TECHNIQUES FOR CAPTURING CO_2 EMISSIONS

Today, power plants that use fossil combustion are the main source of CO_2 emissions among all industries [110]. Extracting CO_2 from flue gases of power plants before separation is an important parameter for carbon management [59]. This process can be performed by three different categories of processes—namely, pre-combustion, post-combustion and oxyfuel combustion processes [111–113].

10.2.1   PRE-COMBUSTION PROCESSES

Pre-combustion processes involve converting coal into clean gases by separating CO_2 from the coal before fuel is placed in the furnace. In a pre-combustion process, coal or natural gas is oxidized before combustion. This is called gasification, and it combines the partial oxidation steam treatment of coal. As described in Equations 10.1 and 10.2, 25–40% of CO_2 is isolated from the product steam during such a process, and excess H_2 is used in gas turbines or fuel cells [114].

For coal, the gasification process must be conducted at low oxygen levels involving CO and H_2 and no other non-combustible gases (Equation 10.1) [115]. In addition, this mixture will experience a water-gas shift (WGS) reaction that forms more hydrogen, whereas CO forms CO_2 (Equation 10.2).

$$COAL \xrightarrow{\text{Gasification}} CO+H_2 \tag{10.1}$$

$$CO+H_2O \xrightarrow{\text{Water gas shift}} H_2O+CO_2 \tag{10.2}$$

There are several gasification methods that can be used for natural gas, such as steam reforming, partial oxidation and autothermal reforming. In the CH_4 steam system, CH_4 is converted to H_2 gas and CO (Equation 10. 3). H_2 is improved by a WGS reaction similar to that which coal undergoes [116].

$$CH_4+H_2O \xrightarrow{\text{reform}} CO+H_2 \tag{10.3}$$

TABLE 10.2

Summary of recent literature about separation methods

	Year	References	Chemical	Physical	Biological
			Separation method		
1.	2008	[43, 44]	✓	✓	✗
2.	2008	[45–48]	✓	✗	✗
4.	2009	[49–53]	✓	✗	✗
5.	2009	[54, 55]	✗	✗	✓
6.	2009	[56]	✓	✓	✗
7.	2010	[57, 58]	✗	✗	✓
8.	2010	[59–61]	✗	✓	✗
9.	2010	[62]	✓	✗	✗
10.	2011	[63–66]	✗	✓	✗
11.	2011	[67]	✗	✗	✓
13.	2012	[68, 69]	✓	✗	✗
14.	2012	[70, 71]	✗	✓	✗
15.	2012	[72, 73]	✗	✗	✓
16.	2013	[74–76]	✗	✓	✗
17.	2013	[77–79]	✗	✗	✓
18.	2013	[80, 81]	✓	✗	✗
19.	2014	[82–84]	✓	✗	✗
20.	2014	[85]	✗	✗	✓
21.	2014	[86, 87]	✗	✓	✗
22.	2015	[88]	✗	✗	✓
23.	2015	[89–91]	✗	✓	✗
24.	2015	[92–94]	✓	✗	✗
25.	2016	[95, 96]	✓	✗	✗
26.	2016	[97–100]	✗	✓	✗
27.	2017	[101]	✓	✗	✗
28.	2017	[102, 103]	✗	✗	✓
29.	2017	[104–106]	✗	✓	✗
30.	2018	[107, 108]	✗	✓	✗
31.	2018	[109]	✓	✗	✗

Partial oxidation reactions use exothermal oxygen and CH_4 (Equation 10.4). Autothermal reforming is a combination of both methods [117]. According to Hoffmann et al. [118], CO_2 trapping efficiency is 80%, and the cost of CO_2 emissions was \$29/1000 kg CO_2 for a planned concept.

$$2CH_4 + O_2 \rightarrow 2CO + 4H_2 \tag{10.4}$$

For the gas hydrate formation process, the operating pressure is shifted from 11 to 1.87 MPa at 278.7 K with the addition of 1 mol% tetrahydrofuran (THF) [119].

The gas composition of the equilibrium gas hydrate is 39.9 mol% CO_2 and 60.1 mol% H_2. Consequently, this $CO_2/H_2/THF$ system is potentially useful for industrial processes with no significant compression costs [114].

10.2.2 Post-Combustion Processes

Post-combustion processes convert power plant exhaust gases into clean gases using chemicals. In the post-combustion process, CO_2 is separated from flue gases comprising NO_x and SO_2. This process is used for retrofitting into existing power plants [115]. CO_2 concentrations are normally quite low, at 7–14% for coal-fired plants and 4% for gas-fired plants. Consequently, the energy penalty and costs to raise CO_2 concentrations are elevated [47, 59]. Membrane post-combustion uses a sweep module with a steam feed of 500 m^3/s with a CO_2 concentration of 13%. This membrane decreases the concentration of CO_2 to 2.1% in treated flue gas and produces a CO_2-enriched permeate. However, the limiting rate for this membrane for CO_2 removal is the gradual permeation of nitrogen across the membrane [120].

Monoethanolamine (MEA) absorption is a chemical absorption method that has been used for more than 60 years. The primary cost of carbon capture using this method is the high demand for regenerating captured media. This is because 40% of the energy output is required to separate CO_2 from the chemical solutions involved [121]. Also, Bounaceur et al. [122] reported that if CO_2 is composed of MEA flue gas absorption, a second separation process is necessary to separate CO_2 and recover the chemical solvents. This was supported by the National Energy Technology Laboratory (NETL), which stated that the cost of the electricity footprint of the host plant and for electricity would increase by 60% and 70%, respectively [123]. A recent study noted that during the post-combustion process, the cost of electricity increases by 32% and 65% in gas-combusting and coal-combusting plants, respectively [124].

The other weakness of post-combustion is that CO_2 must be separated at low partial pressure from the flue gas after being completely burned. This process is costly, is not selective, uses a large number of chemical solvents and is energy-intensive with 40% of power station energy being used to run CCS processes. In order to overcome the drawbacks of this, the chemical solvent used in the process should be changed to amine-2-amino-2-methyl-1-propanol (AMP) due to its higher loading capacity [125, 126]. This is because common chemical solvents such as primary amine and MEA could corrode pipelines and produce minor impurities such as CS_2, resulting in solvent degradation. Therefore, it should be changed to a tertiary amine such as triethanolamine (TEA) or methyl-diethanolamine (MDEA) due to their low heat requirements for CO_2 liberation and minimal production of degradation products [127].

10.2.3 Oxyfuel Combustion

Oxyfuel combustion processes burn coal in the air with high pure O_2 concentrations, ensuring almost pure CO_2 is emitted as exhaust. Then, pure O_2 flows into the energy conversion unit and is mixed with recycled CO_2 gas so that the furnace temperature can be kept low. Lastly, the flue gas of the high-purity steam CO_2 is formed [128].

158 Fossil Free Fuels

Oxyfuels tend to be used with low-sulfur coal and can deliver lower evasion costs than amine when used in combination with SO_2 and non-condensable gases [129].

The presence of nitrogen in exhaust gas will affect the separation process. Therefore, NO_x gas has to be eliminated to produce CO_2-enriched flue gas in this process [130]. The major components of the flue gas are CO_2, water, SO_2 and particulates. SO_2 and particulate compounds are first removed from the flue gas by an electrostatic precipitator and flue gas desulfurization methods to obtain a high-purity CO_2 gas [128]. Large amounts of O_2 are consumed by air separation unit [51] and the CO_2 purification unit [131]. Consequently, the cost is high, and energy is 7% above plant CCS [132, 133]. The oxyfuel process will also produce impurities, such as H_2O, SO_x, NO_x, O_2 and N_2, due to its inflexibility. These impurities need to be reduced before the flue gas can be delivered to pipelines to ensure its safe transportation [131].

10.2.4 COMPARISON AMONG VARIOUS CO_2 REMOVAL COMBUSTION PROCESSES

The comparison results for the three types of combustion processes for CO_2 capture are shown in Table 10.3. Mainly, post-combustion and oxyfuel combustion processes are used for both coal-fueled and gas-fueled plants. Meanwhile, pre-combustion CO_2 capture processes are useful only in coal-fueled plants. Currently, the post-combustion process is the most developed CO_2-capturing process [128, 134]. Gibbins et al. [135] reported the cost of the three combustion processes for CO_2 removal in both coal-fueled and gas-fueled plants.

From the three combustion processes, a few researchers have been developing modified methods, such as temperature swing adsorption (TSA) for post-combustion, chemical looping combustion (CLC) for oxyfuel combustion and integrating gasification combined cycle-syngas (IGCC) for pre-combustion. TSA can potentially improve compressing problems, and it can apply a vacuum to a large volume of a low-pressure gas stream. The working capacity in this method relies on low CO_2/N_2 selectivity and the accessibility of low-grade heat in a power plant as an energy source for regeneration [121]. Chemical looping uses metal oxides as an oxidation source to minimize "unwanted materials." This process uses a dual-fluid bed system. Metal oxide is the first material bed and provides oxygen for fuel reactors. The reduced metal is transferred to the second bed (air reactor) to oxidize the reaction before recycling the fuel reactor [136]. This process provides efficient carbon capture and carbonation technologies that are 99% available directly from coal to CLC. This process has resulted in CLC as carbon energy [137, 138] that is able to capture 99% of CO_2 in a coal-direct chemical looping plant [137, 138]. Integrated gasification combines cycle-syngas for mixtures of CO_2 and H_2 after water gas shift production has led to increased driving forces for mass transfer and flexibility.

Gas-fired power plants significantly reduce emissions of CO_2 compared to coal-fired plants. However, gas-fired power is generally more expensive than coal-fired power. In Table 10.4, the comparison between the cost and efficiency of the coal-fired and gas-fired power sectors are compared, with a focus on the associated capture process.

Power generation is the main source of CO_2 emissions and is a major concern in developing countries. To control and reduce CO_2 emissions from fuel combustion, it

TABLE 10.3

Comparison of three combustion process for dioxide carbon removal

CO$_2$ capture process	Application	Pros	Ref.	Cons	Ref.
Post-combustion process	• Coal-combusting plants. • Gas-combusting plants.	• Gained more acceptability compared to other processes.	[118, 130]	• Low concentration of CO$_2$ causing the energy penalty and costs are elevated	[47, 59, 124]
		• Can be simply retrofitted to the available power plant.	[115]	• The cost of electricity increase.	[132]
Pre-combustion process	• Coal-gasification plants.	• High concentration of CO$_2$. • CO$_2$ capture efficiency is 80%.	[59, 131] [51, 118]	• High operating cost.	[118]
Oxyfuel or O$_2$ /CO$_2$ recycle combustion process	• Coal-combusting plants. • Gas-combusting plants	• High-purity steam CO$_2$. • modified from post-combustion process.	[128] [59]	• High efficiency, cost and energy penalty.	[132, 133]

TABLE 10.4
Cost for different capture processes

Fuel type	Parameter	No capture	Pre-combustion	Post-combustion	Oxyfuel
			Capture processes		
Gas-combusting	Efficiency (%LHV)	55.6	41.5	47.4	44.7
	The initial installation cost (dollar per kW)	500	1180	870	1530
	Electricity cost (cent per kWh)	6.2	9.7	8.0	10
	CO_2 avoided cost (dollar per 10^3kg CO_2)	–	112	58	102
Coal-combusting	Efficiency (%LHV)	44.0	31.5	34.8	35.4
	The initial installation cost (dollar per kW)	1410	1820	1980	2210
	Electricity cost (cent per kWh)	5.4	6.9	7.5	7.8
	CO_2 avoided cost (dollar per 10^3 kg CO_2)	–	23	34	36

is necessary to use CCS. Figure 10.5 shows the mitigation in combustion processes of the chemical loop, TSA and IGCC.

The post-combustion process is the most likely capture method in automobiles that is directly attached to the exhaust to separate CO_2 from the flue gases without any direct interference with the inputs of the internal combustion engine [140]. The applied post-combustion process in automobiles can be in any form of absorption, adsorption or membrane separation.

Currently, CO_2 capture contributes 75% of overall CCS costs. Thus, the electricity production cost of CCS would increase by 50% [126]. Hence, this chapter focuses on holistic CO_2 separation by chemical, physical and biological techniques. There are several choices for CO_2 separation that include absorption, adsorption, membrane processes and cryogenics [141]. All of these options use different chemical and physical approaches [142, 143]. Furthermore, another alternative is biological

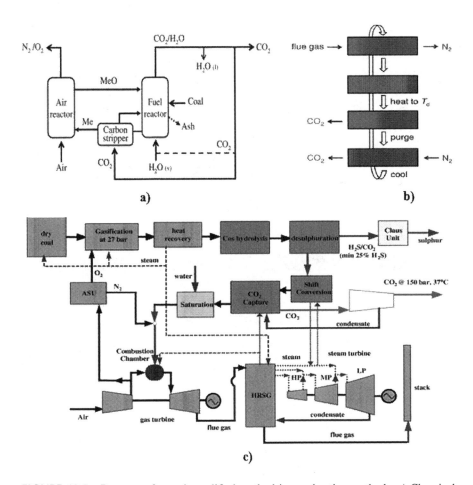

FIGURE 10.5 Processes for each modified method in combustion methods, a) Chemical loop (adapted from [136]), b) TSA (adapted from [121]) and c) IGCC (adapted from [139]).

processes [144]. To achieve environmental and economic sustainability, biological processes are a significant part of the equilibrium of CO_2 concentration [44, 68]. According to Rahman [66], environmental issues are best handled where each citizen has an awareness of environmental matters. The available CO_2 separation and extraction technologies can be divided into three different kinds of methods: chemical, physical and biological. Chemical methods include absorption, chemical adsorption and chemical looping. Physical separation methods include cryogenic condensation, physical absorption and membrane separation. The biological method is biological fixation [145]. Several CO_2 separation and extraction technologies are shown in Figure 10.6.

Several mechanical, chemical and biological technologies for separating CO_2 from emissions have been introduced. Table 10.5 shows a general comparison of different chemical and physical CO_2-separation methods.

10.3 CO_2 CAPTURE PROCESSES

CO_2-capturing techniques can be roughly divided into three categories: physical, chemical and biochemical. In the following subsections, each category is presented and the theory, application and techno-economical evaluation of each process are presented.

10.4 CHEMICAL PROCESS

Chemical methods are widely used for CO_2 removal, of which chemical adsorption, absorption and chemical looping are the most common. In the following section,

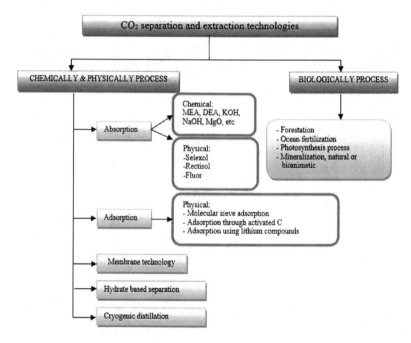

FIGURE 10.6 CO_2 separation and extraction technologies.

TABLE 10.5
General comparison of different separation techniques

Separation Technique	Advantages	Disadvantages	References
Absorption	• Most mature process. • Sorbents regenerate by heating and/or depressurization. • CO_2 absorption efficiency over 90%.	• Efficiency of absorption depends on CO_2 concentration.	[134, 146, 147]
Adsorption	• Reversible process. • Reusable adsorbent. • High adsorption efficiency.	• Adsorbent required higher temperature • High energy for CO_2 desorption.	[59, 148, 149]
Membrane separation	• Effective for conventional • CO_2 separation processes using inorganic or polymeric. • More effective for CO_2 separation compared to liquid absorption process. • Recent apply to industrial processes (separation of air, recovery of hydrogen and carbon capture from natural gas). • Easy installations with high packing density.	• Low fluxes. • Fouling during operation.	[92, 101, 122, 150–152]
Hydrate-based CO_2 separation	• Requires no extra energy consumption.	• Not yet mature enough to be implemented practically.	[119, 123, 153, 154]
Cryogenics distillation	• Established process • Applied for long time in industry for CO_2 recovery.	• Requires low temperature and high pressure. • Very energy-inefficient.	[155–157]

chemical absorption, as well as membrane, hydrate and cryogenic technologies are further discussed.

10.4.1 CHEMICAL ABSORPTION TECHNOLOGIES

Chemical absorption technology is used for lower partial pressures of CO_2. The acidic nature of CO_2 is dealt with using basic solvents for acid–base neutralization reactions for gas streams, such as flue gases. The CO_2 gas will form a weakly bonded intermediate with chemical solvents before the compound is weakened by the application of heat. Then, the solvent is regenerated to form a clean CO_2 stream. Currently, specific solvents are used to remove CO_2 from contaminated natural gases. Many processing plants and industrial plants in chemical and food industries use similar components to recover CO_2 from gas streams. Several alternative methods have been used to separate CO_2 from flue gas mixtures in hydrogen production in oil refineries, ammonia plants and other chemistry works. [158].

The CO_2 trapping technique must be based on the CO_2 pressure in the gas stream, the required range of CO_2 recovery, solvent recovery, gas deficiency sensitivity, the purity of CO_2 product required, the cost of production needed to cover raw costs, erosion and the environmental impact [159]. A chemical absorption technology scheme is depicted in Figure 10.7 [49].

10.4.1.1   Amine Absorption

A liquid sorbent separates CO_2 from the flue gas of exhaust. The absorption rate can be increased by applying heat or depressurizing within the regeneration process [134]. Common sorbents include MEA, diethanolamine (DEA) and methyl diethanolamine (MDEA) [146]. The process of CO_2 absorption using MEA has been widely studied. According to Veawab *et al.* [147], MEA has the highest absorption efficiency (> 90%). Aaron *et al.* [160] mentioned that the most likely way to capture CO_2 is by the absorption of MEAs. One ton of a CO_2/h prototype was tested with a post-combustion technique for coal-fired power plants using 30% MEAs [50].

Currently, adsorption processes with commercial MEA are used to absorb the CO_2 emitted by exhaust gas combustion. The MEA solution is streamed with CO_2 in an absorber to obtain a MEA carbamate. Then, with the CO_2 enriched, the MEA solution is transferred into a heat exchanger where it is heated to produce high-purity CO_2. The obtained solution is then recycled into the absorber [161]. This process is generally not economical, though, as a CO_2 capture plant uses up to 70% of its total cost for the required equipment, solvent size, solvent renewal and intense energy input [162]. Other disadvantages of this process are its low CO_2 loading capacity; high corrosion rates; the degradation of the MEA solvent by SO_2, NO_2, HCl, HF and O_2 in the flue gas with high-temperature regeneration absorbers that increase the rate of absorbent make-up; and high energy consumption [163–165]. This was discussed in detail in Section 10.2.2.

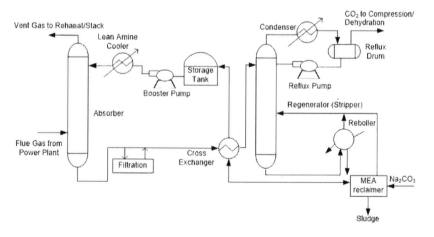

FIGURE 10.7   Chemical absorption system for CO_2 recovery [49].

10.4.1.2 Aqua Ammonia Absorption

According to Resnik *et al.* [164]. And Yeh *et al.*[165] aqua ammonia absorption is an alternative to amine absorption technology to improve the weaknesses of the MEA. Aqueous ammonia is used in this process as a CO_2 sorbent with multi-component control capabilities. Steam gases need to be prepared by oxidizing SO_2 to SO_3 and NO to NO_2. Then, the reaction is carried out with ammonia within wet scrubbers. The recovery of ammonium hydroxide needs thermal input for ammonium bicarbonate and pyrolysis ammonium carbonate. The aqua ammonia process can reduce energy combustion by nearly 60% when compared with the MEA process. The main additives in this process are ammonium compounds of ammonium bicarbonate, ammonium sulfate and ammonium nitrate. Although ammonium bicarbonate is heat-treated for recycling ammonium, it is unreliable for the market. The aqua ammonia process has the potential to promote cheap and abundant sulfur coal combustion. The aqua ammonia process was also implemented in We Energies power plants to extract 35 in 1000 kg CO_2 per day from flue gases [52]. To better understand the situation, the UNIQUAC electrolyte model was used to explain the properties of the NH_3-CO_2-H_2O system [62, 166].

10.4.1.3 Dual Alkali Absorption Approach

Huang *et al.* [167] proposed a modified Solvay binary process using ammonia as a catalyst to support CO_2 reactions with sodium chloride for sodium carbonate production (Equation 10.5).

$$CO_2+NaCl+NH_3+H_2O \rightarrow NaHCO_3 \downarrow +NH_4Cl \qquad (10.5)$$

The sodium carbonate was produced within endothermic reaction of sodium bicarbonate. The reaction between ammonium chloride with lime results in restoration ammonia (Equation 10.6).

$$2NH_4Cl+Ca\left(OH\right)_2 \rightarrow 2NH_3+CaCl_2+2H_2O \qquad (10.6)$$

Nevertheless, some weaknesses are obvious when this process is applied to capture CO_2. This is because limestone is used for ammonia production to reduce the production of CO_2, which makes the carbonation process energy-intensive. According to Yang *et al.* [48], a remarkable choice for CO_2 removal in the dual alkali absorption is done by calcination and carbonation cycles comprised of two mixed bed reactors with limestone as an absorber and regenerator [53]. In the absorber section, $CaCO_3$ is produced by the carbonation of CaO at a temperature range of 600–700°C to produce a concentrate of CO_2. This process is considered an economically competitive technology with efficiencies of over 80% [45, 46, 56].

10.4.2 Membrane Technology

Separation methods using membrane technology are inexpensive when purity gas flow is not the objective. Currently, polymeric membranes were conquered in their industrial applications. Recent research has shown that the use of inorganic membranes has grown rapidly as demand for new applications in high-temperature separations such as membrane reactors has increased. The polymer or inorganic membrane separations have shown better performance than existing CO_2 separation processes [101, 150, 151]. Metallic and polymeric membranes are utilized to convert gasified gases, natural gas or methanol reformations into clean fuels, though inorganic silver alloy membranes are more effective. Separation methods that use metal membranes become more effective as the pressure decreases across the membrane.

According to Yu et al. [92], ceramic hollow fibre membrane shows outstanding potential for real industrial CO_2 capture due to anti-wetting and anti-fouling properties. Brunetti et al. [168] investigated membrane CO_2 separation using adsorption and cryogenics and found that the performance of the technology strongly affected the condition of the stream gas, including CO_2 concentrations and applied pressure. Other researchers [152] developed a multi-permselective mixed matrix membrane by incorporating a multi-purpose filler with ethylene oxide and an amine carrier in the matrix polymer. According to Bounaceur et al. [122], there are several advantages to gas separation membrane processes: i) high separation energy efficiency for balance-based processes, ii) the use of industrial processes (air separation, hydrogen recovery and capture of CO_2 from natural gas) and iii) more packaging density at smaller plants. Some issues regarding recent studies on how to separate amine adsorption technology, aqua ammonia process and dual alkali absorption approaches are presented in Table 10.6.

10.4.3 HYDRATE-BASED SEPARATION

The new hydrate-based separation concept can potentially be used for exhaust gas flow that contains CO_2 by treating it with water in hydrate-forming elevated pressures. When the hydrate forms, the CO_2 is captured. Then, the hydrates are separated and detached, thus releasing the CO_2. The energy consumption of this approach is as low as 6–8%, with a small CO_2 concentration of 0.57kWh/kg [123]. In order to increase CO_2 efficiency, the synthesis rate and hydration pressure reduction can be increased [153]. Linga et al. [119] claimed that the presence of tetrahydrofuran (THF) significantly decreased the pressure formation of the hydrate mixture (CO_2/ N_2) of exhaust gases and the possibility of obtaining CO_2 at an appropriate pressure. THF is a mixed water solvent capable of reacting at low temperatures to produce clathrate hydrate structures. The presence of THF facilities is used as a thermodynamic accelerator for the formation of hydrates. Zhang et al. [154] confirmed the effect of THF on hydrate formation using the isochoric method on the hydrate phase equilibrium.

10.4.4 CRYOGENIC DISTILLATION

The process of separating gases using distillation at very low temperatures and high pressures is called cryogenic distillation. Cryogenic temperatures were acquired in a closed-cycle-operated refrigeration system comprised of compressors, Joule Thompson Valves (JTV), multi-stage heat exchangers and mountings [51]. Since the

TABLE 10.6

Separation method based on absorption processes and membranes technology

Process	Unique features	Reference
Amine absorption technology	• Established method for CO_2 separation.	[134]
	• MEA is the common solvent for CO_2 absorption system with the capable more than 90%.	[50]
	• The most preferable way to catch CO_2 for CCS by absorbing using MEA.	[50]
	• This process is not economical because of the large equipment and intensive energy input required.	[162]
Aqua ammonia process	• The process of ammonia in water avoids the weaknesses of MEA processes.	[164, 165]
	• Up to 60% energy savings compared to MEA process.	
	• It can create potential, which promotes the combustion of high-sulfur coals, which is inexpensive and rich.	[52]
	• Implemented in We Energies power plants, which aim to extract 35 tons of CO_2 per day from flue gases.	[62, 166]
Dual alkali absorption approach	• Solvay dual-alkali method.	[167]
	• Process based on carbonation and calcinations cycle.	[53]
	• CO_2 capture from combustion gases in two fluidized-bed bridges with limestone.	
	• Present high efficiency (> 80%).	[45, 46, 56]
	• Consider as economical.	
	• Competitive technology.	
Membranes separation	• Review about inorganic or polymeric membrane separation processes for conventional CO_2 separation processes.	[101, 150, 151]
	• Advance of the ceramic, metallic and polymeric membranes for membrane diffusion.	[92, 169]
	• The performance of the membranes technology using membranes and others separation method.	[168]

process requires very low temperatures and high pressure, it is an energy-intensive process requiring approximately 600–660 kWh per 1000 kg of CO_2 obtained in liquid form [155]. Some process patents were developed, and many researchers focused on cost optimization [156, 157]. Surveys through simulation and modeling in Aspen HYSYS were conducted to assess what low temperatures were best in yielding CO_2 from oxyfuel combustion at high purity and high pressure [170].

According to Tuinier *et al.* [157], the cryogenics process can also be applied in the post-combustion CO_2 capturing process. They also proposed a cryogenics process using dynamically operated beds. Flue gas comprised of CO_2 and H_2O is separated by differences in condensation and sublimation. The main advantages of this process are the instantaneous separation of CO_2 and H_2O, the use of solvents and low pressure. Table 10.7 presents some issues for the separation method based on cryogenic distillation technology.

TABLE 10.7

Separation method based on cryogenic distillation technology

Process	Issues	References
Cryogenics distillation	• This process requires very low temperatures and a high energy concentration of 600–660 kWh per kg of deposited liquid CO_2.	[155]
	• Many patented works have been introduced and many research are conducted on cost optimization.	[156, 157]
	• Aspen HYSYS has been studied through simulations and models.	[170]
	• The cryogenics process can be applied in post-combustion CO_2 separation with dynamically operated filling bed.	[157]

10.5 PHYSICAL PROCESSES

Physical methods are widely used for CO_2 removal. Physical absorption and cryogenic condensation are among the most common CO_2 separation methods. In the following subsections, physical absorption methods, as well as membrane, hydrate and cryogenic technologies are further discussed.

10.5.1 PHYSICAL ABSORPTION PROCESS

For physical absorption processes, in accordance with Henry's Law, CO_2 is readily absorbed in an organic solvent, which means that pressure and temperature are dependent variables. In physical processes, the acid components of gases are absorbed physically rather chemically. In physical absorption, the disposal of CO_2 depends largely on solubility, partial pressure and the temperature of the CO_2 within the solvents. Physical absorption processes are commonly used to extract acidic gases that consist of CO_2 and H_2S from gaseous mixtures and CO_2 emission in hydrogen (H_2), ammonia (NH_3) and methanol (CH_3OH) production. Physical solvent cleaning of CO_2 uses Selexol (dimethylether of polyethylene glycol) and Rectisol. The Fluor solvent (propylene carbonate) and Purisol (N-methyl-2-pyrollidone) are two physical solvents that are widely used for CO_2 capture. A diagram of the physical absorption approach is described in Figure 10.8.

10.5.1.1 Advantages and Disadvantages among Selexol, Rectisol and Fluor Processes

Table 10.8 shows the pros and cons of physical absorption processes using Selexol, Rectisol and Fluor [148, 151]. Selexol processes are used to sweeten natural gas by removing CO_2 and H_2S gas in bulk. This absorption process operates at very low temperatures (between 0–5°C). Solvents can be used for sulfur removal, CO_2 and water compounds at the same time. The feed gas is dehydrated before it reaches the Selexol unit. Meanwhile, Rectisol treats hydrogen, syngas and city gas flows for impurity removal. Methanol (CH_3OH) is used as a solvent in Rectisol processes due to

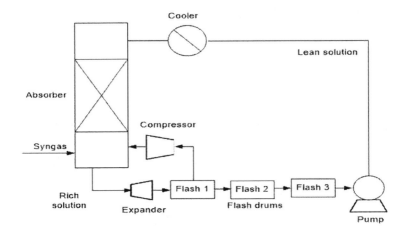

FIGURE 10.8 Physical absorption to capture CO_2 [59].

the high-pressure vapor of methanol. Therefore, the process needs to operate at temperatures within the range of 35–75°C. The Rectisol process is for the purification of heavier components such as ethane [171]. The Rectisol process is one of the most commonly applied commercial CO_2-removal processes used in the natural gas industry. Besides, the Fluor solvent process is used for treating feed gas CO_2 at high partial pressures (> 60 psig). This process is based on propylene carbonate ($C_4H_6O_3$), a physical solvent for removing CO_2 due to its polarity and its high affinity for CO_2 [172].

10.5.2 PHYSICAL ADSORPTION

Physical adsorption largely depends on the thermodynamic potential of materials to be physically or chemically bound from the gas phase into the solid state. The adsorption of CO_2 from flue gas is proceeded by the addition of the adsorbent (zeolite or activated carbon) for removal followed by regeneration (desorption). This process lowers the pressure (pressure swing adsorption [PSA]), raises the temperature (temperature swing adsorption [TSA]) or causes adsorbent (electric swing adsorption [ESA]) or in a hybrid combination of these (PTSA) or cleaning.

10.5.2.1 Molecular Sieve Adsorption

In the sieve adsorption process, molecules are separated on the basis of their mass or size. This method is suitable for different forms of CO_2 capture [161]. A number of studies were done to improve adsorption of CO_2 through chemical modification using a surficial molecular sieve. Adsorption is carried out on large inorganic surface interacting with organic compounds. Interactions among the basic surface and the acidic molecule of CO_2 under water-free conditions produce ammonium carbamates. If the reaction takes place in the presence of water, ammonium bicarbonate and ammonium carbonate are formed [174]. Comparable to the ammonia absorption process without water, the ability to absorb CO_2 is preferable with a molecular weight of 0.5 mol CO_2/mol, whereas 1.0 mol CO_2/mol is preferable in the existence of water.

TABLE 10.8

Advantages and disadvantages of the Selexol, Rectisol and Fluor processes [171–173]

Physical absorption process	Advantages	Disadvantages
Selexol process	• Low temperature of the solvent in the adsorbent due to the exothermic nature of the reaction. • Sweet gas emission from dry adsorption is caused by high-affinity Selexol solvent and water. • Low operating cost of initial plant • Solvent re-generation is by air stripping, it does not require heat reboiler. • Selexol assists in carbon steel formation mainly because of its non-liquid state and also inert chemical properties. • Pperated at low pressure.	• The solvent has a high degree of interaction with heavy hydrocarbons to be produced with CO_2 and causing hydrocarbon vacancy.
Rectisol process	• Solvent (methanol) is not bubbly and completely soluble and reduces losses. • Heat and chemical stability. • Not corrosive. • No problems with degradation. • It is commonly utilized for carbon steel equipment. • The rich solution can be easily revitalized by simplifying pressure without any need for heat inputs in the recharge.	• Cold methanol solvents used to absorb metals such as mercury (Hg) to form amalgam at low temperatures. • Rectisol complex and the need to reduce the temperature of the solvent with the initial installation and operational costs of the separation plant.
Fluor process	• There is no obligation necessary for solvent regeneration. • Fluoro solvents have the ability to dissolve high CO_2 and increase CO_2 loading. • The operation is simple, and the output gas is dry gas. • Propylene carbonate is frozen at -57 ° F. • Modifications to increase CO_2 in feed are low.	• The solvent cycle for the Fluoro solvent process is high. • Fluoro solvents are very costly. • Solvents are highly correlated for heavy hydrocarbons, which is eliminated with CO_2, essentially resulting in loss of hydrocarbons.

Mesoporous molecules such as silica [174, 175], SBA-1 [176], SBA-15 [177], MCM-41 [176, 178, 179] and MCM-48 [167] are better because their pores are wide and easy to reach by the amino group.

According to Chaffee *et al.* [180], there has been much research done to develop new adsorbents to perform vacuum swing adsorptive (VSA) in separating CO_2 from moisture-resistant flue gases that can work at higher temperatures. Theoretically, the adsorption capacity is limited to 1.0 mol molecules of CO_2 for each N atom. Thus, the reaction is equivalent to that which occurs by chemisorption in solutions involving the formation of the chemical bonds. The function of the adsorbent totally differs from the general adsorbent that acts on the principle of dissolution.

10.5.2.2 Adsorption through Activated Carbon

Activated carbon was produced in the form of micro- and mesoporous-sized particles for industrial and technological processes [181]. Anthracites are known to yield high surface areas of activated carbons. Arenillas et al. [182] studied CO_2 capture behaviour of anthracite using steam. The results showed that adsorption capacity decreased rapidly as the adsorption temperature increased. Maximum CO_2 absorption occurs with 65.7 mg CO_2/g at 800°C for 2 hours with a surface area of 540 m²/g. Anthracite has a surface area of 1071 m²/g, capable of adsorbing 40 mg CO_2/g CO_2. The deformation of NH_3 and polyethylaminominic (PE) rapidly improved the adsorption capacity of anthracite at elevated temperatures because of the presence of surficial alkaline nitrogen.

10.5.2.3 Adsorption of Lithium Compounds

Lithium zirconate (Li_2ZrO_3) has an outstanding performance in the adsorption of CO_2 [163]. Chemical reactions of Li_2ZrO_3 for CO_2 removal are expressed in the succeeding reaction:

$$Li_2ZrO_3(s) + CO_2(g) \leftrightarrow Li_2CO_3(s) + ZrO_2(s) \qquad (10.7)$$

Using a temperature swing, the reaction can be reversed within a temperature range of 450–590°C. The formation of carbonates consists of K_2CO_3 and Li_2CO_3, which catalyzes CO_2 adsorption. Zirconate adsorption of binary and titanium lithium is evaluated to capture CO_2 at high temperatures. The combination of carbonate, alkali, carbonate, carbonate, alkali carbonate, ternary alkali carbonate and ternary alkaline carbonate with Li_2ZrO_3 increases the capacity to adsorb CO_2.

Kato et al. [149] and Essaki et al. [183] studied lithium silicate (Li_4SiO_4) for their CO_2 adsorption behaviours. This study concluded that the capacity of Li_4SiO_4 to adsorb CO_2 was greater than that of Li_2ZrO_3. Lithium silicate adsorbs CO_2 at temperatures lower than 720 °C and emits CO_2 above 720 °C. Chemical reactions using Li_4SiO_4 for CO_2 removal are described in the succeeding reaction:

$$Li_4SiO_4 + CO_2(g) \leftrightarrow Li_2SiO_3 + Li_2CO_3 \qquad (10.8)$$

Large capacity, fast absorption, temperature, CO_2 concentration and stability are required for the development of commercial Li_4SiO_4-based CO_2 adsorbents.

10.6 BIOLOGICAL PROCESSES

Biological processes are natural and synthetic ways to trap carbon emitted into the environment by forestation, marine fertilization, photosynthesis, mineral carbonation and hydration.

10.6.1 Forestation

By 1.4 ± 0.7 G photosynthesis in 1000 kg carbon is captured by the ground system [184]. The amount of carbon flux is increased with the reforestation of arid land and desert greening. The capture and storage of CO_2 from the start is positive, and the amount of typical carbon increases. Net capture is zero due to a capture-and-release balance when fully developed. The potential for CO_2 deposition in the global system is about 5–10 G in 1,000 kg of carbon per year [184].

10.6.2 Ocean Fertilization

According to Yamasaki [184], oceans have more CO_2 than land. The ocean comprises about 38,000 G in 1000 kg carbon. Each year, 1.7 ± 0.5 G in 1000 kg carbon is borrowed from the surrounding atmosphere, and the generated carbon by phytoplankton is higher than that of terrestrial plants [184]. Some CO_2 is emitted into the atmosphere again through the respiratory, and the remainder is partially transmitted back to the ocean as inorganic compounds of the phytoplankton's death. This process can be enhanced using sea fertilization by increasing the limiting nutrient to generate phytoplankton production. However, this method has been challenging because of disruptions to the marine ecosystem. Furthermore, sinking organic decompositions can create stronger GHGs, such as methane and N_2O. Therefore, this method needs to be reviewed and improved before it is used.

10.6.3 Photosynthesis Process

Biological carbon storage through the use of photosynthesis reactions can help reduce GHG emissions. Stewart *et al.* [161] studied a combination of solar energy that refers to the transmission of optical fiber and collection systems to encourage the generation of useful byproducts from CO_2 by utilizing biological generators. A photo-bioreactor uses a natural process of photosynthesis to heat up and convert CO_2 into more valuable compounds (e.g., carbohydrates, oxygen-hydrogen). The final product is determined based on the incorporated biological strains. The photosynthetic reaction occurs as follows:

$$6CO_2(aq)+6H_2O(l)+light+heat \leftrightarrow C_6H12O_6(aq)+6O_2(g) \tag{10.9}$$

For photosynthesis, microalgae were used to assure the uniform growth of organisms by distributing light fluxes in the 400–700 nm wavelength range to bioreactors. The effective circulation of light through a photobioreactor will affect CO_2 absorption. Even with more than 50 years of evidence showing the effectiveness of the closed-loop photo-bioreactor systems, their use is still not at the commercial scale. Currently, an open pond system yields approximately 100,000 kg of biomass at an annual cost of approximately USD$10,000 per 1,000,000 kg [161].

10.6.4 MINERALIZATION, NATURAL OR BIOMIMETIC

Other CO_2-trapping processes use the chemical conditions of rocks with silicate compounds such as calcium silicate or magnesium silicate. These silicate rocks react with CO_2 and turn into carbonates according to the following mechanism [182]:

$$\left(Mg,Ca\right)xSiyO_{x+2y} + xCO_2 \leftrightarrow x\left(Mg,Ca\right)CO_3 + ySiO_2 \qquad (10.10)$$

Mineral CO_2 is associated with the storage of CO_2 in solid form, a stable and environmentally friendly carbonate mineral. The carbonate's energy state is 60 to 180 kJ/mol lower than CO_2, which is less than 400 kJ/mol of carbon [182]. Therefore, the formation of carbonate causes long-term CO_2 fixation and avoids any risk of CO_2 leaked from underground storage aquifers. There are two methods to accelerate the reaction rates of CO_2 mineralization: using stronger acids that promote the dissolution of inorganic ions and reacting under high CO_2 pressures. The use of acid can cause environmental problems, but the volume of this anchor is large and high CO_2 pressures consume large amounts of energy.

According to Nelson et al. [185], the silicate compounds were polished repeatedly in the presence of CO_2. Test materials contained natural substances such as synthetic magnesium silicate, wollastonite, lizardite and forsterite. Results showed small fixed amounts of carbon after the intensive milling of lizardite, forsterite, wollastonite and magnesium silicate for 15–20 minutes with CO_2 gas. However, this technology is limited by the power consumed in the grinding of materials. According to Arenillas et al. [182], reaction mineral carbonation rates and efficiencies at adequate reaction states can be significantly improved over physical and chemical activation methods using surface activation. Acidification with sulfuric acid is more effective than physical induction for larger surface areas. Steam activation facilitates CO_2 capture with serpentine. The results showed that the obtained solution of $Mg(OH)_2$ has a 53% carbonation efficiency.

Liu et al. [186] introduced a novel biomimetic method that promotes water treatment lines for the formation of carbonate using enzyme carbonic anhydrase. The enzyme was encapsulated on chitosan-alginate beads. The required cations to produce carbonate were yielded from a simulated saline solution that contained calcium. The results demonstrate that the enzyme carbonic anhydride has the potential to increase precipitation carbonate mineralization with industrial sewage water as a source of cations. According to Haywood et al. [187], CO_2 sequestration within a stable mineral carbonate involves various radical alteration processes and requires a massive source of metal oxides, enough to have a major effect on the iron, manganese, chromium and nickel mining industries.

10.7 THE IMPROVEMENT OF SEPARATION METHODS

An improved amine-based post-combustion system used for the capture of CO_2 was developed by many process developers such as Fluor, Cansolv Technologies and Mitsubishi Heavy Industry (MHI). Fluor Econamine FG plus is a patented degassing of acid contents that demonstrates an availability of over 95% with natural gas power plants. Currently, commercial technology is used to compare other CO_2-recovery technologies. MHI has created a new adsorption process using amine solvents to trap CO_2 from flue gas [188]. Chemical modification membranes have been greatly improved in the most advanced approaches in terms of gas separation efficiency to speed up the marketing of membranes for the hydrogen economy. Therefore, to maintain membrane-system robustness, long-term stability and the performance of the membrane should be considered at elevated temperatures [189].

Ksayer et al. [190] and Rizk et al. [191] developed a cryogenic CO_2-capture process in which the CO_2 heat exchanger acts as a solvent on sand surfaces and is cooled by evaporation from the refrigerant mixture. According to Bansal et al. [192], the performance of CO_2 as a refrigerant is significantly superior to conventional single-level refrigerants. Tuinier et al. [157] investigated a new cryogenic-packed bed that required 22% less energy at temperatures of –105°C to reduce cooling costs. A cryogenic distillation operates at high pressures and low temperatures [193, 194]. The CO_2-capture of the cryogenic capture process is preferable because it can reach 99% efficiency [194, 195].

In recent years, microalgae-based processes have emerged as a promising alternative and a practical system for CO_2 separation from exhaust flue gases have been developed [196]. Lenton et al. [197] reviewed CO_2 removal on biologically based land. Studies have suggested that CO_2 emissions could emit excess CO_2 and reduce global warming and CO_2 concentrations.

10.8   FUTURE CHALLENGES BASED ON EACH SUBSECTION

The continuous emission of GHGs in the form of CO_2 into the atmosphere comes mostly from the transportation and industry sectors. Due to rapid population growth and the escalating demand for transport at present, traffic jams and slow-moving vehicles during traffic back-ups contribute majorly toward CO_2 emissions. Post-combustion, pre-combustion and oxyfuel combustion CO_2 capture processes were investigated and applied, but are mostly inflexible, financially constrained, use a great deal of energy, require large solvent volumes and are only applicable for low-sulfur contexts. Therefore, CO_2 emissions from developed countries still face many challenges due to environmental pollution and health hazards.

Physical separation processes must operate at low temperatures, pressures and CO_2 concentrations to efficiently remove CO_2 from gas streams. Biological separation processes also have a few challenges due to the decomposition of sinking organic compounds that threaten marine life, financial restrictions for bioreactors in underground water and the possible leakage of CO_2 from bioreactors which would affect soil quality and cause water pollution.

10.9   CONCLUDING REMARKS AND RECOMMENDATIONS

In this chapter, a brief overview of traffic pollution from the transportation sector and the separation of pollutants fundamental to CO_2 capture and separation methods was presented. The transportation sector is a major contributor to economic development and is responsible for CO_2 emissions that result in air pollution. Global warming due to the greenhouse effect is becoming a major issue, and CO_2 emissions from the transportation sector are a very broad and disputed issue.

CO_2 can be captured and separated through post-combustion, pre-combustion and oxyfuel combustion methods. These methods have been modified to capture up to 99% CO_2 using chemical looping, TSA and IGCC. Meanwhile, CO_2 separation and extraction technologies are becoming valuable due to innovative conversions to synthetic fuel by chemical, physical and biological methods. In terms of chemical methods, membrane technology is promising due to its low cost and simple operation. Hydrate-based separation and cryogenics distillation are new approaches that can be applied to gases at high pressures and use less energy for the CO_2 separation process. Common physical separation methods use Selexol, Rectisol and Fluor in flue gases at low temperatures and pressures. Physical adsorption has been used to remove CO_2 from gas streams using molecular sieves, activated carbon and lithium elements. In addition, biological separation processes have also been applied using forestation, ocean fertigation, photosynthesis, carbonation and hydration. These methods involve high costs for bioreactor management and maintenance to prevent CO_2 from leaking into the water. Therefore, some improvements in separation methods were investigated, such as the modification of membrane technology, and the development of a cryogenic CO_2 capture process that can operate at low temperatures and high pressures. In summary, future research should consider solid adsorbents for separation processes in order to achieve excellent CO_2 separation at low temperatures with high concentrations of CO_2, reduce energy consumption and encourage long-term stability. Solid adsorbent is inexpensive as it uses carbon nanotubes, alumina-supported metal oxide, zeolite or other substances which do not exceed the customary range of other vehicle-maintenance items.

ACKNOWLEDGMENT

The authors would like to thank the Ministry of Higher Education of Malaysia for their financial support through the research grant (Vote Nos. Q.J130000.2522.15H32 and Q.J130000.2522.19H98), without which this study would not have been possible.

REFERENCES

1. T. Kirschstein and F. Meisel, "GHG-emission models for assessing the eco-friendliness of road and rail freight transports," *Transportation Research Part B: Methodological*, vol. 73, pp. 13–33, 2015.
2. R. O'Driscoll, M. E. J. Stettler, N. Molden, T. Oxley, and H. M. ApSimon, "Real world CO_2 and NOx emissions from 149 Euro 5 and 6 diesel, gasoline and hybrid passenger cars," *Science of the Total Environment*, vol. 621, pp. 282–290, 2018.
3. M. Amann. Final Assessment Report; Baseline Projections of Greenhouse Gases and Air Pollutants in the European Union up to 2030 [Online].
4. J. Gao, H. Hou, Y. Zhai, A. Woodward, S. Vardoulakis, S. Kovats, *et al.*, "Greenhouse gas emissions reduction in different economic sectors: Mitigation measures, health co-benefits, knowledge gaps, and policy implications," *Environmental Pollution (Barking, Essex: 1987)*, vol. 240, pp. 683–698, 2018.

5. M. Sofiev, J. J. Winebrake, L. Johansson, E. W. Carr, M. Prank, J. Soares, *et al.*, "Cleaner fuels for ships provide public health benefits with climate tradeoffs," *Nature Communications*, vol. 9, p. 406, 2018.

6. J. Nègre and P. Delhomme, "Drivers' self-perceptions about being an eco-driver according to their concern for the environment, beliefs on eco-driving, and driving behavior," *Transportation Research Part A: Policy and Practice*, vol. 105, pp. 95–105, 2017/11/01/ 2017.

7. M. Condurat, A. M. Nicuţă, and R. Andrei, "Environmental impact of road transport traffic. a case study for county of iaşi road network," *Procedia Engineering*, vol. 181, pp. 123–130, 2017.

8. D. Adamović, J. Dorić, M. Vojinović Miloradov, S. Adamović, S. Pap, J. Radonić, *et al.*, "The emission of BTEX compounds during movement of passenger car in accordance with the NEDC," *Science of The Total Environment*, vol. 639, pp. 339–349, 10/15/ 2018.

9. B. Edokpolo, Q. J. Yu, and D. Connell, "Health risk assessment of ambient air concentrations of benzene, toluene and xylene (BTX) in service station environments," *International Journal of Environmental Research and Public Health*, vol. 11, pp. 6354–6374, 2014.

10. N. Maizlish, N. J. Linesch, and J. Woodcock, "Health and greenhouse gas mitigation benefits of ambitious expansion of cycling, walking, and transit in California," *Journal of Transport & Health*, vol. 6, pp. 490–500, 2017.

11. N. Mueller, D. Rojas-Rueda, T. Cole-Hunter, A. de Nazelle, E. Dons, R. Gerike, *et al.*, "Health impact assessment of active transportation: A systematic review," *Preventive Medicine*, vol. 76, pp. 103–114, 2015.

12. N. Petrunoff, C. Rissel, and L. M. Wen, "The effect of active travel interventions conducted in work settings on driving to work: a systematic review," *Journal of Transport & Health*, vol. 3, pp. 61–76, 2016.

13. F. A. Rodrigues Filho, J. G. C. Baêta, A. F. Teixeira, R. M. Valle, and J. L. F. de Souza, "E25 stratified torch ignition engine emissions and combustion analysis," *Energy Conversion and Management*, vol. 121, pp. 251–271, 2016/08/01/ 2016.

14. Y.-J. Zhang, Z. Liu, C.-X. Qin, and T.-D. Tan, "The direct and indirect CO_2 rebound effect for private cars in China," *Energy Policy*, vol. 100, pp. 149–161, 2017/01/01/ 2017.

15. Y.-J. Zhang, J.-F. Hao, and J. Song, "The CO_2 emission efficiency, reduction potential and spatial clustering in China's industry: evidence from the regional level," *Applied Energy*, vol. 174, pp. 213–223, 2016.

16. Y.-J. Zhang, H.-R. Peng, Z. Liu, and W. Tan, "Direct energy rebound effect for road passenger transport in China: a dynamic panel quantile regression approach," *Energy Policy*, vol. 87, pp. 303–313, 2015.

17. X. Liu, S. Ma, J. Tian, N. Jia, and G. Li, "A system dynamics approach to scenario analysis for urban passenger transport energy consumption and CO_2 emissions: A case study of Beijing," *Energy Policy*, vol. 85, pp. 253–270, 2015.

18. X. Yin, W. Chen, J. Eom, L. E. Clarke, S. H. Kim, P. L. Patel, *et al.*, "China's transportation energy consumption and CO_2 emissions from a global perspective," *Energy Policy*, vol. 82, pp. 233–248, 2015.

19. ACEA. ACEA Report; Vehicles in use -Europe 2017 [Online].

20. EC. A European strategy for low-emission mobility [Online].

21. INRIX. INRIX Global Traffic Scorecard [Online].

22. EC, "Regulation (EC) No 443/2009 of the European Parliament and of the Council of 23 April 2009," *Official Journal of the European Union*, vol. 140, 2009.

23. DOT. Highway Statistics 2016 [Online].

24. K. McCoy, "Yep, Los Angeles has the world's worst traffic congestion — again," in *USA Today*, ed, 2018.

25. A. Lo and N. Leung, "Los Angeles' notorious traffic problem explained in graphics," in CNN, ed, 2018.
26. J. Deschenes. Air Pollution and the Inner City [Online].
27. Z. Huang, P. Zheng, Y. Ma, X. Li, W. Xu, and W. Zhu, "A simulation study of the impact of the public–private partnership strategy on the performance of transport infrastructure," *SpringerPlus*, vol. 5, p. 958, 2016.
28. T. Y. F. Mailonline. The Great Crawl of China: Thousands of motorists are stranded on Beijing motorway in an incredible FIFTY lane traffic jam as week-long national holiday wraps up [Online].
29. ICCT. CO_2 emissions from new passenger cars in the EU: Car manufacturers' performance in 2016 [Online].
30. P. F. Belgrado, Ľ. Buzna, F. Foiadelli, and M. Longo, "Evaluating the predictability of future energy consumption: application of statistical classification models to data from EV charging points," 2018.
31. E. R. Dennehy and B. P. Ó. Gallachóir, "Ex-post decomposition analysis of passenger car energy demand and associated CO_2 emissions," *Transportation Research Part D: Transport and Environment*, vol. 59, pp. 400–416, 2018.
32. MEP. China Vehicle Environmental Management Annual Report 2017 [Online].
33. ICCT. Automobile production in Canada and implications for Canada's 2025 passenger vehicle greenhouse gas standards [Online].
34. D. Cheng, F. R. Negreiros, E. Apra, and A. Fortunelli, "Computational approaches to the chemical conversion of carbon dioxide," *ChemSusChem*, vol. 6, pp. 944–965, 2013.
35. K. Stangeland, D. Kalai, H. Li, and Z. Yu, "CO_2 methanation: the effect of catalysts and reaction conditions," *Energy Procedia*, vol. 105, pp. 2022–2027, 2017.
36. S. Chu, "Carbon capture and sequestration," ed: American Association for the Advancement of Science, 2009.
37. R. S. Haszeldine, "Carbon capture and storage: how green can black be?," *Science*, vol. 325, pp. 1647–1652, 2009.
38. C. Lastoskie, "Caging carbon dioxide," *Science*, vol. 330, pp. 595–596, 2010.
39. Y. Guo, Z. Yuan, J. Xu, Z. Wang, T. Yuan, W. Zhou, *et al.*, "Metabolic acclimation mechanism in microalgae developed for CO_2 capture from industrial flue gas," *Algal Research*, vol. 26, pp. 225–233, 2017.
40. UNEP, "The role of the transport sector in environmental protection," Department of Economic and Social Affairs, ed: United Nations Environment Programme Commission on Sustainable Development 2001, pp. 16–27.
41. L. Li, W. Conway, R. Burns, M. Maeder, G. Puxty, S. Clifford, *et al.*, "Investigation of metal ion additives on the suppression of ammonia loss and CO_2 absorption kinetics of aqueous ammonia-based CO_2 capture," *International Journal of Greenhouse Gas Control*, vol. 56, pp. 165–172, 2017.
42. P. Subha, B. N. Nair, V. Visakh, C. Sreerenjini, A. P. Mohamed, K. Warrier, *et al.*, "Germanium-incorporated lithium silicate composites as highly efficient low-temperature sorbents for CO_2 capture," *Journal of Materials Chemistry A*, vol. 6, pp. 7913–7921, 2018.
43. J. D. Figueroa, T. Fout, S. Plasynski, H. McIlvried, and R. D. Srivastava, "Advances in CO_2 capture technology—the US Department of Energy's Carbon Sequestration Program," *International journal of greenhouse gas control*, vol. 2, pp. 9–20, 2008.
44. E. Jacob-Lopes, C. H. G. Scoparo, and T. T. Franco, "Rates of CO_2 removal by Aphanothece microscopica Nägeli in tubular photobioreactors," *Chemical Engineering and Processing: Process Intensification*, vol. 47, pp. 1365–1373, 2008.
45. L. Zhen-shan, C. Ning-sheng, and E. Croiset, "Process analysis of CO_2 capture from flue gas using carbonation/calcination cycles," *AIChE Journal*, vol. 54, pp. 1912–1925, 2008.

46. D. Y. Lu, R. W. Hughes, and E. J. Anthony, "Ca-based sorbent looping combustion for CO_2 capture in pilot-scale dual fluidized beds," *Fuel Processing Technology*, vol. 89, pp. 1386–1395, 2008.

47. E. De Visser, C. Hendriks, M. Barrio, M. J. Mølnvik, G. de Koeijer, S. Liljemark, *et al.*, "Dynamis CO_2 quality recommendations," *International Journal of Greenhouse Gas Control*, vol. 2, pp. 478–484, 2008.

48. H. Yang, Z. Xu, M. Fan, R. Gupta, R. B. Slimane, A. E. Bland, *et al.*, "Progress in carbon dioxide separation and capture: A review," *Journal of Environmental Sciences*, vol. 20, pp. 14–27, 2008.

49. K. Smith, U. Ghosh, A. Khan, M. Simioni, K. Endo, X. Zhao, *et al.*, "Recent developments in solvent absorption technologies at the CO_2CRC in Australia," *Energy Procedia*, vol. 1, pp. 1549–1555, 2009.

50. J. N. Knudsen, J. N. Jensen, P.-J. Vilhelmsen, and O. Biede, "Experience with CO_2 capture from coal flue gas in pilot-scale: Testing of different amine solvents," *Energy Procedia*, vol. 1, pp. 783–790, 2009.

51. I. Pfaff and A. Kather, "Comparative thermodynamic analysis and integration issues of CCS steam power plants based on oxy-combustion with cryogenic or membrane based air separation," *Energy Procedia*, vol. 1, pp. 495–502, 2009.

52. F. Kozak, A. Petig, E. Morris, R. Rhudy, and D. Thimsen, "Chilled ammonia process for CO_2 capture," *Energy Procedia*, vol. 1, pp. 1419–1426, 2009.

53. F. Fang, Z.-s. Li, and N.-s. Cai, "CO_2 capture from flue gases using a fluidized bed reactor with limestone," *Korean Journal of Chemical Engineering*, vol. 26, pp. 1414–1421, 2009.

54. R. Thiruvenkatachari, S. Su, H. An, and X. X. Yu, "Post combustion CO_2 capture by carbon fibre monolithic adsorbents," *Progress in Energy and Combustion Science*, vol. 35, pp. 438–455, 2009.

55. I. Eide-Haugmo, O. G. Brakstad, K. A. Hoff, K. R. Sørheim, E. F. da Silva, and H. F. Svendsen, "Environmental impact of amines," *Energy Procedia*, vol. 1, pp. 1297–1304, 2009.

56. V. Manovic, J.-P. Charland, J. Blamey, P. S. Fennell, D. Y. Lu, and E. J. Anthony, "Influence of calcination conditions on carrying capacity of CaO-based sorbent in CO_2 looping cycles," *Fuel*, vol. 88, pp. 1893–1900, 2009.

57. D. Botheju, Y. Li, J. Hovland, T. Risberg, H. A. Haugen, C. Dinamarca, *et al.*, "Biogasification of waste monoethanolamine generated in post combustion CO_2 capture," in *Proceedings of the 2nd Annual Gas Processing Symposium*, 2010, pp. 1–9.

58. D.-J. Kim, Y. Lim, D. Cho, and I. H. Rhee, "Biodegradation of monoethanolamine in aerobic and anoxic conditions," *Korean Journal of Chemical Engineering*, vol. 27, pp. 1521–1526, 2010.

59. A. A. Olajire, "CO_2 capture and separation technologies for end-of-pipe applications–a review," *Energy*, vol. 35, pp. 2610–2628, 2010.

60. D. Zhao, D. Yuan, R. Krishna, J. M. van Baten, and H.-C. Zhou, "Thermosensitive gating effect and selective gas adsorption in a porous coordination nanocage," *Chemical Communications*, vol. 46, pp. 7352–7354, 2010b.

61. D. Zhao, D. Yuan, A. Yakovenko, and H.-C. Zhou, "A NbO-type metal–organic framework derived from a polyyne-coupled di-isophthalate linker formed in situ," *Chemical Communications*, vol. 46, pp. 4196–4198, 2010a.

62. V. Darde, K. Thomsen, W. J. Van Well, and E. H. Stenby, "Chilled ammonia process for CO_2 capture," *International Journal of Greenhouse Gas Control*, vol. 4, pp. 131–136, 2010.

63. G. Qi, Y. Wang, L. Estevez, X. Duan, N. Anako, A. H. A. Park, *et al.*, "High efficiency nanocomposite sorbents for CO_2 capture based on amine-functionalized mesoporous capsules," *Energy and Environmental Science*, vol. 4, pp. 444–452, 2011.

64. A. P. Katsoulidis and M. G. Kanatzidis, "Phloroglucinol based microporous polymeric organic frameworks with-OH functional groups and high CO_2 capture capacity," *Chemistry of Materials*, vol. 23, pp. 1818–1824, 2011.
65. X. Yan, L. Zhang, Y. Zhang, G. Yang, and Z. Yan, "Amine-modified SBA-15: Effect of pore structure on the performance for CO_2 capture," *Industrial and Engineering Chemistry Research*, vol. 50, pp. 3220–3226, 2011.
66. M. F. B. A. Rahman, "Introduction of Pack Test for Participative Environmental Monitoring and Environmental Education for Sustainability in Malaysia," *Journal of Tropical Life Science*, vol. 1, pp. 60–68, 2011.
67. S. E. Wong, E. Y. Lau, H. J. Kulik, J. H. Satcher, C. Valdez, M. Worsely, et al., "Designing small-molecule catalysts for CO_2 capture," *Energy Procedia*, pp. 817–823, 2011.
68. J. Pires, M. Alvim-Ferraz, F. Martins, and M. Simões, "Carbon dioxide capture from flue gases using microalgae: engineering aspects and biorefinery concept," *Renewable and Sustainable Energy Reviews*, vol. 16, pp. 3043–3053, 2012.
69. B. Belaissaoui, Y. Le Moullec, D. Willson, and E. Favre, "Hybrid membrane cryogenic process for post-combustion CO_2 capture," *Journal of Membrane Science*, 415–416, pp. 424–434, 2012.
70. F. Akhtar, L. Andersson, N. Keshavarzi, and L. Bergström, "Colloidal processing and CO_2 capture performance of sacrificially templated zeolite monoliths," *Applied Energy*, vol. 97, pp. 289–296, 2012.
71. P. Shao, M. Dal-Cin, A. Kumar, H. Li, and D. P. Singh, "Design and economics of a hybrid membrane–temperature swing adsorption process for upgrading biogas," *Journal of Membrane Science*, vol. 413, pp. 17–28, 2012.
72. O. G. Brakstad, A. Booth, I. Eide-Haugmo, J. A. Skjæran, K. R. Sørheim, K. Bonaunet, et al., "Seawater biodegradation of alkanolamines used for CO_2-capture from natural gas," *International Journal of Greenhouse Gas Control*, vol. 10, pp. 271–277, 2012.
73. I. Eide-Haugmo, O. G. Brakstad, K. A. Hoff, E. F. da Silva, and H. F. Svendsen, "Marine biodegradability and ecotoxicity of solvents for CO_2-capture of natural gas," *International Journal of Greenhouse Gas Control*, vol. 9, pp. 184–192, 2012.
74. Y. Lin, Q. Yan, C. Kong, and L. Chen, "Polyethyleneimine incorporated metal-organic frameworks adsorbent for highly selective CO_2 capture," *Scientific Reports*, vol. 3, 2013.
75. J. Yu and P. B. Balbuena, "Water effects on postcombustion CO_2 capture in Mg-MOF-74," *Journal of Physical Chemistry C*, vol. 117, pp. 3383–3388, 2013.
76. A. S. González, M. G. Plaza, F. Rubiera, and C. Pevida, "Sustainable biomass-based carbon adsorbents for post-combustion CO_2 capture," *Chemical Engineering Journal*, vol. 230, pp. 456–465, 2013.
77. I. Hauser, A. Einbu, K. Østgaard, H. F. Svendsen, and F. J. Cervantes, "Biodegradation of amine waste generated from post-combustion CO_2 capture in a moving bed biofilm treatment system," *Biotechnology Letters*, vol. 35, pp. 219–224, 2013.
78. S. Wang, J. Hovland, and R. Bakke, "Anaerobic degradation of carbon capture reclaimer MEA waste," *Water Science and Technology*, vol. 67, pp. 2549–2559, 2013a.
79. S. Wang, J. Hovland, and R. Bakke, "Efficiency of the anaerobic digestion of amine wastes," *Biotechnology Letters*, vol. 35, pp. 2051–2060, 2013b.
80. C. A. Scholes, M. T. Ho, D. E. Wiley, G. W. Stevens, and S. E. Kentish, "Cost competitive membrane-cryogenic post-combustion carbon capture," *International Journal of Greenhouse Gas Control*, vol. 17, pp. 341–348, 2013.
81. M. Scholz, B. Frank, F. Stockmeier, S. Falß, and M. Wessling, "Techno-economic analysis of hybrid processes for biogas upgrading," *Industrial and Engineering Chemistry Research*, vol. 52, pp. 16929–16938, 2013.

82. R. Anantharaman, D. Berstad, and S. Roussanaly, "Techno-economic performance of a hybrid membrane - Liquefaction process for post-combustion CO_2 capture," in *Energy Procedia*, 2014, pp. 1244–1247.

83. P. K. Kundu, A. Chakma, and X. Feng, "Effectiveness of membranes and hybrid membrane processes in comparison with absorption using amines for post-combustion CO_2 capture," *International Journal of Greenhouse Gas Control*, vol. 28, pp. 248–256, 2014.

84. L. Zhao, E. Primabudi, and D. Stolten, "Investigation of a Hybrid System for Post-Combustion Capture," *Energy Procedia*, 1756–1772, 2011.

85. S. Wang, J. Hovland, S. Brooks, and R. Bakke, "Detoxifying CO_2 capture reclaimer waste by anaerobic digestion," *Applied Biochemistry and Biotechnology*, vol. 172, pp. 776–783, 2014.

86. M. A. Alkhabbaz, P. Bollini, G. S. Foo, C. Sievers, and C. W. Jones, "Important roles of enthalpic and entropic contributions to CO_2 capture from simulated flue gas and ambient air using mesoporous silica grafted amines," *Journal of the American Chemical Society*, vol. 136, pp. 13170–13173, 2014.

87. Z. H. Xuan, D. S. Zhang, Z. Chang, T. L. Hu, and X. H. Bu, "Targeted structure modulation of 'pillar-layered' metal-organic frameworks for CO_2 capture," *Inorganic Chemistry*, vol. 53, pp. 8985–8990, 2014.

88. T. Wang, J. Hovland, and K. J. Jens, "Amine reclaiming technologies in post-combustion carbon dioxide capture," *Journal of Environmental Sciences*, vol. 27, pp. 276–289, 2015.

89. S. J. Datta, C. Khumnoon, Z. H. Lee, W. K. Moon, S. Docao, T. H. Nguyen, *et al.*, "CO_2 capture from humid flue gases and humid atmosphere using a microporous coppersilicate," *Science*, vol. 350, pp. 302–306, 2015.

90. Z. Chen, K. Adil, L. J. Weseliński, Y. Belmabkhout, and M. Eddaoudi, "A supermolecular building layer approach for gas separation and storage applications: The eea and rtl MOF platforms for CO_2 capture and hydrocarbon separation," *Journal of Materials Chemistry A*, vol. 3, pp. 6276–6281, 2015.

91. X. Wang, Q. Guo, and T. Kong, "Tetraethylenepentamine-modified MCM-41/silica gel with hierarchical mesoporous structure for CO_2 capture," *Chemical Engineering Journal*, vol. 273, pp. 472–480, 2015.

92. X. Yu, L. An, J. Yang, S. T. Tu, and J. Yan, "CO_2 capture using a superhydrophobic ceramic membrane contactor," *Journal of Membrane Science*, vol. 496, pp. 1–12, 2015.

93. L. Cornaglia, J. Múnera, and E. Lombardo, "Recent advances in catalysts, palladium alloys and high temperature WGS membrane reactors: A review," *International Journal of Hydrogen Energy*, vol. 40, pp. 3423–3437, 2015.

94. J. M. Silva, M. A. Soria, and L. M. Madeira, "Challenges and strategies for optimization of glycerol steam reforming process," *Renewable and Sustainable Energy Reviews*, vol. 42, pp. 1187–1213, 2015.

95. Y. W. Fong, C. T. Hsieh, and C. J. Chang, "Spinal gossypiboma," *Formosan Journal of Surgery*, vol. 49, pp. 193–195, 2016.

96. J. Boon, V. Spallina, Y. van Delft, and M. van Sint Annaland, "Comparison of the efficiency of carbon dioxide capture by sorption-enhanced water-gas shift and palladium-based membranes for power and hydrogen production," *International Journal of Greenhouse Gas Control*, vol. 50, pp. 121–134, 2016.

97. M. A. Sakwa-Novak, C. J. Yoo, S. Tan, F. Rashidi, and C. W. Jones, "Poly(ethylenimine)-functionalized monolithic alumina honeycomb adsorbents for CO_2 capture from Air," *ChemSusChem*, vol. 9, pp. 1859–1868, 2016.

98. W. M. Verdegaal, K. Wang, J. P. Sculley, M. Wriedt, and H. C. Zhou, "Evaluation of Metal-Organic Frameworks and Porous Polymer Networks for CO_2-Capture Applications," *ChemSusChem*, vol. 9, pp. 636–643, 2016.

99. B. Li and B. Chen, "A flexible metal-organic framework with double interpenetration for highly selective CO_2 capture at room temperature," *Science China Chemistry*, vol. 59, pp. 965–969, 2016.

100. M. Wei, Q. Yu, T. Mu, L. Hou, Z. Zuo, and J. Peng, "Preparation and characterization of waste ion-exchange resin-based activated carbon for CO_2 capture," *Adsorption*, vol. 22, pp. 385–396, 2016.

101. S. A. Nabavi, G. T. Vladisavljević, A. Wicaksono, S. Georgiadou, and V. Manović, "Production of molecularly imprinted polymer particles with amide-decorated cavities for CO_2 capture using membrane emulsification/suspension polymerisation," *Colloids and Surfaces A: Physicochemical and Engineering Aspects*, vol. 521, pp. 231–238, 2017.

102. W. Zhu, Q. Li, and N. Dai, "CO_2 biofixation of *Actinobacillus succinogenes* through novel amine-functionalized polystyrene microsphere materials," *Applied Biochemistry and Biotechnology*, vol. 181, pp. 584–592, 2017.

103. C. Nwaoha, T. Supap, R. Idem, C. Saiwan, P. Tontiwachwuthikul, M. J. AL-Marri, *et al.*, "Advancement and new perspectives of using formulated reactive amine blends for post-combustion carbon dioxide (CO_2) capture technologies," *Petroleum*, vol. 3, pp. 10–36, 2017.

104. K. Zhang, Z. Qiao, and J. Jiang, "Molecular design of zirconium tetrazolate metal-organic frameworks for CO_2 capture," *Crystal Growth and Design*, vol. 17, pp. 543–549, 2017.

105. S. Dey, A. Bhunia, I. Boldog, and C. Janiak, "A mixed-linker approach toward improving covalent triazine-based frameworks for CO_2 capture and separation," *Microporous and Mesoporous Materials*, vol. 241, pp. 303–315, 2017.

106. P. Billemont, N. Heymans, P. Normand, and G. De Weireld, "IAST predictions vs co-adsorption measurements for CO_2 capture and separation on MIL-100 (Fe)," *Adsorption*, vol. 23, pp. 225–237, 2017.

107. K. M. Elsabawy and A. M. Fallatah, "Fabrication of ultra-performance non-compact graphene/carbon hollow fibers/graphene stationary junction like membrane for CO_2-Capture," *Materials Chemistry and Physics*, vol. 211, pp. 264–269, 2018.

108. T. L. Chitsiga, M. O. Daramola, N. Wagner, and J. M. Ngoy, "Parametric effect of adsorption variables on CO_2 adsorption of amine-grafted polyaspartamide composite adsorbent during post-combustion CO_2 capture: A response surface methodology approach," *International Journal of Oil, Gas and Coal Technology*, vol. 17, pp. 321–336, 2018.

109. K. Maneeintr, K. Photien, and T. Charinpanitkul, "Mixture of MEA/2-MAE for effective CO_2 capture from flue gas stream," *Chemical Engineering Transactions*, vol. 63, pp. 229–234, 2018.

110. A. González-Díaz, A. M. Alcaráz-Calderón, M. O. González-Díaz, Á. Méndez-Aranda, M. Lucquiaud, and J. M. González-Santaló, "Effect of the ambient conditions on gas turbine combined cycle power plants with post-combustion CO_2 capture," *Energy*, vol. 134, pp. 221–233, 2017.

111. C. Yin and J. Yan, "Oxy-fuel combustion of pulverized fuels: Combustion fundamentals and modeling," *Applied Energy*, vol. 162, pp. 742–762, 2016/01/15/ 2016.

112. C. Song, Q. Liu, N. Ji, S. Deng, J. Zhao, Y. Li, *et al.*, "Alternative pathways for efficient CO_2 capture by hybrid processes—A review," *Renewable and Sustainable Energy Reviews*, vol. 82, pp. 215–231, 2018/02/01/ 2018.

113. P. Yuan, Z. Qiu, and J. Liu, "Recent enlightening strategies for co_2 capture: A review," *IOP Conference Series: Earth and Environmental Science*, p. 012046, 2017.

114. H. J. Lee, J. D. Lee, P. Linga, P. Englezos, Y. S. Kim, M. S. Lee, *et al.*, "Gas hydrate formation process for pre-combustion capture of carbon dioxide," *Energy*, vol. 35, pp. 2729–2733, 2010.

115. D. Y. Leung, G. Caramanna, and M. M. Maroto-Valer, "An overview of current status of carbon dioxide capture and storage technologies," *Renewable and Sustainable Energy Reviews*, vol. 39, pp. 426–443, 2014.

116. L. O. Nord, R. Anantharaman, and O. Bolland, "Design and off-design analyses of a pre-combustion CO_2 capture process in a natural gas combined cycle power plant," *International Journal of Greenhouse Gas Control*, vol. 3, pp. 385–392, 2009.

117. T. Riis, E. Hagen, G. Sandrock, P. Vie, and O. Ulleberg, "Hydrogen Production and Storage, International Energy Agency: Paris," ed, 2006.

118. S. Hoffmann, M. Bartlett, M. Finkenrath, A. Evulet, and T. P. Ursin, "Performance and cost analysis of advanced gas turbine cycles with precombustion CO_2 capture," *Journal of Engineering for Gas Turbines and Power*, vol. 131, p. 021701, 2009.

119. P. Linga, R. Kumar, and P. Englezos, "The clathrate hydrate process for post and pre-combustion capture of carbon dioxide," *Journal of Hazardous Materials*, vol. 149, pp. 625–629, 2007.

120. T. C. Merkel, H. Lin, X. Wei, and R. Baker, "Power plant post-combustion carbon dioxide capture: an opportunity for membranes," *Journal of Membrane Science*, vol. 359, pp. 126–139, 2010.

121. J. A. Mason, K. Sumida, Z. R. Herm, R. Krishna, and J. R. Long, "Evaluating metal–organic frameworks for post-combustion carbon dioxide capture via temperature swing adsorption," *Energy & Environmental Science*, vol. 4, pp. 3030–3040, 2011.

122. R. Bounaceur, N. Lape, D. Roizard, C. Vallieres, and E. Favre, "Membrane processes for post-combustion carbon dioxide capture: A parametric study," *Energy*, vol. 31, pp. 2556–2570, 2006.

123. L. C. Elwell and W. S. Grant, "Technology options for capturing CO_2," *Power Magazine*, vol. 150, p. 60, 2006.

124. C. Global, "Institute: Global status of large-scale integrated CCS projects, June 2012 update," Global CCS Institute, Canberra, 2012.

125. K. O. Yoro and P. T. Sekoai, "The potential of CO_2 capture and storage technology in South Africa's coal-fired thermal power plants," *Environments*, vol. 3, p. 24, 2016.

126. P. Feron and C. Hendriks, "CO_2 capture process principles and costs," *Oil & Gas Science and Technology*, vol. 60, pp. 451–459, 2005.

127. F. D. Skinner, K. E. McIntush, and M. C. Murff, "Amine-based gas sweetening and claus sulfur recovery process chemistry and waste stream survey. Topical report," Radian Corp., Austin, TX, 1995.

128. K. A. Mumford, Y. Wu, K. H. Smith, and G. W. Stevens, "Review of solvent based carbon-dioxide capture technologies," *Frontiers of Chemical Science and Engineering*, vol. 9, pp. 125–141, 2015.

129. K. J. Borgert and E. S. Rubin, "Oxyfuel combustion: Technical and economic considerations for the development of carbon capture from pulverized coal power plants," *Energy Procedia*, 2013, pp. 1291–1300.

130. B. J. Buhre, L. K. Elliott, C. Sheng, R. P. Gupta, and T. F. Wall, "Oxy-fuel combustion technology for coal-fired power generation," *Progress in energy and combustion science*, vol. 31, pp. 283–307, 2005.

131. L. Zheng and Y. Tan, "Overview of oxyfuel combustion technology for CO_2 capture. Cornerstone," *The Official Journal of the World Coal Industry*, 2014.

132. GCCSI. CO_2 capture technologies: oxy combustion with CO_2 capture: January 2012. [Online].

133. T. Burdyny and H. Struchtrup, "Hybrid membrane/cryogenic separation of oxygen from air for use in the oxy-fuel process," *Energy*, vol. 35, pp. 1884–1897, 2010.

134. A. S. Bhown and B. C. Freeman, "Analysis and status of post-combustion carbon dioxide capture technologies," *Environmental Science & Technology*, vol. 45, pp. 8624–8632, 2011.

135. J. Gibbins and H. Chalmers, "Carbon capture and storage," *Energy Policy*, vol. 36, pp. 4317–4322, 2008.
136. J. Adanez, A. Abad, F. Garcia-Labiano, P. Gayan, and F. Luis, "Progress in chemical-looping combustion and reforming technologies," *Progress in Energy and Combustion Science*, vol. 38, pp. 215–282, 2012.
137. M. Ishida and H. Jin, "CO_2 recovery in a power plant with chemical looping combustion," *Energy Conversion and Management*, vol. 38, pp. S187–S192, 1997.
138. O. y. Brandvoll and O. Bolland, "Inherent CO_2 capture using chemical looping combustion in a natural gas fired power cycle," in *ASME Turbo Expo 2002: Power for Land, Sea, and Air*,pp. 493–499, 2002.
139. M. Kanniche and C. Bouallou, "CO_2 capture study in advanced integrated gasification combined cycle," *Applied Thermal Engineering*, vol. 27, pp. 2693–2702, 2007.
140. F. A. Rodrigues Filho, T. A. A. Moreira, R. M. Valle, J. G. C. Baêta, M. Pontoppidan, and A. F. Teixeira, "E25 stratified torch ignition engine performance, CO_2 emission and combustion analysis," *Energy Conversion and Management*, vol. 115, pp. 299–307, 2016.
141. R. Steeneveldt, B. Berger, and T. Torp, "CO_2 capture and storage: closing the knowing–doing gap," *Chemical Engineering Research and Design*, vol. 84, pp. 739–763, 2006.
142. J. David, "Economic evaluation of leading technology options for sequestration of carbon dioxide," Massachusetts Institute of Technology, 2000.
143. H. Audus, "Leading options for the capture of CO_2 at power stations," in *Proceedings of the Fifth International Conference on Greenhouse Gas Control Technologies, Cairns, Australia*, 2000, p. 16.
144. H. Berberoglu, P. S. Gomez, and L. Pilon, "Radiation characteristics of Botryococcus braunii, Chlorococcum littorale, and Chlorella sp. used for CO_2 fixation and biofuel production," *Journal of Quantitative Spectroscopy and Radiative Transfer*, vol. 110, pp. 1879–1893, 2009.
145. B. Zhao, Y. Su, W. Tao, L. Li, and Y. Peng, "Post-combustion CO_2 capture by aqueous ammonia: A state-of-the-art review," *International Journal of Greenhouse Gas Control*, vol. 9, pp. 355–371, 2012/07/01/ 2012.
146. C. Hendriks, "Energy conversion: CO_2 removal from coal-fired power plant," Netherlands: Kluwer Academic Publishers, 1995.
147. A. Veawab, A. Aroonwilas, and P. Tontiwachwuthikul, "CO_2 absorption performance of aqueous alkanolamines in packed columns," Fuel Chemistry Division Preprints, vol. 47, pp. 49–50, 2002.
148. J. E. Johnson and A. C. Homme Jr, "SELEXOL Solvent Process reduces lean, high-CO/sub 2/natural gas treating costs," *Energy Prog.;(United States)*, vol. 4, 1984.
149. M. Kato, K. Nakagawa, K. Essaki, Y. Maezawa, S. Takeda, R. Kogo, *et al.*, "Novel CO_2 Absorbents Using Lithium-Containing Oxide," *International Journal of Applied Ceramic Technology*, vol. 2, pp. 467–475, 2005.
150. A. S. Damle and T. P. Dorchak, "Recovery of carbon dioxide in advanced fossil energy conversion processes using a membrane reactor," in First National Conference on Carbon Sequestration, Washington, DC, 2001.
151. Z. Xu, J. Wang, W. Chen, and Y. Xu, "Separation and fixation of carbon dioxide using polymeric membrane contactor," in *Proceedings of the 1st National Conference on Carbon Sequestration*, 2001.
152. X. Li, Y. Cheng, H. Zhang, S. Wang, Z. Jiang, R. Guo, *et al.*, "Efficient CO_2 capture by functionalized graphene oxide nanosheets as fillers to fabricate multi-permselective mixed matrix membranes," *ACS Applied Materials and Interfaces*, vol. 7, pp. 5528–5537, 2015.
153. S. Fan, Y. Wang, and X. Lang, "CO_2 capture in form of clathrate hydrate-problem and practice," in *Proceedings of the 7th International Conference on Gas Hydrates (ICGH 2011)*, UK, 2011.

154. Y. Zhang, M. Yang, Y. Song, L. Jiang, Y. Li, and C. Cheng, "Hydrate phase equilibrium measurements for (THF+ SDS+ CO_2+ N_2) aqueous solution systems in porous media," *Fluid Phase Equilibria*, vol. 370, pp. 12–18, 2014.

155. G. Göttlicher and R. Pruschek, "Comparison of CO_2 removal systems for fossil-fuelled power plant processes," *Energy Conversion and Management*, vol. 38, pp. S173–S178, 1997.

156. C. Hoeger, C. Bence, S. Burt, A. Baxter, and L. Baxter, "Cryogenic CO_2 capture for improved efficiency at reduced cost," in *Proceedings of the AIChE Annual Meeting*, 2010.

157. M. Tuinier, M. van Sint Annaland, G. J. Kramer, and J. Kuipers, "Cryogenic CO_2 capture using dynamically operated packed beds," *Chemical Engineering Science*, vol. 65, pp. 114–119, 2010.

158. S. Anderson and R. Newell, "Prospects for carbon capture and storage technologies," *Annu. Rev. Environ. Resour.*, vol. 29, pp. 109–142, 2004.

159. C. M. White, B. R. Strazisar, E. J. Granite, J. S. Hoffman, and H. W. Pennline, "Separation and capture of CO_2 from large stationary sources and sequestration in geological formations—coalbeds and deep saline aquifers," *Journal of the Air & Waste Management Association*, vol. 53, pp. 645–715, 2003.

160. D. Aaron and C. Tsouris, "Separation of CO_2 from flue gas: a review," *Separation Science and Technology*, vol. 40, pp. 321–348, 2005.

161. C. Stewart and M.-A. Hessami, "A study of methods of carbon dioxide capture and sequestration—the sustainability of a photosynthetic bioreactor approach," *Energy Conversion and Management*, vol. 46, pp. 403–420, 2005.

162. R. Idem, M. Wilson, P. Tontiwachwuthikul, A. Chakma, A. Veawab, A. Aroonwilas, *et al.*, "Pilot plant studies of the CO_2 capture performance of aqueous MEA and mixed MEA/MDEA solvents at the University of Regina CO_2 capture technology development plant and the boundary dam CO_2 capture demonstration plant," *Industrial & Engineering Chemistry Research*, vol. 45, pp. 2414–2420, 2006.

163. D. J. Fauth, E. A. Frommell, J. S. Hoffman, R. P. Reasbeck, and H. W. Pennline, "Eutectic salt promoted lithium zirconate: Novel high temperature sorbent for CO_2 capture," *Fuel Processing Technology*, vol. 86, pp. 1503–1521, 2005.

164. K. P. Resnik, J. T. Yeh, and H. W. Pennline, "Aqua ammonia process for simultaneous removal of CO_2, SO_2 and NOx," *International Journal of Environmental Technology and Management*, vol. 4, pp. 89–104, 2004.

165. J. T. Yeh, K. P. Resnik, K. Rygle, and H. W. Pennline, "Semi-batch absorption and regeneration studies for CO_2 capture by aqueous ammonia," *Fuel Processing Technology*, vol. 86, pp. 1533–1546, 2005.

166. K. Thomsen and P. Rasmussen, "Modeling of vapor–liquid–solid equilibrium in gas–aqueous electrolyte systems," *Chemical Engineering Science*, vol. 54, pp. 1787–1802, 1999.

167. H. Huang, Y. Shi, W. Li, and S. Chang, "Dual alkali approaches for the capture and separation of CO_2," *Energy & Fuels*, vol. 15, pp. 263–268, 2001.

168. A. Brunetti, F. Scura, G. Barbieri, and E. Drioli, "Membrane technologies for CO_2 separation," *Journal of Membrane Science*, vol. 359, pp. 115–125, 2010.

169. W. Yave, A. Car, S. S. Funari, S. P. Nunes, and K.-V. Peinemann, "CO_2–philic polymer membrane with extremely high separation performance," *Macromolecules*, vol. 43, pp. 326–333, 2009.

170. M. T. Besong, M. M. Maroto-Valer, and A. J. Finn, "Study of design parameters affecting the performance of CO_2 purification units in oxy-fuel combustion," *International Journal of Greenhouse Gas Control*, vol. 12, pp. 441–449, 2013.

171. H. Weiss, "Rectisol wash for purification of partial oxidation gases," *Gas Separation & Purification*, vol. 2, pp. 171–176, 1988.

172. B. Li, Y. Duan, D. Luebke, and B. Morreale, "Advances in CO_2 capture technology: A patent review," *Applied Energy*, vol. 102, pp. 1439–1447, 2013.
173. H. Knapp, "Processing Gas Conditioning Conference, Norman, Okla. p," ed: C-1, 1968.
174. G. P. Knowles, J. V. Graham, S. W. Delaney, and A. L. Chaffee, "Aminopropyl-functionalized mesoporous silicas as CO_2 adsorbents," *Fuel Processing Technology*, vol. 86, pp. 1435–1448, 2005.
175. G. P. Knowles, S. W. Delaney, and A. L. Chaffee, "Diethylenetriamine [propyl (silyl)]-functionalized (DT) mesoporous silicas as CO_2 adsorbents," *Industrial & Engineering Chemistry Research*, vol. 45, pp. 2626–2633, 2006.
176. H. Yoshitake, T. Yokoi, and T. Tatsumi, "Adsorption of chromate and arsenate by amino-functionalized MCM-41 and SBA-1," *Chemistry of Materials*, vol. 14, pp. 4603–4610, 2002.
177. M. Gray, Y. Soong, K. Champagne, H. Pennline, J. Baltrus, R. Stevens Jr, *et al.*, "Improved immobilized carbon dioxide capture sorbents," *Fuel Processing Technology*, vol. 86, pp. 1449–1455, 2005.
178. C. Song, "Global challenges and strategies for control, conversion and utilization of CO_2 for sustainable development involving energy, catalysis, adsorption and chemical processing," *Catalysis Today*, vol. 115, pp. 2–32, 2006.
179. X. Xu, C. Song, B. G. Miller, and A. W. Scaroni, "Adsorption separation of carbon dioxide from flue gas of natural gas-fired boiler by a novel nanoporous "molecular basket" adsorbent," *Fuel Processing Technology*, vol. 86, pp. 1457–1472, 2005.
180. A. L. Chaffee, G. P. Knowles, Z. Liang, J. Zhang, P. Xiao, and P. A. Webley, "CO_2 capture by adsorption: Materials and process development," *International Journal of Greenhouse Gas Control*, vol. 1, pp. 11–18, 2007.
181. R. Saxena, V. K. Singh, and E. A. Kumar, "Carbon dioxide capture and sequestration by adsorption on activated carbon," in *Energy Procedia*, 2014, pp. 320–329.
182. A. Arenillas, F. Rubiera, J. Parra, C. Ania, and J. Pis, "Surface modification of low cost carbons for their application in the environmental protection," *Applied Surface Science*, vol. 252, pp. 619–624, 2005.
183. K. Essaki, K. Nakagawa, M. Kato, and H. Uemoto, "CO_2 absorption by lithium silicate at room temperature," *Journal of Chemical Engineering of Japan*, vol. 37, pp. 772–777, 2004.
184. A. Yamasaki, "An overview of CO_2 mitigation options for global warming—emphasizing CO_2 sequestration options," *Journal of Chemical Engineering of Japan*, vol. 36, pp. 361–375, 2003.
185. M. G. Nelson, "Carbon dioxide sequestration by mechanochemical carbonation of mineral silicates," University of Utah (US)2004.
186. N. Liu, G. M. Bond, A. Abel, B. J. McPherson, and J. Stringer, "Biomimetic sequestration of CO_2 in carbonate form: Role of produced waters and other brines," *Fuel Processing Technology*, vol. 86, pp. 1615–1625, 2005.
187. H. Haywood, J. Eyre, and H. Scholes, "Carbon dioxide sequestration as stable carbonate minerals–environmental barriers," *Environmental Geology*, vol. 41, pp. 11–16, 2001.
188. BP, "CO_2 capture project technical report DEFC26-01NT41145," National Energy Technology Laboratory, 2005.
189. L. Shao, B. T. Low, T.-S. Chung, and A. R. Greenberg, "Polymeric membranes for the hydrogen economy: contemporary approaches and prospects for the future," *Journal of Membrane Science*, vol. 327, pp. 18–31, 2009.
190. E. B. L. Ksayer, M. Younes, and D. Clodic, "Concentration of brine solution used for low-temperature air cooling," *International Journal of Refrigeration*, vol. 35, pp. 418–423, 2012.

191. J. Rizk, M. Nemer, and D. Clodic, "A real column design exergy optimization of a cryogenic air separation unit," *Energy*, vol. 37, pp. 417–429, 2012.

192. P. Bansal, "A review–Status of CO_2 as a low temperature refrigerant: Fundamentals and R&D opportunities," *Applied Thermal Engineering*, vol. 41, pp. 18–29, 2012.

193. M. Zhang and Y. Guo, "Rate based modeling of absorption and regeneration for CO_2 capture by aqueous ammonia solution," *Applied Energy*, vol. 111, pp. 142–152, 2013.

194. A. Chikukwa, N. Enaasen, H. M. Kvamsdal, and M. Hillestad, "Dynamic modeling of post-combustion CO_2 capture using amines–a review," *Energy Procedia*, vol. 23, pp. 82–91, 2012.

195. A. Sarkar, W. Pan, D. Suh, E. D. Huckaby, and X. Sun, "Multiphase flow simulations of a moving fluidized bed regenerator in a carbon capture unit," *Powder Technology*, vol. 265, pp. 35–46, 2014.

196. D. Matovic, "Biochar as a viable carbon sequestration option: Global and Canadian perspective," *Energy*, vol. 36, 2011.

197. T. M. Lenton, "The potential for land-based biological CO_2 removal to lower future atmospheric CO_2 concentration," *Carbon Management*, vol. 1, pp. 145–160, 2010.

11 Algal Biofuel: A Promising Alternative for Fossil Fuel

Hoofar Shokravi, Zahra Shokravi, Md. Maniruzzaman A. Aziz, and Hooman Shokravi

CONTENTS

11.0 INTRODUCTION

Climate change is receiving significant attention all around the world. The main reason for climate change is the greenhouse effect caused by the emission of greenhouse gases (GHGs). GHGs include fluorinated gases (F-gases), nitrous oxides (N_2O), methane (CH_4) and carbon dioxide (CO_2), etc. These gasses block heat escaping from the atmosphere, resulting in global warming. The burning of fossil fuels is the main reason for the concentration of CO_2 in the atmosphere [1, 2]. On the other hand, air pollution is an environmental concern that occurs close to big cities. Air pollutants are accounted for as being one of the main reasons for respiratory problems and cardiovascular diseases. Air pollutants include volatile organic compounds (VOC), sulfur dioxide (SO_2) nitrogen oxides (NO_x) and fine particulate matter (PM2.5), etc. [3,4]

In order to mitigate climate change, a combination of strategies such as using renewable energies, increasing energy efficiency and reducing climate-damaging F-gases, as well as promoting sustainable mobility and city planning, proposes to reduce or retard the accumulation of GHG in the Earth's atmosphere [5]. Developing and using renewable and sustainable energy sources is the most promising way to mitigate the damages caused by GHG emissions and air pollution into the environment. Renewable energy is produced by using natural resources that can be constantly replaced and will never run out. Renewable energies can be divided into several classes including solar, ocean, wind, hydropower and geothermal, as well as hybrid/enabling technologies and bioenergy [6].

In 1995 only 5% of total energy consumption in the EU was from renewables. In 2013, renewable energy accounted for 12% of energy consumption, of which 7.7% came from biomass and renewable waste. In January 2014, the European Commission announced their long-term strategy to engage with climate-change mitigation. They have targeted to boost the use of renewable energy into 27% by 2030—from the 12% of 2013 [7]. The data for the consumption of the renewable energies (See Figure 11.1) shows that the EU has the highest share of renewable energy consumption equal with 38.6% of total production, where Asia and Oceania stand in second place with 30.9% and America—including North, South and Central American countries—has the third place [8].

Most of the bioenergy is sourced by solid biomass, where the biofuel, biogas and municipal solid waste have smaller shares. Figure 11.2 shows information and quantitative data regarding the inland bioenergy consumption in the EU in 2016 [9].

Biofuels are classified into four classes of first generation, second generation, third generation and fourth generation [10]. First-generation biofuels include the biodiesel that is produced from vegetable oils and bioethanol, which is directly obtained from the fragmentation of the edible feedstock of conventional sugar and starch crops. Biodiesel and bioethanol are compatible with conventional diesel fuel, and they can be used as complete alternatives or as an admixture blended in any proportion [11]. Though biodiesel offers no improvement in ignition quality, it reduces the amount of soot in consumption. Containing an ester moiety in the formation of biodiesel in the absence of aromatic species leads to low soot production in biodiesel fuels [12].

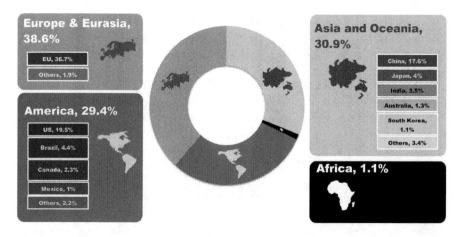

FIGURE 11.1 Consumption of renewable energy based on continents (data source: [8]).

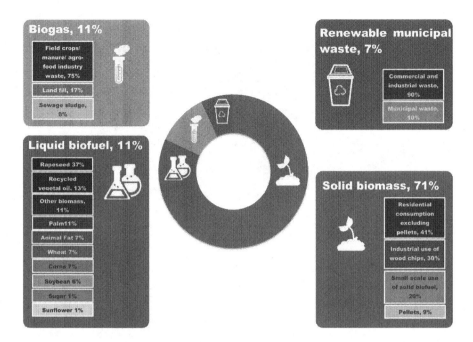

FIGURE 11.2 Quantitative data regarding the inland bioenergy consumption in EU in 2016 (data source: [9]).

Second-generation biofuel technology is from nonfood plants and typically made from energy crops, residues and wood. The second-generation biofuels are gaining wide attention for the production of biofuels using lignocellulosic materials. The idea behind them is to integrate different conversion techniques to produce a wide variety of products by taking advantage of all the fractions in the biomass [13]. In order to promote renewable energies and reduce emissions, in 2007, the EU set an ambitious goal for the year 2020 to cut GHG emissions to 20% of the 1990 level. Furthermore, it was agreed to increase the share of combustion from renewable energies from 9.8% in 2010 into 20% in 2020 [14]. The stringent regulation formulated by EU forces coal-based power plants to use biomass fuel such as wood pellets. A wood pellet is a lignocellulosic byproduct from the wood industry—i.e., from sawdust and shavings—and is typically produced by the compression of dried biomass particles. Since 2008 the export of wood pellets from the United States to the EU has increased sixfold, and the southern United States has become the world's largest exporter of wood pellets for industrial biomass use. A report by the Natural Resources Defense Council stipulated that 15 million acres of unprotected forests in the Southeastern United States are at risk of clearance due to wood exports [15]. Figure 11.3 shows a wetland clear-cut in North Carolina due to biomass exploitation.

Palm oil is one of the energy corps that is widely used for second-generation biofuel due to the higher productivity rate compared to other oilseeds, such as soybean and rapeseed. Across Southeast Asia, the expanding palm oil plantation has been destroying forests in the region and imperiling endangered species in tropical

FIGURE 11.3   Wetland clearcut in North Carolina linked to the second generation biomass industry [15].

forests. The widespread fire burning was reported in southern Kalimantan (Borneo) and western Sumatra in Indonesia to develop palm oil plantation [16]. The devastating fire was the main source of the 2015 haze crisis in Malaysia, Singapore and Indonesia. It was reported that more than 100,000 people were likely to have died due to the haze of the agricultural fire in the affected countries [17].

The aim of using renewable energies is to mitigate climate change, but the gap between legislation and practice could result in bioenergy use becoming counterproductive. Currently, there is no limitation regarding the land use for biomass harvesting in either the Kyoto Protocol, or the existing climate legislation in Europe and the United States. In 2014, the European Commission published a working document on the sustainability of biomass that included information to reduce negative impacts on the environment. It was stated in the report that the GHG emission due to the change in land use have not been accounted in the life cycle analysis (LCA) due to the assumption of the immediate uptake via plant regrowth [18]. However, concerning the "land use, land use change and forestry emissions (LULUCF)" the Parliament and Council of European Commission, adopted a decision on LULUCF as a first step toward the inclusion of the related sector in the EU's climate policy [19]. Figure 11.4

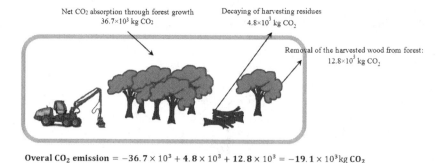

Net CO_2 absorption through forest growth
36.7×10^3 kg CO_2

Decaying of harvesting residues
4.8×10^3 kg CO_2

Removal of the harvested wood from forest:
12.8×10^3 kg CO_2

Overal CO_2 emission $= -36.7 \times 10^3 + 4.8 \times 10^3 + 12.8 \times 10^3 = -19.1 \times 10^3$ kg CO_2

FIGURE 11.4   A simple example for LCA of a "system" consists of a one-hectare stand of trees.

shows that a "system" consists of a one-hectare stand of trees and the associated life cycle analysis (LCA).

In order to deal with the concerns regarding the forest-based bioenergy and to ensure optimal bioenergy use, the extension of sustainability and GHG saving criteria to biomass fuels is hoped to be reviewed by the European Parliament in January 2018 [20]. With all the skepticism over the sustainability of the first and second generation of biofuels, it must be considered that bioenergy offers one of the most efficient ways to transition from coal or fossil fuel combustion into renewables. Sustainable biomass is the ideal solution for decarbonizing and mitigating global warming; hence, to address this issue, producing the third generation of biofuels from aquatic inedible microorganisms was developed in recent years [21]. The advantages and disadvantages of biofuel generation are presented in Table 11.1.

The current researches and development in the field of biofuels dealt with developing alternative biosources to alleviate the disadvantages of the first and second generation in land usage and competing with food crops. Hence, third- and fourth-generation biofuels were introduced that were derived from algae biomass [25]. Algae represent a large group of heterogeneous photosynthesizing organisms that belong to a different phylogenetic group with approximately 30,000 known species. Algae are defined as thallophytes that can be divided into two basic types of prokaryotic and eukaryotic photosynthesizing organisms. They can be macroscopic or microscopic, unicellular—e.g., chlorella, diatoms, coccoid, palmelloid, colonial and filamentous—or multicellular—e.g., giant kelp-attached or free-living, motile or immotile (based on the presence of flagella), terrestrial or aquatic and aerial or subaerial [27]. Algae exhibit a wide range of reproduction strategies from simple vegetative to complex forms of sexual reproduction [24]. Algae can be autotrophic—i.e., require only inorganic compounds for their growth—or heterotrophic—i.e., non-photosynthetic and require an external source of nutrients and energy [28].

Most of the research to date mainly focuses on second-generation biofuel, whereas the number of the papers written on the third and fourth generation is low. There is only one review paper on fourth-generation biofuel that just focuses on the metabolic engineering of biofuel production. In light of the above, this chapter intends to conduct a thorough survey of the cultivation process, with emphasis on the microalgae feedstock of biofuel and its associated genetic modification methods.

11.1   ALGAL BIOMASS FEEDSTOCK

Marine algae are one of the most favorable resources that are widely used in different industries, including human food, aquaculture animal feed, biofuel, pharmaceuticals and cosmetics. The first interest for the cultivation of algae perhaps dates from World War II as a potential source of food and antibiotics [29]. Third and fourth generation biofuels are energetically viable and environmentally sustainable resources that eventually could replace petroleum-based fuel for combustion in power plants, engines and boilers [86]. Open-pond cultivation is a promising method of producing algae for biomass targets due to several advantages, such as not requiring arable land and having no need to compromise food supplies [30].

TABLE 11.1

Advantages and disadvantages of biofuel generations

Biofuels	Biomass source	Pros	Cons	References
First generation	Edible feedstock	• Renewable source. • Environmentally friendly. • Easy conversion to biofuels.	• Compete with food crops. • Rise of food price due to the compete. • Land scarcity.	[22]
Second generation	Nonedible feedstock	• Renewable source. • Easy conversion to biofuels. • Do not compete with food crops. • Effective land utilization. • Environmentally friendly.	• Land and water use competition. • Required sophisticated downstream processing technology. • High production cost. • Uncertain long-term supply.	[22]
Third and fourth generation	Microalgae and macroalgae	• Renewable source, environmentally friendly. • No conflict with food or land usage. • High productivity tendencies within a short period. • Less financial input for cultivation. • No expenditure on fertilizer. • 3D cultivation. • No land and fresh water limit. • Ecological impact such as CO_2 sequestration. • High photosynthesis efficiency. • No roots hence no absorbing of nutrition. • Can endure harsh environmental condition. • Ability to grow throughout the year. • Higher tolerance to high carbon dioxide content. • No requirement of herbicides or pesticides. • Microalgae can produce a wide variety of biofuels such as diesel, gasoline, bioethanol and etc.	• Insufficient for commercialization. • Possible ecological impact such as marine eutrophication. • Possible marine algae bloom. • No regulation available for marine cultivation. • Higher cultivation cost. • Solar energy for cultivation is only available during daytime. • Harvesting of microalgae is also a challenging task. • Commercial scale of microalgae biomass production need large amount of CO_2 as carbon source. • The limited light penetration in photo-bioreactors. • Demand for biofuels is not high due to low market price of fossil fuel.	[22–26]

The use of algae feedstock as an alternative for the first and second generation of biofuels is in urgent demand. Algae can produce valuable organic materials such as carbohydrates, proteins and lipids using photosynthesis and recycling greenhouse gases (GHG). Simply, algae can be divided into two major types of microalgae and macroalgae. Several species of microalgae and macroalgae are used for the production of biofuel. Below, further detail about algal feedstock is presented.

Microalgae are among the oldest forms of life on earth, and the number of documented microalgae species is around 4400, but because of a high percentage of undiscovered species and the continuous discovering of new microalgae, the exact numbers are not clearly determined and, according to estimations, the actual number of microalgae ranges between 70,000 to over a million species [31,32]. Microalgae comprise a large and very heterogenic group of eukaryotes and cyanobacteria, which are commonly present in aquatic ecosystems. They are stress-tolerated microorganisms that adapt well to extreme environmental conditions, including salinity, drought, photo-oxidation, anaerobiosis, osmotic pressure and temperature. Photosynthesis of sunlight can be considered the main biochemical mechanism of light-energy conversion into organic matters on the earth [33]. Photosynthesis takes place in specialized intracellular structures of the chloroplast. The chloroplasts consist of internal photosynthetically active membrane systems named thylakoids [34]. In eukaryotic microalgae, chloroplasts are enclosed by limiting membranes, whereas in prokaryotic microalgae, thylakoid membranes are not separated from cytosol [35].

Macroalgae, also known as seaweed, are an aquatic plant that is harvested across the world to produce foods, cosmetics and fertilizers. Macroalgae is a fast-growing biomass that can grow large within a short period of time. Marine microalgae are harvested from wild stocks, managed stands or cultivated beds. Large-scale cultivation of macroalgae in coastal areas is not only economically favorable but also environmentally beneficial, playing an important role in ecosystem balance and environmental sustainability [36]. The photosynthetic efficiency of macroalgae is three to four times that of terrestrial plants, hence better performance in carbon absorption and sequestration. The iodine production from macroalgae is stated to be acting as an atmospheric cooling agent [37]. Moreover due to the algal bio-absorption and neutralization properties of macroalgae it can be used to scavenge toxic metals, such as selenium, arsenic and molybdenum from waters, as well as serving as a bio-filter to remove nitrogen and phosphorus produced by marine animals [38]. Figure 11.5 shows the global harvesting quantity of macroalgae in the world.

Macroalgae can be classified based on their pigmentation into three main classes of brown macroalgae—i.e., phaeophyceae—red macroalgae—i.e., rhodophyceae—and green macroalgae—i.e., hlorophyceae. [39]. In recent years macroalgae have attracted attention as a feedstock for biofuel due to the higher content of carbohydrate as a carbon source compared to terrestrial biosources [40]. The higher carbon content of macroalgae makes it suitable for producing biofuels such as biobutanol, biogas and bioethanol [41]. Algae is made up of three main compositions of protein, lipids and carbohydrates and the content of each composition is varied from one to the other [42]. Microalgae and macroalgae differ significantly in terms of composition, where microalgae has a higher lipid content (up to 80%), but the carbohydrate content of macroalgae is high. Hence, the production process of biofuel from each

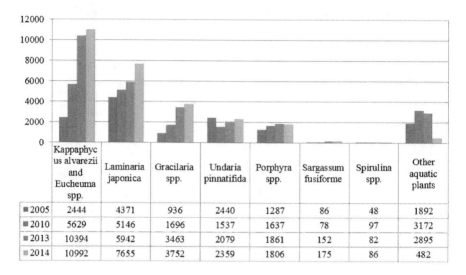

	Kappaphyc us alvarezii and Eucheuma spp.	Laminaria japonica	Gracilaria spp.	Undaria pinnatifida	Porphyra spp.	Sargassum fusiforme	Spirulina spp.	Other aquatic plants
■ 2005	2444	4371	936	2440	1287	86	48	1892
■ 2010	5629	5146	1696	1537	1637	78	97	3172
■ 2013	10394	5942	3463	2079	1861	152	82	2895
■ 2014	10992	7655	3752	2359	1806	175	86	482

FIGURE 11.5   The macroalgae harvest in the world.

alga is substantially different from the other due to the different composition of each feedstock. The composition profile of some well-known micro-and macro algae are shown in Table 11.2 [42].

Green macroalgae contains glucans as the main carbohydrate component that is easily fermentable. Unlike green macroalgae, the existing carbohydrate component

TABLE 11.2
Chemical composition of some algae (% of dry matter) [42]

Type of algae	Compositions (%)		
	Protein	Carbohydrate	Lipid
Microalgae			
Chlamydomonas rheinhardii	48	17	21
Chlorella vulgaris	41–58	12–17	10–22
Porphyridium creuntum	28–39	40–57	9–14
Prymnesium parvum	28–45	25–33	22–39
Scenedesmus dimorphus	8–18	21–52	16–40
Scenedesmus obliquus	50–56	10–17	12–14
Macroalgae			
Eucheuma cottonii	9–10	26	1
Gelidium amansii	20	66	0.2
Laminaria japonica	8	51	1
Sargassum ilicifolium	8–9	32–33	2
Ulva lactuca	17	59	3–4
Undaria pinnitifida	24	43	3–4

in red macroalgae and brown macroalgae, *i.e.,* alginate and mannitol—3,6-anhydro-L-galactose—are not easily fermentable [43]. Finding or engineering the appropriate microbes to be capable of converting all types of carbohydrates into biofuels is one of the challenges of microbiological biotechnologies in recent years. The harvesting of brown algae is twofold that of red algae, whereas the green algae have the least value among all macroalgae and the lowest harvest [44]. Hence several research studies are conducted to enhance the hydrocarbon utilization of brown and red macroalgae through microbiological biotechnologies. There is a limited number of microbes for the modification of alginate for biofuel production. Hence, extensive studies are conducted to use bioengineering tools to modify the structure of microbes toward facilitating the production of algal biofuels. Bacterium *Sphingomonas* sp. A1 is one of the well-known microbes used to assimilate alginate. However, this bacterium is unable to assimilate mannitol to extract the carbohydrates [45]. Takeda et al. [46] introduced two gene encoded version of *Sphingomonas* sp. A1 to optimize the condition of algal biofuel production. Bioengineered *E. coli* is another example that was very advantageous and improved 40% in the production of biofuel from macroalgae compared to its plasmid-based counterpart [47,48]. *S. cerevisiae* is a unicellular eukaryote yeast with a well-characterized genetic system. Hence it has none of the genes required to assimilate alginate and mannitol. Bioengineered *S. cerevisiae* is able to metabolize alginate and mannitol for biofuel production [49].

Considerable scientific effort is dedicated to improving algal biology in order to enhance biomass production. Using metabolic engineering and genetic editing are two main tools for enhancing algal biology, including the optimization of nutrient absorption and culture conditions as well as increasing the yield and productivity [50].

11.2   GENETIC MODIFICATION AND METABOLIC ENGINEERING

A number of laboratory-based studies have searched for ways to improve the photosynthesis efficiency of microalgae with the help of genetic modification methods. The green algae possess three modifiable compartments (nucleus, mitochondria and chloroplast), of which the chloroplast has received considerable attention because of its capability of increasing growth rates and biomass productivity through genetic engineering [51]. As the Calvin cycle plays an important role in carbon assimilation, searching a breakthrough in the regulation of this cycle is one area that has been subjected to many studies. A number of genetic investigations have been done to increase the light utilization efficiency in them. The reduction of the chlorophyll antenna size of photosystems is one of the areas which has been a focus of the investigations. Light intensity and availability is a major concern for maximizing growth rates [96]. Extremely high flux can cause photoinhibition, while low light intensity can cause reduced photosynthetic activity, then chloroplast engineering is a promising approach for obtaining maximum feedstock production under different light conditions throughout the year [51].

Microalgae have developed effective light-harvesting antennae complexes (LHCs) over the long periods of their evolution to trap light energy. Under high light irradiation, photosystems I and II can be affected, leading to inhibition. In order to increase photosynthetic efficiency and subsequently raise biomass productivity, mutant species with reduced antenna size, namely TLH (truncated light-harvesting), were constructed. A number of photosynthesis regulation genes were discovered which promote light utilization in microalgae [52]. Tetali *et al.* [53] identified a new gene (*tla1*) which is associated with the regulation size of antenna. After the characterization of this new mutant more investigations were carried out; Kirst *et al.* [54] showed that the light antenna of *C. reinhardtii* could be modified to alter chlorophyll content and subsequently adjust photosynthetic productivity.

Genetic engineering of RuBisCO is another approach in improving biomass yield; despite the crucial position of RuBisCO in carbon assimilation, slow catalytic rate and confused specificity make it the target of many investigations. Alternative approaches have been applied for three other Calvin cycle enzymes to promote the photosynthetic capacity and increase biomass yield. Previous researches have shown that three non-regulated Calvin-cycle enzymes including fructose 1,6-bisphosphate aldolase (aldolase), sedoheptulose 1,7-bisphosphatase (SBPase) and transketolase (TK) decrease photosynthesis rate. In addition, the efficiency of microalgal photosynthesis can also be promoted by several abiotic factors, including light availability, temperature and carbon availability, and nutrients increase [55]. Introducing the fructose-1,6-/sedohe ptulose-1,7-bisphosphatase (FBP/SBPase) gene from the cyanobacterial species into Euglena chloroplasts leads to enhanced biomass production [56].

Microalgae with the aid of sunlight convert simple inorganic nutrients to feed-stock, consisting of primary metabolites and secondary metabolites, unlike primary metabolites that are involved in the overall survival of microalgae, secondary metabolites are only synthetized under specific ecological conditions or certain phases of their life [57]. Different species of microalgae accumulate various ratios of secondary metabolites including carbohydrates, proteins and lipids (in forms of oils like triacylglycerol TAG or fatty acid) which can be used in the production of third- or fourth-generation biofuels and several other biocompounds. These rates for protein and carbohydrate are around half of their dry weight, and for lipid, it is up to $2 \sim 40\%$ of the dry cell weight, which can be seen in plasma membrane or as energy source or storage materials. By optimizing the culture condition this can be increased to 30–50% oil [55].

Lipid biosynthesis in microalgae mainly takes place through both fatty acid synthesis and TAG synthesis, which occur in the chloroplast and the endoplasmic reticulum, respectively. The fatty acid and TAG biosynthetic pathways have been fully characterized in microalgae. Fatty acid synthesis is performed by two different enzymatic systems including acetyl-CoA carboxylase (ACCase) and fatty acid synthase (FAS). The first step for the biosynthesis of fatty acid is synthesized under the catalysis of ACCase, which transforms acetyl-CoA to malonyl-CoA. The second step is catalyzed by FAS complex. The malonyl moiety is transferred to acyl carrier protein (ACP) and makes malonyl-ACP, which is added to another acyl-ACP to form an acyl chain with two carbons longer. Further reactions lead to a saturated and unsaturated acyl chain with acyl carrier protein. When the chain reaches the appropriate length, acyl carrier protein is removed from fatty acid, yielding the complete fatty acid. Furthermore, the synthesis of TAG is performed by four enzymes, including glycerol-3- phosphate dehydrogenase (GPDH), lysophosphatidic acyltransferase (LPAAT), diacylglycerol acyltransferase (DGAT) and glycerol-3-phosphate acyltransferase (GPAT). Therefore, the overexpression of these genes has been used as a technique to promote lipid content [58–60].

In the last decades, a large number of random mutagenesis was done with the purpose of generating species with a higher oil content. In this process, after generating random mutants, the best-performing ones were selected and then characterized. One of the most successful studies about random mutagenesis was done by de Jaeger *et al.*, which resulted in a starchless mutant with a higher TAG content in *A. obliquus*. Nowadays, the administration of molecular biology and genetic engineering techniques on microalgal investigations has opened a new horizon for the modification of main genes, and regulators are involved in fatty acid and TAG biosynthetic pathways. Overexpression of acetyl-CoA carboxylase (ACCase) in oleaginous microalga *Phaeodactylum tricornutum* (PtACC2) into the choloroplast led to altered fatty acid profile and an increased lipid content of up to 40% dry weight [12]. Other attempts have been made to bring the microalgal oil profile toward producing medium-chain fatty acids (MCFAs) by inserting the Chaetoceros ketoacyl-ACP synthase III enzymes gene into Synechococcus [61]. Overexpression of GPAT in marine diatom, *Phaeodactylum tricornutum*, enhances polyunsaturated fatty acids rate [62]. Introducing lysophosphatidic acyltransferase gene (*c-lpaat*) and glycerol-3-phosphate dehydrogenase gene (*c-gpd1* into the genomic DNA led to increase in fatty acid content, furthermore, the insertion of new genes induces a change in fatty acid composition [60]. Under natural conditions, microalgae have high potential to respond to environmental stresses with changing metabolic mechanisms. Therefore, abiotic stresses can be used as tools for changing microalgal metabolism in favor of biofuel feedstock production. Many investigations have been conducted in this area.

Light intensity is a critical factor in the enhancing of lipid accumulation. Adequate light irradiation not only affects the performances of microalgal photosynthesis but could be used to induce elevated lipid accumulation [57]. Efficient starch and lipid production in microalgae occur only up to a point, which is generally similar to the specific photosynthesis saturation of that particular microalgae species. Extremely high flux above saturation level leads to photoinhibition, while low light intensity causes reduced photosynthetic efficiency, which results in a negative influence on lipid production [63,64]. Adequate levels of CO_2 concentrations in microalgal cultivation can increase photosynthetic efficiency and, as a result, increase bioproducts like lipid accumulation. Lower levels of CO_2 concentrations in the growth medium decrease the metabolism of microalgae, leading to reduced lipid content. Therefore, elevated CO_2 supply flue gas (rich in CO_2) is considered to be a promising approach in decreasing the costs of microalgae production; on the other hand, the growth of microalgae will be limited by high amounts of CO_2 because of the pH reduction in culture which comes from the converting of unutilized CO_2 to carbonic acid (H_2CO_3). Therefore, in order to achieve maximum biomass and lipid production, an optimal CO_2 level is required that differs among species [63,65].

Temperature is another crucial parameter that affects microalgal lipid and fatty acid contents. Like the other environmental factors, the rates of lipid and fatty acids can be boosted by adjusting the optimal temperature in culture. The optimal temperature at which the highest biomass productivity can be obtained varies from species to species. Xing *et al.* [66] investigated the molecular mechanism involved in the regulations of FA metabolism under low and high-temperature stress in oleaginous microalgae in *Auxenochlorella protothecoides* UTEX 2341. Physiological research indicates that imposing nitrogen starvation can reduce growth rate and photosynthetic pigments. On the other hand, it enhances lipid accumulation. Transcriptome analysis of *Scenedesmus acutus* under nitrogen starvation characterize the molecular responses involving in lipid production [67].

11.3 STRAINS SELECTION

In terms of oil production, diatoms and green algae are the most promising microalgae for biodiesel production [55]. D'Ippolito *et al.* [68] evaluated the potential of 17 diatom species for biomass production and lipids yield. The two best-performing diatoms, *Thalassiosira weissflogii* and *Cyclotella cryptica*, were selected as good candidates for biofuel production. El-Sheekh *et al.* [69] investigated different characteristics of green algae *Scenedesmus obliquus* and introduced it as a potential candidate for biodiesel production, the green microalga *Scenedesmus obliquus* can tolerate extreme environmental conditions, including salinity, various light intensity and pH conditions, imposing solitary stress up to 100% can promote biomass productivity, esterified fatty acid (EFA) rate and EFA production. Biomass productivity also can be enhanced with increasing of light intensity. In terms of biodiesel characterization, all biodiesel properties of the biodiesel S. *obliquus* except viscosity are in agreement with the European standards for fuel diesel (EN 590:1999). The iodine value of biodiesel of S. *obliquus* was 70 g iodine/100 g oil.

Song *et al.* [70] evaluated the capacity of seven algal species for biodiesel production by analyzing their different characteristics, including fatty acid profiles, biodiesel characteristics, growth rate, biomass productivity and lipid accumulation; among these species, five best-performing ones (*Selenastrum capricornutum, Chlorella vulgaris, Scenedesmus obliqnus, Phaeodactylum tricornutum* and *Isochrysis sphacrica*) were selected because of their higher growth potential and lipid accumulation; among these, seven species *P. tricornutum*, with lipid content of 61.43 %, lipid productivity of 26.75 mg $L^{-1}d^{-1}$, fatty acid composition of C16–C18 (74.50%), C14:0 (11.68%) and C16:1 (22.34%) as well as suitable biodiesel properties of higher cetane number (55.10), lower iodine number (99.2 gI2/100 g) were the most promising species for biodiesel production. A challenge for selecting the most suitable genera as feedstock for biodiesel production is the screening of microalgal species that can maintain a high growth rate with high lipid content. Nannochloropsis is considered to be a potential oleaginous model microalga because of the great photosynthetic efficiency, high lipid productivity, well-established genetic toolbox and relatively mature technology for outdoor cultivation systems on a large scale [71].

11.4 CULTIVATION STRATEGIES

Cultivation condition plays an important role in the growth and content of algal cells to achieve highest productivity in a cost-effective way. Abiotic, biotic and hydrodynamic factors are influential parameters in the cultivation of microalgae. The abiotic factors deals with energy sources, nutrients, etc., where the biotic factors are the pathogens that could be microalgae competitors for nutritional and environmental resources, such as bacteria, fungi and viruses [72]. On the other hand, hydrodynamic parameters relate to the operational factors, such as turbulence degree, velocity and depth that are influential in bioreactor design [73]. The most important parameters in medium composition and cultivation condition are discussed in the following subsections.

Increasing the photosynthetic efficiency, enhancing the light penetration into dense microalgae culture, and reducing photoinhibition are some of the interesting strategies for improving the quality of the obtained algal biomass in fourth-generation biofuels [50]. Several researches are conducted for finding the target genes to enhance the photosynthesis efficiency and light intake of microalgae [74]. One of the solutions for enhancing the light penetration into dense microalgae cultures is by the truncation of the chlorophyll antenna of chloroplast [75]. On the other hand, the photosynthesis efficiency can be improved by expansion of the absorbing spectrum range of microalgae in photosynthesis [76].

Previous studies stipulated that change in metabolism conditions—i.e., environmental and nutritional factors—can lead to a significant increase in the lipid or carbohydrate content of microalgae [77]. Lipid content is the most attractive factor for efficiency of microalgae in biodiesel production. The lipid content for different microalgae species can reach 70% of the dry cell weight, but it is associated with lower productivity [78]. On the other hand, the carbohydrate maximization in algal feedstock is a favorable option in ethanol production [79]. There are four different metabolism conditions for microalgae cultivation: photoautotrophic, heterotrophic, mixotrophic and photoheterotrophic. Strain selection plays an important role in boosting the quality—i.e., lipid content—and quantity—i.e., dry weight of biomass or chlorophyll content—of algal biofuel. Hence, comprehensive analysis of the microalgae strain in phototroph, heterotroph or mixotroph species is required to reach the optimum cultivation condition and maximum productivity [80]. On the other hand, the carbohydrate maximization in algal feedstock is a favorable option in ethanol production [79]. The mode of cultivation plays an important role in determining the growth style of microalgae, and thus has a direct effect on the quality and quantity of biomass yield. There are three different metabolism conditions for microalgae cultivation: photoautotrophic, heterotrophic and mixotrophic. In photoautotrophic cultivation, microalgae with the aid of sunlight convert simple inorganic nutrients to feedstock through photosynthesis process. There are two types of photobioreactors for photoautotrophic cultivating of algae, which are the open cultivation system—i.e., having direct contact with the environment such as tanks, artificial ponds, raceways and thin-layer platforms—and the closed system [81]. The photobioreactors are designed based on the requirement of the microalgae-based process. Hence, the designing processes must satisfy the environmental and nutritional consideration of the associated microalgae. Heterotrophic cultivation is the more flexible growth style, which microalgae utilize the organic compounds—e.g., glucose, acetate, wastewater, crop flours and others— as energy sources. The lipid accumulation of the heterotrophic cultivation is at par or higher compared to the one in photoautotrophic mode [80]. However, due to the need for a bioreactor and organic compounds, the cost of heterotrophic cultivation is higher than the photoautotrophic mode.

Mixotrophy is the mode that microalgae can drive the combination of photo-autotrophic and heterotrophic cultivation with the possibility of using organic and inorganic carbon [82]. Inorganic carbon is fixed through photosynthesis affected by illumination, whereas the organic carbon is sourced from aerobic respiration [80]. Table 11.3 shows different strategies for the cultivation of microalgae, as well as the advantages and disadvantages of each class.

11.4.1 CARBON SOURCE

Carbon dioxide is an essential nutrient for the cultivation of algae for biofuel production. In order to produce 1000 kg of algal biomass, at least 1830 kg of carbon dioxide is required [90]; purchasing CO_2 contributes to nearly 50% of the production cost and the algae culture for biofuel is not feasible without having access to free resources of carbon dioxide. Potentially the CO_2 is emitted by transportation vehicles and by industrial and residential heating equipment. However, the concentration of the emitted CO_2 is a major limitation in the applicability of the emitted gases in algae production. The flue gases produced during power generation in coal-fired power plants and cement manufacture are the main source of concentrated carbon dioxide used for algae cultivation.

The concentration of the CO_2 in atmosphere is around 0.039% by volume. However, there is no method for high-productive algae cultivation in normal atmosphere concentration [91]. Hence, some strategies are required for cheaply capturing and concentrating the carbon dioxide from the atmosphere for use in algal culture. Many algae and cyanobacteria are known to facilitate the concentration of carbon dioxide from the culture medium into the cell. But the absorption rate from the atmosphere into a culture medium is slower than the speed for providing the required nutrients for the growth of algae [92]. For algae with growing capability in highly alkaline conditions—*e.g., Spirulina sp.*—the feeding of dissolved inorganic carbon—*i.e.,* bicarbonate and carbonate—can be a substituted for direct purging of CO_2. Another solution for alleviating the insufficiency of the concentrated source of heterotrophic cultivation of the microalgae is using organic carbon. Meyer *et al.* [93] presented the economic analysis of biodiesel production that shows the higher cost of using heterotrophic cultivation due to using bioreactors.

11.4.2 ILLUMINATION

The effect of temperature and illumination, as well as their cross-interactions, have an important role in the metabolic regulation and growth of algae [94]. Especially for photoautotrophic cultivation, illumination plays an important role in boosting the productivity of microalgae. Hence, the selection of appropriate source of light, wavelength and intensity for the target microalgae is an essential factor in large-scale and low-cost production [80]. The efficient light wavelength for microalgae photosynthesis is ranged between 400–700nm, whereas the wavelength between 400–500nm—*i.e.,* blue light—improves the overall growth efficiency and the wavelength between

TABLE 11.3
Variation of culture methods for microalgae

Culture method	Advantages	Disadvantages

Photoautotrophic

Light   Initial microalgae

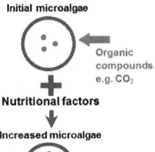

Organic compounds e.g. CO$_2$

Nutritional factors

Increased microalgae

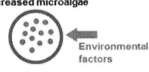

Environmental factors

- Low energy consumption.
- Low production cost.
- Generates high-value light-induced components [83].
- High dry weight of biomass [80].
- Environment-friendly due to CO$_2$ absorption.

- Low productivity.
- The technological subsystem for implementation is not fully developed [83].

Heterotrophic

Dark   Initial microalgae

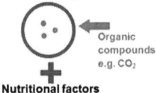

Organic compounds e.g. CO$_2$

Nutritional factors

Increased microalgae

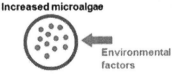

Environmental factors

- No need for to light supply for metabolism [84].
- Higher cell densities are achievable [84,85].
- Industrial scale can be in conventional fermenters to reach high controllability [87].
- High conversion ratio of the input energy to ATP [88]

- CO$_2$ emission [82].
- Limited types of strains are adaptable to this system.
- Expensive additive compounds of organic substrate.
- Prone to contamination.
- Unable to generate light-induced metabolite [89].

(Continued)

TABLE 11.3   (CONTINUED)
Variation of culture methods for microalgae

Culture method	Advantages	Disadvantages
Mixotrophic Dark Light 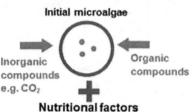 Inorganic compounds e.g. CO₂ — Organic compounds **+ Nutritional factors** Increased microalgae Environmental factors	• High productivity [80]. • Reduced risk of photoinhibition. • Producing light-induced metabolite with high value components [83]. • High lipid productivity [80].	• Prone to contamination. • The technological subsystem for implementation is not fully developed [83]. • Limited types strains are adapted. • Low dry weight of biomass [80]. • CO_2 emission during the metabolism.
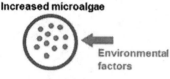 **Photoheterotrophic** Light Initial microalgae  Organic compounds **+ Nutritional factors** Increased microalgae  Environmental factors	• High productivity [80]. • Reduced risk of photoinhibition.	• Prone to contamination. • High initial installation cost. • Light requirement [78].

600–700nm—*i.e.,* red light though the narrow spectrum illumination can enhance the productivity of microalgae. The cost of the associated light sources is high compared to sunlight or other wide-spectrum light sources. Light intensity is another important parameter for lipid accumulation and fatty acid composition. The reaction of microalgae for different light intensity can be defined by the three phases of photolimitation, photosaturation and photoinhibition that the growth rate in the associated light intensity is increased, independent and declined, respectively. The photosaturation limit for microalgae is around 6500 lux and exposure to intense light in photoinhibition phase can lead to oxidative stress, reduced growth or death of algae [94,95]. However, the influence light intensity and light wavelength on metabolic regulatory and growth rate of algae are species-dependent and even strain-specific.

11.4.3 TEMPERATURE

The primary influence of temperature can be associated with an improvement in the yield and reaction rate, as well as enhancing the structural components of algae cells—*i.e.,* mainly protein and lipid—as the consequence of the main influence, temperature is also attributed to physiological-biochemical changes—*e.g.,* cell permeability, cell size and composition and enzyme reactions [94]. Growth reduction of microalgae in low temperature is due to a decrease in enzyme activity, the degree of membrane fluidity and electron transfer in electron transport chains. On the other hand, excessive temperature leads to denaturation and degradation of some proteins resulting in the growth reduction of algae. However, the sensitivity of the algae to temperature variation is species-dependent and even strain-specific.

11.4.4 AERATION

Optimization of aeration is one of the key parameters in increasing productivity and lipid content of microalgae species [97]. The improvement in growth of microalgae cells through aeration is attributed to various factors, such as mass transfer, toxic level control of dissolved oxygen, deficiency-avoidance of CO_2, controlling CO_2 inhibitory level, gradient-reduction of the nutrients, avoidance cell sedimentation, emergence of dead zones, clumping of cells, bioreactor fouling and optimization of the light/dark cycle [98]. However, excessive aeration should be avoided because of increased running costs as well as resulting cell damage in the microalgae culture [97]. Various parameters must be considered in aeration systems, such as light penetration and distribution, hydrodynamics and mass transfer that are closely interrelated. In order to design a well-balanced process, with appropriate interaction between the incorporated factors in the aeration system, a deep knowledge of fluid dynamics and mass transfer is necessary.

11.4.5 CONTAMINATION

Contaminants significantly reduce the yields and increase the risk of culture crashes that result in poor control over the entire process. The main contaminants in microalgae culture can be grouped within five classes: grazers, photosynthetic organisms, fungi, bacteria and viruses [99]. Due to the lower controllability of open ponds, a wide variety of biological contaminants co-cultivates with the target microalgae species, and competes for their nutritional and environmental resources. Theses contaminants may reduce the yield, and pose health concerns about food or a safety threat for the microalgae. Among all, bacteria and some macroalgae species are the most hazardous type of biological contamination in algal cultivation.

11.5 WATER FOOTPRINT AND LAND USAGE

Access to the required water and land for algae production is an important param-eter that should be engaged beside other factors, such as energy consumption and GHG emission. Ensuring safe and clean water is an overriding global challenge due to water pollution and water scarcity. The water footprint is a concept introduced by researchers to relate the biofuels to water pollution and scarcity into biofuel production analysis the sustainability of the process from a water perspective. In order to quantify the water consumption of the biomass feedstock production, three independent components of green—*i.e.*, consumption of rainfall; blue—*i.e.*, con-sumption of surface and groundwater—and gray water footprint—*i.e.*, impaired water quality—are required to be known [100]. Zhang *et al.* [101] presented total direct water footprints for three terrestrial plants—*i.e.,* sweet sorghum, cassava and Jatropha curcas—and microalgae. It was stipulated that the direct green water footprint of the microalgae was almost a quarter of the average direct green water footprint of the three terrestrial plants [102]. Figure 11.6 shows the schematic repre-sentation of the water footprint and computing the water demand in algae biomass cultivation.

The harvesting is the most important phase in the life cycle—*i.e.*, cultivation, harvest, treatment and energy extraction—of biofuel production. Recycling the dis-charged water from the harvesting phase can reduce 90.2% of water consumption in biofuel production [103]. In addition to freshwater, seawater and wastewater also can be used for algae production. The use of undiluted wastewater is a promising strategy to save freshwater resources. In addition, the application of wastewater for the culti-vation of algae can potentially make biofuel production economically attractive and environmentally sustainable [104]. Open ponds have a higher footprint compared to closed systems. It is estimated that the water footprint for microalgae production in an open pond can reach 5.5 ha-ft/acre/year [105]. Yang *et al.* [106] conducted a case study to quantify the water footprint in the production of biodiesel from microalgae. It was stipulated that for producing of 1 kg microalgal biodiesel, 3726 kg water, 0.33 kg nitrogen and 0.71 kg phosphate are required without considering recycling of the applied freshwater [102].

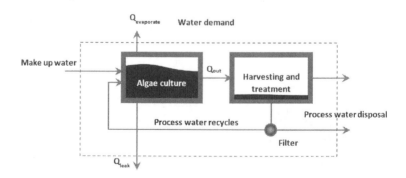

FIGURE 11.6 Schematic representation of the water footprint and computing the water demand in algae biomass cultivation [102].

Land is needed for growing plants as well as for freshwater. Land use is a challenging parameter for biomass feedstock in first- and second-generation biofuel. Change in land use by expanding the cultivation area of biomass has a direct effect on GHG emissions. Lardon *et al.* [107] estimated that the land use by algal biomass is negligible compared to terrestrial biofuel resources due to considerably low land use by algal biomass, high biomass yield and algae oil content compared to other biomass feedstocks. The GHS emission caused by land use change is varied by different factors, such as facility locations—*i.e.*, forestland, cropland, grassland or marginal/barren—or environmental conditions—*e.g.*, tropical, subtropical and temperate [108]. In general, the land use of microalgal biofuels is estimated to be ranged from 20 to 200 m^2/GJ net energy [109].

11.6 CONCLUSION

Algal biofuels are widely accepted as a sustainable and environmentally friendly fuel source. Algae are an easily available and accessible feedstock that do not have any of the disadvantages of first and second generation biofuels. Recently, a large number of studies were conducted on genetic modification and the nutrient optimization of microalgae to enhance the biomass and lipid yields. Microalgae have different cultivation strategies that are mixotrophic, autotrophic, heterotrophic and photohetrotrophic modes.

Nutritional requirements are the most important factor in the cultivation of microalgae. Low-cost nutrient sources from residual waste are considered to be commercially viable nutrient sources. As a conclusion, the challenges in the production of algal biofuel need to be carefully addressed in order to ensure that biofuel production runs on a cost-effective and environmentally friendly platform.

ACKNOWLEDGMENT

The authors would like to thank the Ministry of Higher Education of Malaysia for their financial support through the research grant (Vote No. Q.J130000.2522.15H32 and Q.J130000.2522.19H98), without which this study would not have been possible.

REFERENCES

1. T. Kirschstein, and F. Meisel, "GHG-emission models for assessing the eco-friendliness of road and rail freight transports", *Transp. Res. Part B Methodol.*, vol. 73, pp. 13–33, 2015.
2. R. O'Driscoll, M. E. J. Stettler, N. Molden, T. Oxley, and H. M. ApSimon, "Real world CO_2 and NOx emissions from 149 Euro 5 and 6 diesel, gasoline and hybrid passenger cars", *Sci. Total Environ.*, vol. 621, pp. 282–290, 2018.
3. J. Nègre, and P. Delhomme, "Drivers' self-perceptions about being an eco-driver according to their concern for the environment, beliefs on eco-driving, and driving behavior", *Transp. Res. Part A Policy Pract.*, vol. 105, pp. 95–105, 2017.
4. M. Condurat, A. M. Nicuţă, and R. Andrei, "Environmental impact of road transport traffic. A case study for county of Iaşi Road Network", *Procedia Eng.*, vol. 181, pp. 123–130, 2017.

5. BMU, *Mitigating Greenhouse Gas Emissions*. Available at: https://www.international -climate-initiative.com/en/issues/mitigation/, 2018.

6. ARENA, *What Is Renewable Energy?* Australian Government. Available at: https://ar ena.gov.au/about/what-is-renewable-energy/, 2018.

7. M. Albani, A. Bühner-Blaschke, N. Denis, and A. Granskog, *"Bioenergy in Europe: A new beginning–or the end of the road"*, McKinsey Company, pp. 47–52, 2014.

8. B. Dudley, *BP Statistical Review of World Energy June 2017.* British Petroleum (BP). Available at: https://www.bp.com/content/dam/bp/en/corporate/pdf/energy-economi cs/statistical-review-2017/bp-statistical-review-of-world-energy-2017-full-report.pdf, 2017.

9. AEBIOM, *What Are the Biomass Sources Consumed in the EU-28?.* European Biomass Day, European Biomass Association (AEBIOM). Available at: http://www.europeanb ioenergyday.eu/bioenergy-facts/bioenergy-in-europe/what-are-the-volumes-of-bioma ss-used-in-the-eu28-to-produce-energy/, 2018.

10. K. Ullah, V. K. Sharma, M. Ahmad, P. Lv, J. Krahl, and Z. Wang, "The insight views of advanced technologies and its application in bio-origin fuel synthesis from lignocel-lulose biomasses waste, a review", *Renew. Sustain. Energy Rev.*, 2017.

11. M. Lapuerta, J. Barba, A. D. Sediako, M. R. Kholghy, and M. J. Thomson, "Morphological analysis of soot agglomerates from biodiesel surrogates in a coflow burner", *J. Aerosol Sci.*, vol. 111, pp. 65–74, 2017.

12. D.-W. Li et al., "Constitutive and chloroplast targeted expression of Acetyl-CoA car-boxylase in oleaginous microalgae elevates fatty acid biosynthesis", *Mar. Biotechnol.*, vol. 20, no. 5, pp. 566–572pp. 1–7, 2018.

13. D. L. Aguilar et al., "Operational strategies for enzymatic hydrolysis in a biorefin-ery", In: Kumar S., Sani R. K. (Eds.), *Biorefining of Biomass to Biofuels.* Springer, pp. 223–248, 2018.

14. EC, 2020 *Climate & Energy Package*. European Commission. Availabe at: https://ec .europa.eu/clima/policies/strategies/2020_en, 2018.

15. A. Colette, *Uncovering the Truth: Investigating the Destruction of Precious Wetland Forests*. Dogwood Alliance, 2015.

16. L. Allen, "Is Indonesia's fire crisis connected to the palm oil in our snack food?", *The Guardian*, 2015.

17. P. Jacobson, "SE Asian governments dismiss finding that 2015 haze killed 100,300", *Mongabay*, 2016.

18. EC, *State of Play on the Sustainability of Solid and Gaseous Biomass Used for Electricity, Heating and Cooling in the EU, SWD(2014) 259 Final*. EUROPEAN COMMISSION. Available at: https://ec.europa.eu/energy/en/topics/renewable-energy/ biomass, 2014.

19. EC, "Decision No 529/2013/EU of the European Parliament and of the Council", *Official Journal of the European Union*. The European Parliament and of the Council. Available at: https://eur-lex.europa.eu/LexUriServ/LexUriServ.do?uri=OJ:L:2013: 165:0080:0097:EN:PDF, 2013.

20. UmweltBundesamt, "Scientific Opinion Paper: What Issues Need To Be Addressed in the RED II Draft to Ensure Environmental Integrity and Net Climate Benefit of Bioenergy Use?" German Environment Agency. Available at: https://www.umweltbu ndesamt.de/sites/default/files/medien/376/dokumente/scientific_opinion_paper_red _ii_and_lulucf.pdf, 2018.

21. F. Darvishi, Z. Fathi, M. Ariana, and H. Moradi, "Yarrowia lipolytica as a workhorse for biofuel production", *Biochem. Eng. J.*, vol. 127, pp. 87–96, 2017.

22. W.-H. Leong, J.-W. Lim, M.-K. Lam, Y. Uemura, and Y.-C. Ho, "Third generation biofuels: A nutritional perspective in enhancing microbial lipid production", *Renew. Sustain. Energy Rev.*, vol. 91, pp. 950–961, 2018.

23. R. P. John, and G. S. Anisha, "Macroalgae and their potential for biofuel", *Plant Sci. Rev.*, vol. 2011, pp. 151–162, 2012.
24. S. Chinnasamy, P. H. Rao, S. Bhaskar, R. Rengasamy, and M. Singh, "Algae: A novel biomass feedstock for biofuels", In: Arora R. (Ed.), *Microbial Biotechnology: Energy and Environment.* CABI UK, pp. 224–239, 2012.
25. S. Behera, R. Singh, R. Arora, N. K. Sharma, M. Shukla, and S. Kumar, "Scope of algae as third generation biofuels", *Front. Bioeng. Biotechnol.*, vol. 2, p. 90, 2015.
26. J. Milano et al., "Microalgae biofuels as an alternative to fossil fuel for power generation", *Renew. Sustain. Energy Rev.*, vol. 58, pp. 180–197, 2016.
27. D. Verma, A. Singla, B. Lal, and P. M. Sarma, "Conversion of biomass-generated syngas into next-generation liquid transport fuels through microbial intervention: Potential and current status", *Curr. Sci.*, vol. 110, no. 3, pp. 329–336, 2016.
28. L. Brennan, and P. Owende, "Biofuels from microalgae—A review of technologies for production, processing, and extractions of biofuels and co-products", *Renew. Sustain. Energy Rev.*, vol. 14, no. 2, pp. 557–577, 2010.
29. C. F. Murphy, and D. T. Allen, "Energy-water nexus for mass cultivation of algae", *Environ. Sci. Technol.*, vol. 45, no. 13, pp. 5861–5868, 2011.
30. A. Ozkan, K. Kinney, L. Katz, and H. Berberoglu, "Reduction of water and energy requirement of algae cultivation using an algae biofilm photobioreactor", *Bioresour. Technol.*, vol. 114, pp. 542–548, 2012.
31. J. Brodie et al., "The algal revolution", *Trends Plant Sci.*, vol. 22, no. 8, pp. 726–738, 2017.
32. P. Neofotis et al., "Characterization and classification of highly productive microalgae strains discovered for biofuel and bioproduct generation", *Algal Res.*, vol. 15, pp. 164–178, 2016.
33. S. Flori et al., "Plastid thylakoid architecture optimizes photosynthesis in diatoms", *Nat. Commun.*, vol. 8, p. 15885, 2017.
34. B. Serive et al., "Community analysis of pigment patterns from 37 microalgae strains reveals new carotenoids and porphyrins characteristic of distinct strains and taxonomic groups", *PLOS ONE*, vol. 12, no. 2, p. e0171872, 2017.
35. U. C. Vothknecht, and P. Westhoff, "Biogenesis and origin of thylakoid membranes", *Biochim. Biophys. Acta (BBA)-Molecular Cell Res.*, vol. 1541, no. 1–2, pp. 91–101, 2001.
36. C. L. Johansson, N. A. Paul, R. de Nys, and D. A. Roberts, "Simultaneous biosorption of selenium, arsenic and molybdenum with modified algal-based biochars", *J. Environ. Manage.*, vol. 165, pp. 117–123, 2016.
37. J. S. Rowbotham, P. W. Dyer, H. C. Greenwell, and M. K. Theodorou, "Thermochemical processing of macroalgae: A late bloomer in the development of third-generation biofuels?", *Biofuels*, vol. 3, no. 4, pp. 441–461, 2012.
38. A. K. Zeraatkar, H. Ahmadzadeh, A. F. Talebi, N. R. Moheimani, and M. P. McHenry, "Potential use of algae for heavy metal bioremediation, a critical review", *J. Environ. Manage.*, vol. 181, pp. 817–831, 2016.
39. FAO, *The State of World Fisheries and Aquaculture 2016 (SOFIA), SOFIA 2016.* Food and Agriculture Organization of the United Nations. Available at: http://www.fao.org/documents/card/en/c/2c8bcf47-2214-4aeb-95b0-62ddef8a982a, p. number204, 2016.
40. S. Kawai, and K. Murata, "Biofuel production based on carbohydrates from both brown and red macroalgae: Recent developments in key biotechnologies", *Int. J. Mol. Sci.*, vol. 17, no. 2, p. 145, 2016.
41. M. Suutari, E. Leskinen, K. Fagerstedt, J. Kuparinen, P. Kuuppo, and J. Blomster, "Macroalgae in biofuel production", *Phycol. Res.*, vol. 63, no. 1, pp. 1–18, 2015.
42. S. A. Jambo, R. Abdulla, S. H. Mohd Azhar, H. Marbawi, J. A. Gansau, and P. Ravindra, "A review on third generation bioethanol feedstock", *Renew. Sustain. Energy Rev.*, vol. 65, pp. 756–769, 2016.

43. R. Jiang, K. N. Ingle, and A. Golberg, "Macroalgae (seaweed) for liquid transportation biofuel production: What is next?", *Algal Res.*, vol. 14, pp. 48–57, 2016.
44. F. C. Ertem, P. Neubauer, and S. Junne, "Environmental life cycle assessment of biogas production from marine macroalgal feedstock for the substitution of energy crops", *J. Clean. Prod.*, vol. 140, pp. 977–985, 2017.
45. S. Kawai et al., "Bacterial pyruvate production from alginate, a promising carbon source from marine brown macroalgae", *J. Biosci. Bioeng.*, vol. 117, no. 3, pp. 269–274, 2014.
46. H. Takeda, F. Yoneyama, S. Kawai, W. Hashimoto, and K. Murata, "Bioethanol production from marine biomass alginate by metabolically engineered bacteria", *Energy Environ. Sci.*, vol. 4, no. 7, pp. 2575–2581, 2011.
47. A. J. Wargacki et al., "An engineered microbial platform for direct biofuel production from brown macroalgae", *Science*, vol. 335, no. 6066, pp. 308–313, 2012.
48. C. N. S. Santos, D. D. Regitsky, and Y. Yoshikuni, "Implementation of stable and complex biological systems through recombinase-assisted genome engineering", *Nat. Commun.*, vol. 4, p. 2503, 2013.
49. M. Enquist-Newman et al., "Efficient ethanol production from brown macroalgae sugars by a synthetic yeast platform", *Nature*, vol. 505, no. 7482, p. 239, 2014.
50. P. Tandon, and Q. Jin, "Microalgae culture enhancement through key microbial approaches", *Renew. Sustain. Energy Rev.*, vol. 80, pp. 1089–1099, 2017.
51. M. A. Scranton, J. T. Ostrand, F. J. Fields, and S. P. Mayfield, "Chlamydomonas as a model for biofuels and bio-products production", *Plant J.*, vol. 82, no. 3, pp. 523–531, 2015.
52. A. E. Gomma, S. K. Lee, S. M. Sun, S. H. Yang, and G. Chung, "Improvement in oil production by increasing Malonyl-CoA and Glycerol-3-Phosphate Pools in *Scenedesmus quadricauda*", *Indian J. Microbiol.*, vol. 55, no. 4, pp. 447–455, 2015.
53. S. D. Tetali, M. Mitra, and A. Melis, "Development of the light-harvesting chlorophyll antenna in the green alga *Chlamydomonas reinhardtii* is regulated by the novel Tla1 gene", *Planta*, vol. 225, no. 4, pp. 813–829, 2007.
54. H. Kirst, J. G. García-Cerdán, A. S. Zurbriggen, and A. Melis, "Assembly of the light-harvesting chlorophyll antenna in the green alga *Chlamydomonas reinhardtii* requires expression of the TLA2-CpFTSY gene", *Plant Physiol.*, vol. 158, pp. 930–945, 2012.
55. A. Demirbas, and M. F. Demirbas, "Importance of algae oil as a source of biodiesel", *Energy Convers. Manag.*, vol. 52, no. 1, pp. 163–170, 2011.
56. T. Ogawa et al., "Enhancement of photosynthetic capacity in *Euglena gracilis* by expression of cyanobacterial fructose-1, 6-/sedoheptulose-1, 7-bisphosphatase leads to increases in biomass and wax ester production", *Biotechnol. Biofuels*, vol. 8, no. 1, p. 80, 2015.
57. L. Barsanti, and P. Gualtieri, "Is exploitation of microalgae economically and energetically sustainable?", *Algal Res.*, vol. 31, pp. 107–115, 2018.
58. A. J. Klok, P. P. Lamers, D. E. Martens, R. B. Draaisma, and R. H. Wijffels, "Edible oils from microalgae: Insights in TAG accumulation", *Trends Biotechnol.*, vol. 32, no. 10, pp. 521–528, 2014.
59. L. Kirchner et al., "Identification, characterization, and expression of diacylglycerol acyltransferase type-1 from *Chlorella vulgaris*", *Algal Res.*, vol. 13, pp. 167–181, 2016.
60. C. Wang, Y. Li, J. Lu, X. Deng, H. Li, and Z. Hu, "Effect of overexpression of LPAAT and GPD1 on lipid synthesis and composition in green microalga *Chlamydomonas reinhardtii*", *J. Appl. Phycol.*,vol. 32, no. 10, pp. 1711–1719, 2018.
61. H. Gu, R. E. Jinkerson, F. K. Davies, L. A. Sisson, P. E. Schneider, and M. C. Posewitz, "Modulation of medium-chain fatty acid synthesis in *Synechococcus* sp. PCC 7002 by replacing FabH with a Chaetoceros ketoacyl-ACP synthase", *Front. Plant Sci.*, vol. 7, p. 690, 2016.

62. Y.-F. Niu et al., "Molecular characterization of a glycerol-3-phosphate acyltransferase reveals key features essential for triacylglycerol production in Phaeodactylum tricornutum", *Biotechnol. Biofuels*, vol. 9, no. 1, p. 60, 2016.
63. C. Zuñiga et al., "Genome-scale metabolic model for the green alga chlorella vulgaris UTEX 395 accurately predicts phenotypes under autotrophic, heterotrophic, and mixotrophic growth conditions", *Plant Physiol.*, vol. 172, no. 1, pp. 589–602, 2016.
64. M. Vitova, K. Bisova, S. Kawano, and V. Zachleder, "Accumulation of energy reserves in algae: From cell cycles to biotechnological applications", *Biotechnol. Adv.*, vol. 33, no. 6, pp. 1204–1218, 2015.
65. G. Sibi, V. Shetty, and K. Mokashi, "Enhanced lipid productivity approaches in microalgae as an alternate for fossil fuels–A review", *J. Energy Inst.*, vol. 89, no. 3, pp. 330–334, 2016.
66. G. Xing et al., "Integrated analyses of transcriptome, proteome and fatty acid profilings of the oleaginous microalga Auxenochlorella protothecoides UTEX 2341 reveal differential reprogramming of fatty acid metabolism in response to low and high temperatures", *Algal Res.*, vol. 33, pp. 16–27, 2018.
67. A. Sirikhachornkit, A. Suttangkakul, S. Vuttipongchaikij, and P. Juntawong, "De novo transcriptome analysis and gene expression profiling of an oleaginous microalga Scenedesmus acutus TISTR8540 during nitrogen deprivation-induced lipid accumulation", *Sci. Rep.*, vol. 8, no. 1, p. 3668, 2018.
68. G. d'Ippolito et al., "Potential of lipid metabolism in marine diatoms for biofuel production", *Biotechnol. Biofuels*, vol. 8, no. 1, p. 28, 2015.
69. M. El-Sheekh, A. E.-F. Abomohra, H. Eladel, M. Battah, and S. Mohammed, "Screening of different species of Scenedesmus isolated from Egyptian freshwater habitats for biodiesel production", *Renew. Energy*, vol. 129, pp. 114–120, 2018.
70. M. Song, H. Pei, W. Hu, and G. Ma, "Evaluation of the potential of 10 microalgal strains for biodiesel production", *Bioresour. Technol.*, vol. 141, pp. 245–251, 2013.
71. X. Ma, Q. Zhu, Y. Chen, and Y.-G. Liu, "CRISPR/Cas9 platforms for genome editing in plants: Developments and applications", *Mol. Plant*, vol. 9, no. 7, pp. 961–974, 2016.
72. P. L. Kashyap, A. K. Srivastava, S. P. Tiwari, and S. Kumar, *Microbes for Climate Resilient Agriculture*. John Wiley & Sons, 2018.
73. J. C. M. Pires, M. C. M. Alvim-Ferraz, and F. G. Martins, "Photobioreactor design for microalgae production through computational fluid dynamics: A review", *Renew. Sustain. Energy Rev.*, vol. 79, pp. 248–254, 2017.
74. S. Qin, H. Lin, and P. Jiang, "Advances in genetic engineering of marine algae", *Biotechnol. Adv.*, vol. 30, no. 6, pp. 1602–1613, 2012.
75. G. Buitrón, J. et al., "Biohydrogen production from microalgae", In Gupta, V. K., Tuohy, M. G. (Eds.), *Microalgae-Based Biofuels and Bioproducts*. Elsevier, pp. 209–234, 2018.
76. B. M. Wolf, D. M. Niedzwiedzki, N. C. M. Magdaong, R. Roth, U. Goodenough, and R. E. Blankenship, "Characterization of a newly isolated freshwater Eustigmatophyte alga capable of utilizing far-red light as its sole light source", *Photosynth. Res.*, vol. 135, no. 1–3, pp. 177–189, 2018.
77. C.-H. Hsieh, and W.-T. Wu, "Cultivation of microalgae for oil production with a cultivation strategy of urea limitation", *Bioresour. Technol.*, vol. 100, no. 17, pp. 3921–3926, 2009.
78. C.-Y. Chen, K.-L. Yeh, R. Aisyah, D.-J. Lee, and J.-S. Chang, "Cultivation, photobioreactor design and harvesting of microalgae for biodiesel production: A critical review", *Bioresour. Technol.*, vol. 102, no. 1, pp. 71–81, 2011.
79. C. Simas-Rodrigues, H. D. M. Villela, A. P. Martins, L. G. Marques, P. Colepicolo, and A. P. Tonon, "Microalgae for economic applications: Advantages and perspectives for bioethanol", *J. Exp. Bot.*, vol. 66, no. 14, pp. 4097–4108, 2015.

80. D. Wang et al., "Nannochloropsis genomes reveal evolution of microalgal oleaginous traits", *PLOS Genet.*, vol. 10, no. 1, 2014.

81. F. G. Acién et al., "Photobioreactors for the production of microalgae", In: C. Gonzalez-Fernandez, and R. Muñoz, eds. *Microalgae-Based Biofuels and Bioproducts.* Woodhead Publishing, pp. 1–44, 2017.

82. J. Lowrey, M. S. Brooks, and P. J. McGinn, "Heterotrophic and mixotrophic cultivation of microalgae for biodiesel production in agricultural wastewaters and associated challenges—A critical review", *J. Appl. Phycol.*, vol. 27, no. 4, pp. 1485–1498, 2015.

83. R. Muñoz, and C. Gonzalez-Fernandez, *Microalgae-Based Biofuels and Bioproducts: From Feedstock Cultivation to End-Products.* Woodhead Publishing, 2017.

84. F. Chen, Y. Zhang, and S. Guo, "Growth and phycocyanin formation of Spirulina platensis in photoheterotrophic culture", *Biotechnol. Lett.*, vol. 18, no. 5, pp. 603–608, 1996.

85. D. Morales-Sánchez, O. A. Martinez-Rodriguez, J. Kyndt, and A. Martinez, "Heterotrophic growth of microalgae: Metabolic aspects", *World J. Microbiol. Biotechnol.*, vol. 31, no. 1, pp. 1–9, 2015.

86. M. A. Scaife, A. Merkx-Jacques, D. L. Woodhall, and R. E. Armenta, "Algal biofuels in Canada: Status and potential", *Renew. Sustain. Energy Rev.*, vol. 44, pp. 620–642, 2015.

87. O. Perez-Garcia, F. M. E. Escalante, L. E. de-Bashan, and Y. Bashan, "Heterotrophic cultures of microalgae: Metabolism and potential products", *Water Res.*, vol. 45, no. 1, pp. 11–36, 2011.

88. J. Hu, D. Nagarajan, Q. Zhang, J. S. Chang, and D. J. Lee, "Heterotrophic cultivation of microalgae for pigment production: A review", *Biotechnol. Adv.*, vol. 36, no. 1, pp. 54–67, 2018.

89. X. B. Tan, M. K. Lam, Y. Uemura, J. W. Lim, C. Y. Wong, and K. T. Lee, "Cultivation of microalgae for biodiesel production: A review on upstream and downstream processing", *Chin. J. Chem. Eng.*, vol. 26, no. 1, pp. 17–30, 2018.

90. Y. Chisti, "Constraints to commercialization of algal fuels", *J. Biotechnol.*, vol. 167, no. 3, pp. 201–214, 2013.

91. A. Kumar et al., "Enhanced CO_2 fixation and biofuel production via microalgae: Recent developments and future directions", *Trends Biotechnol.*, vol. 28, no. 7, pp. 371–380, 2010.

92. F. Bumbak, S. Cook, V. Zachleder, S. Hauser, and K. Kovar, "Best practices in heterotrophic high-cell-density microalgal processes: Achievements, potential and possible limitations", *Appl. Microbiol. Biotechnol.*, vol. 91, no. 1, p. 31, 2011.

93. M. Meyer, and H. Griffiths, "Origins and diversity of eukaryotic CO_2-concentrating mechanisms: Lessons for the future", *J. Exp. Bot.*, vol. 64, no. 3, pp. 769–786, 2013.

94. G. Gacheva, and L. Gigova, "Biological activity of microalgae can be enhanced by manipulating the cultivation temperature and irradiance", *Open Life Sci.*, vol. 9, no. 12, pp. 1168–1181, 2014.

95. Y. S. Shin, H. I. Choi, J. W. Choi, J. S. Lee, Y. J. Sung, and S. J. Sim, "Multilateral approach on enhancing economic viability of lipid production from microalgae: A review", *Bioresour. Technol.*, vol. 258, pp. 335–344, 2018.

96. V. da Silva Ferreira, and C. Sant'Anna, "Impact of culture conditions on the chlorophyll content of microalgae for biotechnological applications", *World J. Microbiol. Biotechnol.*, vol. 33, no. 1, p. 20, 2017.

97. C. Ji, J. Wang, and T. Liu, "Aeration strategy for biofilm cultivation of the microalga Scenedesmus dimorphus", *Biotechnol. Lett.*, vol. 37, no. 10, pp. 1953–1958, 2015.

98. X. Guo, L. Yao, and Q. Huang, "Aeration and mass transfer optimization in a rectangular airlift loop photobioreactor for the production of microalgae", *Bioresour. Technol.*, vol. 190, pp. 189–195, 2015.

99. T. P. Lam, T. M. Lee, C. Y. Chen, and J. S. Chang, "Strategies to control biological contaminants during microalgal cultivation in open ponds", *Bioresour. Technol.*, vol. 252, pp. 180–187, 2018.
100. P. W. Gerbens-Leenes, "Green, blue and grey bioenergy water footprints, a comparison of feedstocks for bioenergy supply in 2040", *Environ. Process.*, vol. 5, no. 1, pp. 167–180, 2018.
101. Y. Zhang, A. Kendall, and J. Yuan, "A comparison of on-site nutrient and energy recycling technologies in algal oil production", *Resour. Conserv. Recycl.*, vol. 88, pp. 13–20, 2014.
102. B. Abdullah et al., "Fourth generation biofuel: A review on risks and mitigation strategies", *Renew. Sustain. Energy Rev.*, vol. 107, pp. 37–50, 2019.
103. P.-Z. Feng, L.-D. Zhu, X.-X. Qin, and Z.-H. Li, "Water footprint of biodiesel production from microalgae cultivated in photobioreactors", *J. Environ. Eng.*, vol. 142, no. 12, p. 4016067, 2016.
104. D. De Francisci, Y. Su, A. Iital, and I. Angelidaki, "Evaluation of microalgae production coupled with wastewater treatment", *Environ. Technol.*, vol. 39, no. 5, pp. 581–592, 2018.
105. B. G. Subhadra, and M. Edwards, "Coproduct market analysis and water footprint of simulated commercial algal biorefineries", *Appl. Energy*, vol. 88, no. 10, pp. 3515–3523, 2011.
106. J. Yang, M. Xu, X. Zhang, Q. Hu, M. Sommerfeld, and Y. Chen, "Life-cycle analysis on biodiesel production from microalgae: Water footprint and nutrients balance", *Bioresour. Technol.*, vol. 102, no. 1, pp. 159–165, 2011.
107. L. Lardon, A. Helias, B. Sialve, J.-P. Steyer, and O. Bernard, "Life-cycle assessment of biodiesel production from microalgae", *Environ. Sci. Technol.*, vol. 43, no. 17, pp. 6475–6481, 2009.
108. R. M. Handler, R. Shi, and D. R. Shonnard, "Land use change implications for large-scale cultivation of algae feedstocks in the United States Gulf Coast", *J. Clean. Prod.*, vol. 153, pp. 15–25, 2017.
109. P. W. Gerbens-Leenes, L. Xu, G. J. De Vries, and A. Y. Hoekstra, "The blue water footprint and land use of biofuels from algae", *Water Resour. Res.*, vol. 50, no. 11, pp. 8549–8563, 2014.

12 The Fourth-Generation Biofuel: A Systematic Review on Nearly Two Decades of Research from 2008 to 2019

*Zahra Shokravi, Hoofar Shokravi,
Md. Maniruzzaman A. Aziz, and Hooman Shokravi*

CONTENTS

12.0 INTRODUCTION

Recently, biofuels have attracted considerable scientific and public attention as a promising alternative to fossil fuels. Biofuels are made from biomass through processes such as chemical, biochemical or hybrid conversions [1]. Biofuels do not create additional emissions apart from those produced during the production and transportation stages [2]. Hence, biofuels avoid the environmental drawbacks associated with the consumption of fossil fuels. Biofuels are usually classified into four classes of first-, second-, third- and fourth-generation biofuel [3]. The definition and highlights of each class are presented in the following.

First-generation biofuel includes biodiesel and ethanol produced from oil, sugar and starch crops, respectively. First-generation bioethanol is mainly produced from food crop feedstock, thus it is competing for agricultural areas used for food production. Hence, it is believed that the efficiency of the ethanol produced from the first generation feedstock to achieve targets for the substitution of fossil fuels in reducing global warming and to reach economic growth is limited [4]. The cumulative impact of these concerns has increased pressure on the energy sector in producing ethanol from non-edible feedstock. In this regard, the second-generation biofuel was developed to overcome the deficiencies of the first-generation biofuel in terms of food security, biodiversity and sustainability.

Second-generation biofuels differ in feedstock which, in this case, comes from nonfood plants such as agricultural crops, residues and wood (so-called lignocellulosic biomass). Biorefinery is gaining wide attention for converting lignocellulosic biomass to biofuels. It is a concept analogous to petroleum refineries, which produce various output products from crude oil. The idea behind designing a biorefinery is to integrate different conversion techniques to produce a wide variety of products by taking advantage of all the fractions in the biomass [5]. Biorefinery integrates producing bioethanol or other biochemical sustainably from lignocellulosic biomass. Palm oil is one of the energy crops widely used for second-generation biofuel due to its high rate of productivity—per unit of planted area—compared to other oilseeds such as soybean and rapeseed. Deforestation and the associated health risks are the key concerns faced by governments as a result of the over-reliance on palm oil [6, 7].

In order to deal with the concerns relating to forest-based bioenergy and to ensure optimal bioenergy use, biofuel production from aquatic inedible microorganisms, the so-called third generation of biofuel was developed in recent years [8]. Third-generation biofuel is mainly derived from algae biomass [9]. Algae represent a large group of heterogeneous photosynthesizing organisms that belong to different phylogenetic groups with approximately 30,000 known species. Algae are defined as thallophytes—*i.e.,* plants lacking roots, stems and leaves—and can be divided into two basic types of prokaryotic and eukaryotic photosynthesizing organisms. They can be unicellular—*e.g.,* chlorella, diatoms, coccoid, palmelloid, colonial and filamentous—or multicellular—*e.g.,* giant kelp-attached or free-living, motile or immotile (based on the presence of flagella), terrestrial or aquatic, macroscopic or microscopic and aerial or subaerial [10]. Algae exhibit a wide range of reproduction strategies, from simple vegetative to complex forms of sexual reproduction [11]. Algae can be autotrophic—i.e., require only inorganic compounds for their growth—or heterotrophic—*i.e.,* non-photosynthetic and require an external source of nutrients and energy—or mixotrophic growth conditions [12, 13].

All microalgae species are not equally attractive or feasible for biodiesel production. Chlorophyceae is the most promising microalgae species due to relatively high growth rate, high lipid content as well as its relatively easy cultivation. Bacillariophyceae, Eustigmatophyte, Chrysophyceae, Haptophyceae and Cyanophyceas are other microalgae classifications that are important for biofuel production [14]. Lipid contents of microalgae vary significantly for different microalgae strains. Thought implementation of biochemical strategies improve the lipid content but simultaneously decreases the growth rate in microalgae species [15].

For instance, *Scenedesmus obtusus* microalgae strains with a higher lipid content have a low growth rate when cultivated under biochemical stress [16].

While microalgal biofuel production is a well-established practice in small-scale systems, its industrial-scale production is not economically viable. This is mainly due to the low lipid content and growth rate of the microalgae strains [17]. Genetic engineering offers a wide range of options to enhance the lack of industrially competent strains by several approaches, such as transcription and targeted expression of key proteins involved in microalgal lipogenesis [18, 19]. Production of biofuel from genetically engineered algae is discussed under the FGB term [20]. The review studying the first, second, third and fourth generations of biofuels is presented in Table 12.1.

The presented review of the available literature on biofuels shows that most of the research to date mainly focuses on second-generation biofuel, whereas there are also several papers about the first and third generations of biofuels. Nonetheless, there are only two review papers for the FGB, and their focus is mainly on metabolic engineering and risk-related concerns of the FGB, respectively. In the paper by Lü *et al.* [38], a survey is carried out on metabolic engineering of algae for the production of FGB. The findings of the research mainly focused on FGB production by introducing a cell factory research paradigm. In further research by the co-authors of this chapter (Abdullah *et al.* [21]) in 2019, the health and environmental risk of FGB production was reviewed and the associated mitigation strategies were identified. Although some specific aspects of FGB production, notably risks and metabolic engineering, were covered by the aforementioned review articles, some other important features such as cultivation, harvest, environmental impact, sustainability and techno-economy were obscured or lacking in their works. In light of the above, this paper intends to conduct a thorough review of the papers published for FGB, with a focus on aspects other than microbiological ones. To this end, the main keywords involved in FGB were selected by consultancy with two experts. After carrying out three steps of literature search, eligibility identification and inclusion, the selected papers were included into the meta-analysis. Then, the included papers were classified after a full-text reading into 16 classes that was later narrowed into five groupings of strain selection, genetic modification, cultivation and harvesting, environment and sustainability and industrialization and economy. The available papers in each group were discussed and reviewed and the associated meta-analysis was presented. The extensive diversity of the studies in terms of scope, discipline and research area, as well as the limited space of the present review obstructs the authors from including all the references used in the meta-analysis.

12.1   RETRIEVAL AND ANALYSIS OF THE ARTICLES

In order to include all available relevant published papers on FGB, three main steps were followed: (1) literature search, (2) selecting the eligible papers and (3) data extraction and summarizing of the results. Relevant studies were retrieved using a multi-step process shown in Figure 12.1.

Selecting a suitable database was the first challenge to overcome, in the literature search step. Web of Science (WOS), Google Scholar and Scopus are the most widespread databases, which are frequently used for basic bibliometric searching

TABLE 12.1

Distribution of the review papers based on biofuel generation and publish year

No.	Year	Title	Ref.	Generation			
				First	Second	Third	Fourth
1	2019	Fourth generation biofuel: A review on risks and mitigation strategies	Abdullah *et al.* [21]	✗	✗	✗	✓
2	2018	Heterogeneous sulfur-free hydrodeoxygenation catalysts for selectively upgrading the renewable bio-oils to second generation biofuels	Li *et al.* [22]	✗	✓	✗	✗
3	2018	Third generation biofuels: A nutritional perspective in enhancing microbial lipid production	Leong *et al.* [23]	✗	✗	✓	✗
4	2017	Biodiversity impacts of bioenergy production: Microalgae vs. first generation biofuels	Correa *et al.* [24]	✓	✗	✗	✗
5	2017	Efficiency of second-generation biofuel crop subsidy schemes: Spatial heterogeneity and policy design	Andrée *et al.* [25]	✗	✓	✗	✗
6	2017	Current status and strategies for second generation biofuel production using microbial systems	Bhatia *et al.* [26]	✗	✓	✗	✗
7	2016	Guidelines for emergy evaluation of first, second and third generation biofuels	Saladini *et al.* [27]	✓	✓	✓	✗
8	2015	Microbial conversion of pyrolytic products to biofuels: a novel and sustainable approach toward second-generation biofuels	Islam *et al.* [28]	✗	✓	✗	✗
9	2015	Combustion pathways of biofuel model compounds: A review of recent research and current challenges pertaining to first-, second-, and third-generation biofuels	Hayes *et al.* [29]	✗	✗	✓	✗

(Continued)

TABLE 12.1 (CONTINUED)
Distribution of the review papers based on biofuel generation and publish year

No.	Year	Title	Ref.	First	Second	Third	Fourth
					Gen	eration	
10	2014	The impact of the rebound effect of the use of first generation biofuels in the EU on greenhouse gas emissions: A critical review	Smeets *et al.* [30]	✓	✗	✗	✗
11	2014	Reviews of science for science librarians: Second-generation biofuels feedstock: Crop wastes, energy grasses, and forest byproducts	Stankus [31]	✗	✓	✗	✗
12	2013	Technoeconomic assessment of second-generation biofuel pathways: Challenges and solutions	Brown [32]	✗	✓	✗	✗
13	2013	Second generation biofuels and food crops: Co-products or competitors?	Thompson *et al.* [33]	✗	✓	✗	✗
14	2013	Third-generation biofuels: Current and future research on microalgal lipid biotechnology	Li-Beisson *et al.* [34]	✗	✗	✓	✗
15	2012	Thermochemical processing of macroalgae: A late bloomer in the development of third-generation biofuels?	Rowbotham *et al.* [35]	✗	✗	✓	✗
16	2011	Thermochemical conversion of biomass to second generation biofuels through integrated process design-A review	Damartzis *et al.* [36]	✗	✓	✗	✗
17	2011	Bioconversion of synthesis gas to second generation biofuels: A review	Mohammadi *et al.* [37]	✗	✓	✗	✗
18	2011	Metabolic engineering of algae for fourth generation biofuels production	Lü *et al.* [38]	✗	✗	✗	✓
19	2010	First generation biofuels compete	Martin [39]	✓	✗	✗	✗

(Continued)

TABLE 12.1   (CONTINUED)
Distribution of the review papers based on biofuel generation and publish year

No.	Year	Title	Ref.	First	Second	Third	Fourth
					Generation		
20	2010	Technologies in second-generation biofuel production	Hromádko *et al.* [40]	✗	✓	✗	✗
21	2010	Production of first and second generation biofuels: A comprehensive review	Naik *et al.* [41]	✓	✓	✗	✗
22	2010	An overview of second generation biofuel technologies	Sims *et al.* [42]	✗	✓	✗	✗
23	2008	Third generation biofuels via direct cellulose fermentation	Carere *et al.* [43]	✗	✗	✓	✗
24	2006	The second-generation biofuels	Jęczmionek *et al.* [44]	✗	✓	✗	✗

of scientific literature [45]. For many years WOS from Thomson Reuters (ISI) was the only standard source to extract the publication and citation data. However, in 2004 Elsevier Science introduced the Scopus database as an alternative to ISI-WOS [46]. On the other hand, Google Scholar was released in the same year by Google to provide accessibility to the full text and meta-data of scholarly literature for researchers [47]. Many researchers have examined these databases from different aspects to be able to answer questions about their efficacy and effectiveness [48–50]. Different results were achieved that sometimes support or undermine one another. The results show that the advantages of each database significantly depend on the particular area of research, the explicit topic to be analyzed and the period of analysis [51]. Comparison of these databases has not identified any clear winner, but it was reported that the results are case-sensitive and the requirement of the study is the dominant factor in determining a suitable database [45, 50]. Using a database with the capability of filtering within a particular dataset field in the manuscript, such as the abstract, article title, keywords or full-text, was a requisite in this study to achieve more reliable classification results and reduce the possibility of error in the inclusion phase. Such capability is only available in the Scopus database. Hence, the Scopus database was nominated to retrieve information on FGB. The main keywords for the literature search were selected in consultation with two experts. The literature search was accomplished using some keywords within the Scopus database with the search end dated March 2019. A total of 566 records were identified from searching the keywords. The title and abstract were screened and the articles were removed if they did not explicitly involve genetic modification on algae, if they failed to cover any steps in relation to FGB production, or overlapping with other biofuel generations (i.e., first-, second- and third-generation biofuels) without focusing directly on

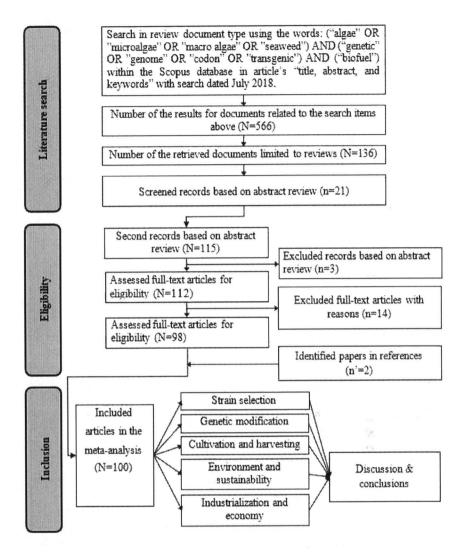

FIGURE 12.1 Study flowcharts for the identification, screening, eligibility and inclusion of articles.

any specific point of FGB. Finally, 466 articles were excluded, leaving a total of 115 scholarly papers in the final step. The eligible articles were selected to be included in the continuation phase of the study.

In the eligibility phase, 115 articles that potentially met the inclusion criteria through reading titles and abstracts were selected before being critically read in full. The non-English papers, as well as textbooks, book chapters, doctoral theses, editorial notes, unpublished working papers and conference papers were excluded, yielding a total number of 100.

For the last stage of the review section, the selected papers were read thoroughly and the information was retrieved after consultation with experts. The selected papers were categorized based on their specific research purposes and key findings. Accordingly, the included studies were classified into five groups. The following categories were identified: strain selection, genetic modification, cultivation and harvesting, environment and sustainability and industrialization and economy. (See Figure 12.1). Finally, findings of the review are discussed and, in the light of these findings, suggestions and recommendations for future studies were proposed.

12.2 FOURTH GENERATION BIOFUEL (FGB)

The FGB is achieved by enhancing the quality and productivity of microalgae using genetic modification. A microalga comprises molecules of lipids, proteins, carbohydrates and nucleic acids [52]. The lipid content in the biochemical composition of microalgae can be in any form of polar or neutral lipids. Polar lipids constitute the structural lipid by bounding into the organelle membranes or in bilayer structure of cell membrane. Neutral lipids are storage lipids, which may include triglycerides (TAGs), diglycerides (DAGs), monoglycerides (MAGs) and free fatty acids (FFAs) [14]. Lipids contain the highest energy level (37.6 kJ g^{-1}) among biochemical components of microalgae; hence, achieving a higher lipid content from algal biomass plays a key role in the efficiency of FGB production [53].

Three different strategies are generally used for the overproduction of lipids in microalgae strains: biochemical, genetic and transcription methods [54]. The efficiency of the selected strategy is widely dependent on the appropriateness of the microalgae species for the purpose [53]. Biochemical strategies describe the act of applying environmental stresses during the cultivation process and mainly deal with controlling the nutrient, salinity, mineral, chemical and physical factors of microalgae cultivation [55]. On the other hand, advancements in the genetic engineering of microalgae enable further improvement of production efficiency. Understanding of the metabolism pathway is considered to be a basis for fully utilizing the microalgae's potential as a lipid source. Genetic engineering is composed of a variety of strategies in FGB for genome editing, such as gene knockdown, knockout and genetic modifications [56]. Transcription factor engineering deals with the regulation of metabolic pathways within whole cells rather than at the single pathway level. It is carried out by controlling the abundance or function of various enzymes relevant to the production of a desired constituent in microalgae [57].

12.2.1 STRAIN SELECTION

Microalgae strain selection is an important factor in the overall success of algal biofuel production. Strain selection is based on the idea that certain microalgae are better suited than others for a particular task [58]. Growth rate and condition, digestibility and chemical composition are among the most important parameters to have a significant influence on the applicability of a microalgae strain in biofuel production. Lipid is the main composition in microalgae that is used for biodiesel production. To date, biodiesel is the only alternative fuel that can be used directly in the combustion of existing engines. Biodiesel can be produced from the transesterification of the accumulated lipid in microalgae cells [59]. Biodiesel is compatible with conventional diesel fuel and they can be used as a complete alternative or as an admixture blended in any proportion [60]. Though biodiesel offers no improvement in ignition quality, it reduces the amount of soot in consumption. The containing of an ester moiety in the formation of biodiesel and in absence of aromatic species cause low soot production in biodiesel fuels [61].

The carbohydrate fraction of microalgae mainly comes from starch in the plastids and cellulose/polysaccharides in the cell wall [62]. The absence of lignin and low hemicellulose content makes microalgal carbohydrate a more attractive choice for biofuel production [63]. The composition and metabolism of carbohydrates differ significantly from species to species, and therefore great care must be taken while selecting the appropriate strain for biofuel feedstock. The high productivity and sugar composition are the most important parameters to increase the efficiency and potential of biofuel production. Some algae strains such as *Chlorella*, *Dunaliella*, *Chlamydomonas* and *Scenedesmus* are capable of accumulating more than 50% starch (dry cell weight). The cellulose and starch components are the most feasible feedstock for production of bioethanol and biobutanol.

The energy content of protein (16.7 kJ g^{-1}) and carbohydrate (15.7 kJ g^{-1}) compounds is lower compared to that of lipids. In contrast to lipids and carbohydrates, proteins are not widely used to synthesize biofuel due to the difficulty in deaminating protein hydrolysates [55]. The chemical content of some of the most widely used microalgae in FGB production is shown in Table 12.2. Adopting genetic modification of microalgae strains is a solution to increase the amount of starch, lipid and hydrocarbons released by the algae [64]. Though genetic engineering is a common practice

TABLE 12.2
Properties of some of the most widely used microalgae in biofuel production [14, 67]

Microalgae strain	Lipids (%)	Proteins (%)	Carbohydrates (%)
Chlorella vulgaris	41–58	51–58	12–17
Chlorella sorokiniana	22–24	40.5	26.8
Chlorella pyrenoidosa	2	57	26
Chlorella protothecoides	40–60	10–28	11–15
Chlorella minutissima	14–57	47.89	8.06
Botryococcus braunii	25	–	–
Scenedesmus obliquus	30–50	10–45	20–40
Dunaliella tertiolecta	11–16	20–29	12.2–14
Dunaliella salina	6–25	57	32
Scenedesmus dimorphus	16–40	8–18	21–52
Tetraselmis suecica	15–23	–	–
Haematococcus pluvialis	25	–	–
Scenedesmus quadricauda	1.9	40–47	12
Phaeodactylum tricornutum	18–57	30	8.4
Thalassiosira pseudonana	20	–	–
Spirulina platensis	4–9	46–63	8–14

for enhancing the productivity and lipid accumulation of microalgae strains, these methods possess some limitations. Genetic modification cannot be used for all species due to a lack of available genomic data, the complexity of transgenesis, as well as the complications in establishing the delicate balance between metabolic and energy storage pathways [65]. Almost 20 whole-algal genomes are available that allow for the modification of the targeted genomes for enhancing lipid metabolism [66].

Cyanobacteria are among the first microorganisms to have lasted for a few billion years. They play a significant role in the average human's everyday life as an important atmospheric oxygen source [13]. Cyanobacteria have a wide range of potential applications, such as nutrition source, agricultural biofertilizer and wastewater treatment [68]. Cyanobacteria can be of freshwater strains belonging to the genera *Anabaena* and *Plectonema*, or of marine cyanobacteria belonging to the genus *Synechococcus*, *Gloeothece*, *Oscillatoria* and *Trichodesmium*. Cyanobacteria present a diverse range of morphology including unicellular—i.e., *Gloeothece*, *Synechococcus*; filamentous—i.e., *Plectonema*, *Oscillatoria*, *Anabaena* and *Nostoc*—and surface-attached strains—i.e, *Oscillatoria*, *Phormidium* and *Lyngbya* [69]. A variety of chemical types can be synthesized in cyanobacteria including alcohols and fatty alcohols, fatty acids, hydrocarbons and industrial enzymes [70]. Figure 12.2 shows the chemical types and biofuels produced from cyanobacteria.

The selection of a suitable microalgae strain is a vital step in biofuel production. The behavior of selected strains under biochemical, genetic and transcription strategies are of great importance. For instance, using biochemical stress in the cultivation stage may improve the lipid content, but it simultaneously decreases the growth rate of the microalgae strain. All microalgae species cannot be modified using genetic engineering due to several constraints, such as the availability of the genomic data, the complexity of transgenesis and the difficulty of maintaining a balance between metabolic and energy storage pathways. Cyanobacteria are of particular interest in the production of a variety of biofuels, such as ethanol, biogas, hydrogen and biodiesel, due to their simple genetic structure, low nutrient demand and broad environmental tolerance. The cultivation and harvesting of microalgae, with a focus on genetically modified species, will be reviewed in the following section.

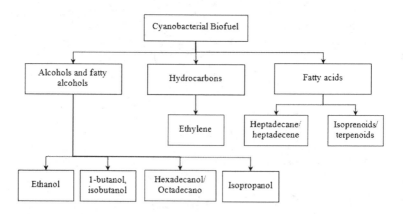

FIGURE 12.2 The chemical type and the biofuels produced from cyanobacteria.

12.2.2 CULTIVATION AND HARVESTING OF GENETICALLY MODIFIED MICROALGAE

Microalgae are photosynthetic microorganisms that need light and CO_2 as energy and carbon sources, respectively. Microalgae grow under different culture conditions of autotrophic, mixotrophic or heterotrophic [13, 71]. Autotrophic microalgae cultivation is a sunlight-driven system that converts CO_2 into lipids and other valuable components. Autotrophic cultivation is not a suitable choice for transgenic microalgae due to the possible threat to the environment and health of leaving species [72]. The limitation of the autotrophic method can be overcome by using heterotrophic cultivation. Heterotrophic cultivation takes advantage of using organic carbon sources such as glucose and acetate [73]. Tabernero *et al.* [74] conducted a comprehensive evaluation into the economic potential of implementing the heterotrophic cultivation of microalgae. It was stipulated that the heterotrophic cultivation of microalgae is not viable for industrial production of biofuel due to the large consumption of organic compounds and use of photobioreactors (PBR). In order to adopt genetic modification in heterotrophic cultivation, it must be determined that strain modification should target up or downstream metabolism [66].

The mixotrophic mode offers a unique opportunity to combine the advantages of autotrophic and heterotrophic cultivation to deal with the above issues [75]. In the mixotrophic method, microalgae can drive the autotrophic mode by receiving inorganic carbon from photosynthesis, as well as heterotrophic form by using organic carbon sources such as glucose, glycerol, acetate, maltose, sucrose and fructose. Isleten-Hosoglu *et al.* [76] observed that the lipid accumulation in the heterotrophic cultivation of the *Chlorella saccharophila* was three times higher than that in the autotrophic condition. Wang *et al.* [77] indicated that the biomass productivity of *Dunaliella salina*, *Nannochloropsis oculata*, *Spirulina platensis*, *Chlorella sorokiniana* and *Scenedesmus obliquus* in mixotrophic culture were 2.2, 1.4, 3.8, 2.4–4.2 and 8.7 times that under photoautotrophic condition.

The photosynthetic products in microalgae are accumulated in the form of cell structural or storage components. The product accumulation in microalgae can be optimized by nutrient deprivation. Nitrogen (N) and phosphorus (P) are two main macronutrients for the growth and reproduction of algae that are typically consumed in relatively large amounts. Ample reports have discussed the effect of N and P limitation on microalgae production [55,71,78]. It is stated that the productivity and lipid accumulation under nutrient depletion varies based on strain type and culture condition [73]. Přibyl *et al.* [79] tested ten different *Chlorella* genuses under specific conditions, including N deficiency in autotrophic cultures, to evaluate the lipid accumulation growth. The results showed that *Chlorella vulgaris* (strain CCALA 256) had the highest lipid accumulation and biomass productivity, whereas *Chlorella sorokiniana* had the lowest productivity rate. The N and P depletion in *Chlorella pyrenoidosa* cultivation increases the productivity and lipid accumulation 50.32% and 34.29% of dry cell weight (DCW), respectively [80]. Mata *et al.* [81] analyzed the influence of adding N into *Dunaliella tertiolecta* culture medium. It was shown that a tenfold increase of N can increase the lipid accumulation three times that of the rate in a standard culture medium. Yeesang and Cheirsilp [73, 82] cultured four *Botryococcus* spp. to assess the effect of N deficiency on lipid accumulation. It was shown that a higher rate of lipid accumulation was achieved in the N-depleted condition compared to the one in an N-rich medium. Finally, Roleda *et al.* [83] cultivated six different diatoms of *Thalassiosira pseudonana, Odontella aurita, Nannochloropsis oculata, Isochrysis galbana, Chromulina ochromonoides* and *Dunaliella tertiolecta* in an N-depleted culture. Only N. *oculata* had a significant increase in lipid content compared to the nutrient-rich medium. The number of research articles discussing the effect of P concentration on lipid accumulation is limited [73]. Xin *et al.* [84] studied the influence of P concentrations on growth and lipid accumulation of *Scenedesmus* sp. The study showed that lipid content increased by 53% under P starvation. Chen *et al.* [85] determined the nutritional requirements of *Dunaliella tertiolecta.* Deprivation of P had a lesser effect on lipid accumulation compared to N. Furthermore, the starvation of iron or cobalt improved the accumulation of neutral lipids on D. *tertiolecta.* Other nutrient particles in the production of microalgae include macronutrient Ca, Na, K and Mg, as well as microelements such as Zn, B, Co, Mn, Fe and Mo [86]. However, the N and P content in the medium has a much higher impact on lipid accumulation compared to the other micro- and macro-nutrients [85]. Many microalgae strains do not produce large amounts of lipids during logarithmic growth. In order to improve the productivity of the structural lipid, environmental stresses such as a lack of N, high salinities and a high amount of irradiance are applied to slow down their proliferation and start producing lipids [87]. Overexpression of the lipid synthesis pathway genes is a solution to slow down microalgal proliferation. While the overexpression of the genes involved in fatty acid synthesis has had little successes, some interesting results have been achieved through the overexpression of genes involved in TAG assembly [88]. On the other hand, inhibiting lipid catabolism may also result in reducing proliferation and biomass productivity. In Table 12.3, some strategies for the genetic modification of the lipid metabolism are presented.

Light intensity is another important parameter for lipid accumulation and fatty acid composition. The reaction of microalgae for different light intensities can be defined by three phases of photolimitation, photosaturation and photo-inhibition that the growth rate in the associated light intensity is increased, independent and declined, respectively [58, 94]. However, the influence of light intensity and light wavelength on metabolic regulatory and growth rate of algae are species-dependent and even strain-specific [58]. Atta *et al.* [95] investigated the effect of the light quality and the intensity on lipid content and biomass productivity of the *Chlorella vulgaris.* It was observed that the maximum lipid accumulation was achieved under the blue light with an intensity of 200 μmol m^{-2} s^{-1}. Teo *et al.* [96] studied the effect of light wavelengths on productivity and lipid content of *Tetraselmis* sp. and *Nannochloropsis* sp.

Genetic engineering tools are widely adapted to increase photosynthetic efficiency to effectively capture light energy [97]. On the other hand, several studies have been conducted to genetically modify the accumulation through impairing photosynthetiec machinery [66]. Reducing the number of light-harvesting antenna complex (LHC) pigments or lowering the chlorophyll antenna size is used in some studies to overcome the light-saturation effect [98]. Genetic modification could reduce the production cost in FGB by 50% or more—*e.g.*, astaxanthin, fucoxanthin,

TABLE 12.3
Genome editing approaches used for increasing lipid

Algae strain	Gene/platform involved	Nutrient condition	Result	Reference
Nannochloropsis oceanica CCMP1779	Overexpressing DGTT in	Nutrient-replete medium	increased production of TAG	[89]
Synechococcus elongatus PCC 7942	bicA	N starvation	Accelerated development of biosolar cell factories	[90]
Phaeodactylum tricornutum	Overexpression of Phaeodactylum tricornutum ME (PtME)	N-deprivation	Increased neutral lipid content	[91]
Chlamydomonas reinhardtii	Overexpressing a Dof-type transcription factor	Nutrient deficiency	Increased total lipids an higher proportion of specific fatty acids	[92]
Phaeodactylum tricornutum Pt4	Co-expression of a yeast diacylglycerol acyltransferase (ScDGA1) and a plant oleosin (AtOLEO3)	N stress	Increase in TAG levels	[93]

carotenoid and polyunsaturated fatty acids contents can be doubled or tripled by genetic engineering. Applying safe genetic engineering techniques such as mutagenesis or self-cloning for the production of industrially valuable algal products may decrease biosafety concerns [99].

Algal harvesting is the process of separating or detaching of algae from its growth medium [100]. Microalgae harvesting methods intensely depend on the type of target microalgae, the density and cell size of the algae, the characteristics of the final product and the reusability of the culture medium for the next cultivation cycle [101]. The harvesting methods can be from any of the chemical, biological, mechanical or electrical power classes [102]. Harvesting process of microalgae can be divided into two main stages, including thickening and dewatering [100]. The thickening stage is aimed at increasing the solid concentration of microalgal suspension and reducing the processing volume. The obtained sludge from the thickening process is still quite slushy and needs further processing to proceed into the drying stage, which is carried out at the dewatering stage. There are a wide variety of thickening and dewatering methods for solid–liquid separation, including coagulation, flocculation, centrifugation, filtration, flotation or a combination of these methods [103]. One of the concerns in the harvesting stage of the genetically modified (GM) strains relates to the residual water from the harvesting phase. Discharged water from the harvesting process may contain plasmid or chromosomal DNA from the transgene GM algae that may result in horizontal gene transfer [21]. To deal with this problem, several treatment methods are advised that are reviewed in Abdullah *et al.* [21] and Beacham *et al.* [104].

Genetic modification is a viable solution to improve the yield of valuable products in microalgae at a reduced cost. Several approaches are available for improvising the lipid content in microalgae. These enhancement strategies can be generally divided into four groups that include improving photosynthetic efficiencies, biomass productivity, diversity and the ability to thrive in diverse ecosystems. Among the aforementioned fields, enhancing photosynthetic efficiencies and biomass productivity are the most researched. Different stress conditions, including the addition or depletion of nutrients, light and salinity level are used for the cultivation of the genetically modified strains. However, health and environmental risks are the main concerns in the cultivation and harvesting of GM strains. Further discussion on the biosafety risks of the FGB is presented in the following subsection.

12.2.3 ENVIRONMENT AND SUSTAINABILITY

It is well known that the emission of greenhouse gasses (GHGs), such as fluorinated gases (F-gases), nitrous oxides (N_2O), methane (CH_4) and carbon dioxide (CO_2) are the main causes of climate change. These gasses block escaping heat from the atmosphere, resulting in global warming. The consumption of fossil fuels is the main cause of GHG emissions, but biofuels are a promising alternative to meet humanity's needs for clean energy in transportation, power generation and heating. The GHG emission attributed to biofuel combustion is much lower than that of conventional fuels, hence it can be engaged with climate change mitigation strategies [105]. Biofuels can be used as a complete substitute or admixture for fossil fuels [60].

12.2.3.1 Environmental Effect

Microalgae-derived biofuel is one of the most promising renewable energy sources, not only due to its lower GHG emissions but also owing to its sequestration of CO_2 [106]. The CO_2 sequestration in microalgae is 10–50 times more than many terrestrial plants, while a higher concentration of CO_2 results in a higher yield of lipids. However, the low concentration of atmospheric CO_2 is still a challenge for algal sequestration by cultivation. Algal-based wastewater treatment offers an efficient and cost-effective tool to eliminate organic and inorganic contaminant wastes from wastewater. The wastewater from these sources can be divided into organic and inorganic compounds. Carbon-containing biodegradable substances are the main constituent of organic waste, while inorganic waste contains nitrate, phosphate and heavy metal, etc. [107]. Various successful studies have been reported for municipal, agricultural and industrial wastewater treatment through the cultivation of microalgae [108–110]. Most municipal wastewater is from domestic sources rich in phosphate (PO_4-), ammonia (NH_3) and other essential nutrients and metals. Different microalgae species have been used for wastewater treatment. Komolafe *et al.* [111] grew *Desmodesmus* sp. and a mixed culture dominated by *Oscillatoria* and *Arthrospira* on domestic wastewater. Higher biomass concentration and lipid were achieved in monoculture and mix-culture conditions. Eighty-two percent of N and 61% orthophosphate were removed from the wastewater during the culturing period. Wang et al. [112] evaluated the cultivation of *Chlorella* sp. on municipal wastewater. The highest P and NH4–N removal was obtained in the wastewater before and after primary settling scenarios (90.6%, 82.4%), respectively. It was stated that the metal ions, especially Mg, Al, Fe, Mn and Ca were removed efficiently. Khalid *et al.* [113] assessed the growth rate and nutrient removal efficiency of *Characium* sp. cultured in agricultural wastewater. Findings showed that the macroalgae *Characium* sp. can remove 80.0% and 89.9% of total N and of P, respectively. Kamyab *et al.* [114] evaluated the lipid production in *Chlorella Pyrenoidosa* microalgae cultivated on palm oil mill effluent agricultural wastewater. The maximum removal of organic carbon was reported in the culture with continuous illumination (65%). In recent years, a variety of research has been conducted studying industry wastewater treatment by microalgae, including the wastewater from the paper industry effluent [115], the textile industry effluent [116] and the petroleum industry [117].

While the algal biofuels are generally believed to be more environmentally friendly than fossil fuels, they still face significant obstacles in large-scale cultivation and commercialization. Among the constraints involved in the commercialization of FGB, the potential health and environmental risks in the mass production of microalgae have been mostly neglected [118]. Hewett *et al.* [119] investigated the risks related to the health and environment of the modified algae for biofuel production. It was stated that the environmental risk issues associated with the application of synthetic biology in algal biofuel production were in competition with native species, changes in natural habitat, toxicity and horizontal gene transfer. The assessment was obtained from interviews with professionals and scientists engaged in synthetic biology. Competitive growth of native and wild species was the topic of several studies to evaluate the potential risks posed by open-pond cultivation of modified microalgae. The United States Environmental Protection Agency (US EPA) [120] conducted an experiment aimed at understanding the risk attributed to the possible release of genetically engineered *Acutodesmus dimorphus* into the surrounding environment during the open-pond cultivation period. The findings showed that the GM algae were not able to outcompete native strains and the effect of uncontained culturing modified algae on native species and their composition was negligible. Russo *et al.* [121] carried out a competitive growth experiment of *Chlamydomonas reinhardtii* strains (CC-124) and a mutant cell of (CC-4333). It was observed that, due to the slow transition phase, the mutant cells outcompete wild strains; hence, they pose an insignificant risk in an escape case. Aquatic invasion has a negative impact on biodiversity and some ecosystem functions. Algal blooms by the toxic dinoflagellate *Alexandrium minutumalong* on the French coast in 1985 is one of the examples of environmental change and depletion cases that contributed to the enrichment of waterborne nutrients in marine waters. Several poisoning cases were reported as a result of the produced neurotoxins by algae species [122]. Horizontal or lateral gene transfer is defined as a transferring mechanism of genetic information of an organism from one genome to another in a non-genealogical manner [123]. Snow and Smith [124] addressed the potential environmental risk associated with horizontal gene transfer of genetically engineered algae and cyanobacteria. The risks related to human health fall mainly into four groups of antibiotic resistance, allergies, pathogenicity or toxicity and carcinogens. Further discussion on the health-related risks of algal feedstock can be found in [119, 125,126].

12.2.3.2 Sustainability

Microalgae are a more sustainable choice for biofuel production due to the shorter life cycle of the algal feedstock compared to the terrestrial crops. The lifecycle range of algae biomass ranges from one to four days, while food crops need between 90 and 180 days for growth [127]. Several studies were conducted to identify the constraints on commercial production of the FGB. The production cost of FGB is expensive compared to fossil-based fuels. Hence, it is required to increase competence by reducing the expenditure on raw material procurement and increasing production efficiency [128]. A substantial amount of expenditure on raw material is spent to purchase carbon dioxide. Hence, culturing microalgae feedstock is not economically feasible without access to free carbon dioxide sources. At least 1830 kg of carbon dioxide is needed to produce 1000 kg of microalgae [129]. Low-concentration carbon dioxide emissions cannot be directly used in microalgae cultivation. Providing the required amount of carbon dioxide is one of the main constraints in large-scale cultivation of microalgae. The flue gases from coal-fueled power plant, or cement manufacture factory is of the concentrated source of carbon for algae production [128].

Nowadays, water pollution and groundwater recharge deficiency are overriding global concerns and causing damage to the ecosystem. Zhang *et al.* [130] analyzed the water footprint of microalgae and terrestrial plants. It was observed that a green water footprint in the production of the microalgae was almost a quarter of the average green water footprint in three terrestrial plants. Furthermore, through the implementation of recycling and reusing of the discharged water from the harvesting phase, up to 90.2% of the discharged water can be returned into the production line [131]. Crofcheck *et al.* [132] stated that the media used to exploit biomass can be recycled back into the process and reused up to four times. Providing an appropriate amount of freshwater for the large-scale FGB production poses a significant challenge. However, besides the growth rate, there are some other parameters for assessing the sustainability of a product, which include the environmental, energetic and economic impacts of the whole to achieve a concept of sustainability.

Adopting a system to almost fully recycle the nutrients in microalgae cultivation could improve the efficiency in terms of economic performance and minimize material input [133]. Anaerobic digestion (AD) is an emerging biological process that has been widely used for wastewater treatment and nutrient recycling. AD can be used for large-scale recycling of different types of organic wastes [134]. Crofcheck *et al.* [132] investigated the effect of using nutrient-containing residue of algae biomass harvesting to enrich the cultivation substrate. Algal digestate in AD proved to be a sufficient replacement for an N and P source to replenish nutrients. Barbera *et al.* [135] used the aqueous phase from flash hydrolysis of *Scenedesmus obliquus* as a cultivation medium. It was found that the nutrient recycling from flash hydrolysis media was a feasible choice for algae cultivation. Furthermore, it was shown that *Scenedesmus* sp. was able to use the recovered recycled nutrients. Zhang *et al.*

[136] compared the efficiency of different nutrient recycling techniques used in algal biofuel production. Two AD and hydrothermal liquefaction (HTL) methods were shortlisted for further analysis. The results showed that AD had better performance compared with HTL in terms of maturity of the technology as well as nutrients and energy recycling. Talbot [137] conducted a comparative study to evaluate nutrient recovery in rapid and conventional HTL for microalgae cultivation. It was observed that rapid HTL had better performance in nutrient recovery by replacing almost 50% of the recycled P and N.

Kenny *et al.* [138] conducted a model of microalgal growth mechanism to obtain the optimal condition for the industrial production of algal biofuel. It was observed that open-pond cultivation is the only choice for massive production. Accordingly, it was found that genetic modification is the necessary requirement for economically viable biofuel production. The open-pond growing of microalgae has been widely practiced worldwide [18,139]. However, using open-pond for culturing genetic strains poses a significant risk to the environment. Matsuwaki *et al.* [90] investigated the spread of genes into the surrounding environment by detecting the gene of *Pseudochoricystis ellipsoidea* MBIC 11204 in vessels installed at variable distances from a culturing open-pond. It was observed that the related genes of the experimented microalgae were found up to 150 m from the cultivation pond. Wind was an important factor in spreading genes into the surrounding area. The sustainability of the FGB in terms of economic and environmental impacts are still the main drawbacks of biofuel production.

Several studies have been conducted aiming to elevate the sustainability level of the FGB production chain. Compliance to the GHG emissions' reduction targets, security of the water and food supplies and environmental conservation are the main concerns in sustainable FGB production. Several successful reports are published on combining microalgae cultivation with municipal, agricultural and industrial wastewater treatment. Microalgae can consume nutrients, heavy metals and CO_2 for growth. The water footprint of microalgae is much less than that of other biofuels. Furthermore, the recycling of the cultivation medium can significantly reduce the water footprint in microalgae production. Several studies have emerged looking at the impact of GM microalgae on ecosystem and environment sustainability. The health and environmental risks of GM microalgae are still the key concerns in the FGB production phase.

12.2.4 INDUSTRIALIZATION AND ECONOMY

Biofuel is considered to be the main alternative to fossil fuel as the future's green combustible source. However, biofuel production does not economically compete with petroleum-based fuel [140]. Cost reduction is a step closer toward achieving an economically sustainable production of FGB. FGB productivity is a function of algae growth rate and the oil content. A fast growth rate enables low cultivation cost, higher yield and a lower risk of contamination per harvest. Hence, bulk biomass culturing of relatively fast-growing microalgae can be an outstanding alternative to current, unsustainable production methods. In addition, the available fatty acids in microalgae are not equally favorable for algal biofuel production. The microalgae oil consists of four different fatty acids: free fatty acids (FFA), monounsaturated fatty acid (MUFA), polyunsaturated fatty acids (PUFA) and saturated fatty acid (SUFA). The existing types of fatty acid in microalgae determine the quality of the potential extracted biodiesel [141]. Finally, the choice of an easier harvest is preferred, owing to the saving in production costs.

Cost analysis of biofuel production is a case-sensitive tool due to the variation in the applied technologies and the specific microalgae strain used for biofuel extraction. Furthermore, a shortage of the published information from the commercial producers and lack of facilities for bulk cultivation made the cost estimation of algal biomass more elusive [142]. The available cost-evaluations generally have used pilot experiments that are unreliable and overly optimistic when dealing with mass production cost analysis. Hence, the economic output of such experiments must be interpreted with care and require substantiation with other studies [143]. Wide varieties of algal strain have the potential for industrial production of high-value products with pharmaceutical, nutraceutical, cosmetic, feed additives, food colorant-stabilizer and biofuel production. Among them, microalgae biofuel production needs the largest production scale to become economically feasible and offset the capital investment involved in cultivation, harvesting and extraction facilities [144]. The production price for the two-stage cultivation of *Haematococcus pluvialis* is $103.6 /kg in small facilities. This price can be reduced by using a fast-growing strain within one stage in a larger tubular reactor facility into $27.6/kg. The price to produce *Spirulina* biomass in open-pond is $5.0/kg. The price could be even cheaper in very large new facilities in countries like India and China that have low wages and market-friendly policy environments [142].

Sunlight, nutrient and water medium are the main components in microalgae cultivation that are available by open-pond cultivation. Hence the open-pond is the most cost-efficient cultivation system. Nonetheless, outdoor cultivation is vulnerable to contamination and climate conditions [127]. Davis *et al.* [145] compared the capital and operational costs of open-pond and PBR. It was reported that the installation cost of PBR is much higher than an open-pond system. However, the consumed water in PBR is just 30% of the water required for open-pond cultivation. It was observed that the depreciation of the capital cost contributes to 80% of the overall cost of the product in PBR. However, the capital cost in open-pond was evenly allocated without apparent lasting effect [140]. The involvement of the capital cost for open-ponds and PBRs in the overall price of the product is shown in Figure 12.3.

The produced biofuel from microalgae is an important factor in the economic viability of a product. Microalgae biomass is a valuable product that is used for the production of a wide variety of biofuels, such as biohydrogen, biodiesel, ethanol and butanol. In Figure 12.4, various biofuels from microalgae, as well as the processing approach of raw materials, are presented. Biodiesel is the most widely used biofuel in diesel engines. Low sulfur, aromatic content and a higher flash point have made biodiesel a promising substitute for gasoline in the transportation sector. Depending on the scenario used for the production of biodiesel, the cost ranged between $1.68/L^{-1} and $75/L^{-1} [146]. It was stated that the production cost of biodiesel can be reduced by large-scale outdoor cultivation into the range of $0.42 L^{-1} to 0.97 L^{-1} [146]. Biohydrogen is denoted as the fuel of the future due to its high conversion efficiency

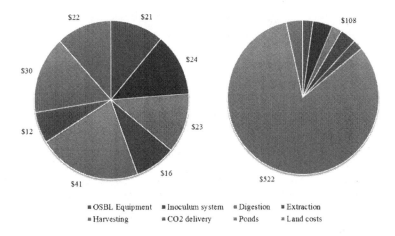

FIGURE 12.3 The cost distribution for open-ponds and PBRs [140].

and nonpolluting nature [147]. The target commercial cost of hydrogen ($0.30 /kg) corresponds to the equivalent energy content for gasoline priced at $2.5 /GJ in a competitive market [148]. Show *et al.* [149] scaled up a pilot reactor to conduct a techno-economic analysis of biohydrogen production. It was observed that the construction cost was considerably lower than the one stated in the literature. The economic analysis showed that biohydron can be produced at a cost between $10/GJ and $20/GJ.

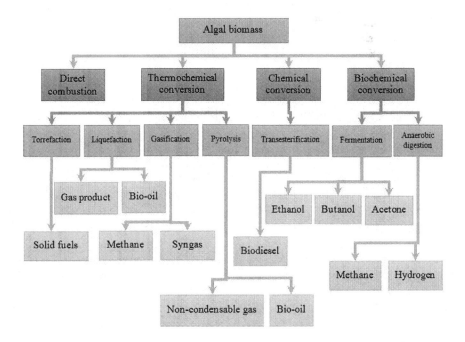

FIGURE 12.4 The flow of algal biofuel including the processes and main products.

Bioethanol is an alcohol-based adjunct for gasoline that reduces the emissions in internal combustion engines. In an industrial-scale production conducted by Algenol, the production cost of bioethanol was estimated to be almost $0.79/L [150].

Combining algae cultivation with wastewater treatment has been recognized as a promising pathway for cost-effective and sustainable biofuel production. It is estimated that the purchase of nutrients and carbon dioxide in the commercial-scale cultivation of microalgae accounted for almost 10–30% of the total production costs [151]. However, there are still several challenges that must be overcome for an industrial take-up of the technology. The incorporation of large-scale algae cultivation systems within urban wastewater treatment plants and the variation of the composition and nutrient ratio, as well as the contamination impacts, are the concerns that hamper using wastewater as a nutrient source of algae growth [151].

In the period of 1978 to 1996, the U.S. Department of Energy (DOE) funded a program for the development of algal biofuel for transportation purposes. In Japan, the government financed a large research project entitled "Biological CO_2 Fixation and Utilization" from 1990 to 1999 [144]. In the 2000s, several startups were established with a focus on algal biofuel production. Increasing the demand for algal biofuel driven by environmental concerns has convinced the governments to invest in microalgae production projects and take algal biofuel into account in their energy policies. An example is the EnAlgae project by the EU to finance home-grown microalgal biomass production [152]. The years 2008 and 2009 were very important in the history of algal biofuel, where it drew serious attention from nearly 150 commercial sectors worldwide [153]. In July 2009, Exxon Mobil Corporation announced a USD$600 million investment in algae-based biofuels [154]. In August 2009, BP reserved USD$10 million to bring large-scale algae biofuels to commercialization [154]. Sapphire Energy, Inc. constructed an algae cultivation farm in New Mexico, consisting of 21 acres for research and development (R&D) and to bridge the transition from lab to field [144]. Although the program yielded some success, none of the projects have proved economical on a large scale.

Using microalgae as biomass is a promising approach for biofuel production due to the fast growth rate and high lipid contents. However, algal biofuel production is too costly to compete with fossil fuel. Cost of algal biofuel production depends on the applied technologies and the specific microalgae strain used for biofuel extraction; hence, the cost estimation of algal biomass is an elusive issue. The available cost analysis for biofuels are based on pilot-scale data, and the obtained results are unreliable and overly optimistic when dealing with mass production. It is even worse for GM algal biomass, where a limited number of publications are available to deal with the mass production of modified strains. Furthermore, the environmental and health risks of GM algal biomass are topics that are not covered in much depth in the literature, and the economic aspects of such risks are also radically unknown. The environmental and economic opportunity of the algae-based biofuel's by-products contributes to the reduced waste amount and more sustainable resource consumption. In GM biomass, the strains may contain plasmid or chromosomal DNA from the GM algae that may increase the risk of horizontal gene transfer. Treatment or disposal of the by-products and discharged water from the cultivation and harvesting process of GM microalgae and their economic impact are the topics that need to be addressed.

12.2.5 GENETIC MODIFICATION

Enhancement of productivity and lipid accumulation is the easiest way to reduce the cost, nutrient consumption and water footprint. Genome editing methods are widely used to increase the productivity and lipid content of microalgae. Presently, three types of genome editing tools of zinc-finger nuclease (ZFN), transcription activator-like effector nucleases (TALEN), and clustered regularly interspaced palindromic sequences (CRISPR/Cas9) are widely used for genome editing of microalgae strains [144]. The first genome editing experiment in microalgae was reported on *Chlamydomonas reinhardtii* using ZFN [155]. Genome editing of *Phaeodactylum tricornutum* (with TALEN [156,157] and with CRISPR/Cas9 [158]), *Nannochloropsis oceanica* (with CRISPR/Cas9 [159]), *Chlamydomonas reinhardtii* (with TALEN [160] and CRISPR/Cas9 [161]) has been successfully demonstrated. Some examples of genome editing in microalgae are shown in Table 12.4. Using CRISPR/Cas9 has significantly improved the potential for the modification of microalgal metabolism. However, CRISPR/Cas9 is in its infancy for microalgae due to the relatively small number of published studies [66]. The mechanisms of genome editing have been reviewed elsewhere [162, 163].

The genetic modification strategies for stress mitigation of cyanobacteria have been successfully employed so far. The first exogenous DNA transfer in *Anacystis nidulans* in the 1970s opened the way to use recombinant DNA technology for the construction of genetically modified cyanobacteria. The most important genetically engineered methods to enhance production in stress-related conditions of outdoor cultivation include sunlight irradiance, medium temperature, salt content and chemical or solvent composition [164]. Ruffing [164] discussed the genetic tools and strategies to physiologically enhance cyanobacteria for large-scale production. The availability of extensive genomic information and a well-established genetic model make cyanobacteria a favored target for genetic engineering [70]. However, by increasing the titers, the toxic compounds formed in cyanobacteria reduce productivity and slow down growth [165]. Further details about strain selection and genetic engineering of cyanobacteria could be found in review papers by Kitchener and Grunden [165], Xie *et al.* [70] and Mazard *et al.* [125].

Recently, four modern genome editing technologies have revolutionized the field of genetic engineering. These include: ZFN, TALEN, meganucleases and CRISPR. Compared to other gene-editing technologies, CRISPR has gained both a lot of attention and widespread use due to its design simplicity, transfection efficiency and improved performance in multiplexed mutations. However, CRISPR is in its infancy for microalgae due to the relatively small number of published studies. More research should be done to perfect the conclusion about CRISPR. In the next section, the classification and meta-analysis of the bibliographic information on the FGB are presented.

12.3 CLASSIFICATION OF THE PUBLICATIONS BASED ON THEIR APPLICATION AREA

Published papers on FGB are diverse and scattered over different disciplines and a large variety of scientific journals. In order to integrate the findings from a wide range of distinct, disparate sources and contribute them to a body of knowledge on

TABLE 12.4
Genetic engineering in microalgae [144, 163, 166]

Species	Delivery/ nuclease	Gene/marker	Comment	Reference
Chlamydomonas reinhardtii CCMP2561	Electropora-tionCRISPR/ Cas9	Hygro, mGFP,FKB12, and Gluc	First application of CRISPR/Cas9 in microalgae	[167]
Chlamydomonas reinhardtii strain CC4350	Glass beadsZFN	COP3 encodes light-gated cation channel: channelrhodopsin-1 (ChR1)	First application of CRISPR/Cas9 in diatoms	[155]
Chlamydomonas reinhardtii CC124	Electropora-tionCRISPR/ Cas9	MAA7, CpSRP43 and ChlM	First application of CRISPR/Cas9 in diatoms	[168]
Chlamydomonas reinhardtii CC4349	Electropora-tionCRISPR/ Cas9	ZEP and CpFTSY	Production of two-gene knockout mutant	[169]
Chlamydomonas reinhardtii CC400	Glass bead-scRISPR/dCas9	PEPC1 and RFP	CRISPRi in Chlamydomonas	[170]
Phaeodactylum tricornutum CCMP2561	Bombardment-TALEN	UGPase/NAT gene	Triacylglycerol accumulation was increased	[171]
Nannochloropsis gaditana CCMP1894	Electropora-tionCas9 Editor line	ZnCys TF BSD	Flow cytometry, Western blotting, PCR were used for mutation detection	[172]
Synechococcus elongates UTEX 2973	Conjugation-CRISPR/Cas9	nblA	Genome editing using CRISPR/ Cas9-based nucleases	[173]
Synechococcus elongates PCC 7942	Conjugation-CRISPR/Cas9	glgc/Gm^R gene	Increase of succinate Production	[156]

the FGB a thorough review is conducted. The main search keywords were selected in consultation with two experts. After carrying out three steps of literature search, eligibility identification and inclusion, the selected papers were included into the meta-analysis. The paper's research area in the Scopus database was extracted, and the results showed that the papers belong to one of five categories shown in Table 12.5. A large variety of research is conducted on assessing and analyzing genomic data and discussing emerging approaches for genetic modification and the metabolic engineering of microalgae in the "biochemistry, genetics, immunology, microbiology and molecular biology" application area. The scattering of the FGB papers in

TABLE 12.5
Distribution of the papers based on research areas

Research area	No.	(%)
Biochemistry, genetics, immunology, microbiology and molecular biology	76	35%
Chemical and other engineering disciplines (except genetic engineering)	53	24%
Agricultural, environmental and biological sciences	48	22%
Energy	22	10%
Others	18	8%
Total	217	100%

the aforementioned area is nearly 35% of the papers. The engineering applications are in the second place and the most papers in this category are explicitly in the chemical engineering area. "Agricultural, environmental and biological sciences" and "energy" applications have a 22% and 10% share of the content, respectively. Even though the number of journals in the energy and environmental discipline is lower compared to the genetic, microbiology and engineering aspects, the average of the papers in the those journals is much higher. For example, "Renewable and Sustainable Energy Reviews" in the field of "Energy" has 8% of all papers in the literature (See Table 12.6). Hence, it is very important to study the paper within the application area and their source title, simultaneously.

The presented results in Table 12.6 show that the selected scholarly papers are from 64 international journals. Based on the results, "Renewable and Sustainable Energy Reviews" and "Bioresource Technology" have the first place over the journals presented in the table by 8%, while "Current Opinion in Biotechnology" was ranked second with six papers and a percentage of 6%. In third place, "Algal Research," "Applied Microbiology and Biotechnology," "Biotechnology Advances," "Biotechnology Journal," and "Frontiers in Microbiology" are five journals with three articles in each. The distribution of the papers based on the journal is presented in Table 12.6. The large variety of papers is mainly due to the high distribution of the FGB in many fields.

The distribution of papers based on the research area and publication year is shown in Figure 12.5. In general, the first paper in FGB was published in 2008 in the field of genetic modification and cultivation and harvesting. Except for a reduction during the period of 2014–2016, the research in genetic modification is shown to have an upward trend until 2017, where it reached a peak of 14 papers. Regarding the scientific papers on cultivation and harvesting, the first paper was also published in 2008 and, except for 2013–2015, had an upward trend until 2017, when it reached a peak of 12 papers. In the case of strain selection and related papers, these reached a peak of 12 in 2013 and 2017. The research on "environment and sustainability" and "industrialization and economy" have been constantly fluctuating from one to five. The study shows that the environmental and economic aspects of FGB production have attracted the least attention among research areas.

In order to investigate the reason for the drop in the number of publications in the period of 2013–2015, a study was conducted. Governments play a significant

TABLE 12.6

Distribution of the papers based on the published journal

Name of journal	Application area	Number	Percentage
Renewable and Sustainable Energy Reviews	Energy	8	8%
Bioresource Technology	Energy; Environmental Science; Chemical Engineering	8	8%
Current Opinion in Biotechnology	Engineering; Biochemistry, Genetics and Molecular Biology; Chemical Engineering	6	6%
Algal Research	Agricultural and Biological Sciences	3	3%
Applied Microbiology and Biotechnology	Biochemistry, Genetics and Molecular Biology; Immunology and Microbiology	3	3%
Biotechnology Advances	Biochemistry, Genetics and Molecular Biology; Immunology and Microbiology; Chemical Engineering:	3	3%
Biotechnology Journal	Immunology and Microbiology; Biochemistry, Genetics and Molecular Biology	3	3%
Frontiers in Microbiology	Medicine; Immunology and Microbiology:	3	3%
International Journal of Environmental Science and Technology	Agricultural and Biological Sciences; Environmental Science:	2	2%
Journal of Applied Microbiology	Biochemistry, Genetics and Molecular Biology; Immunology and Microbiology:	2	2%
Journal of Biotechnology	Biochemistry, Genetics and Molecular Biology; Immunology and Microbiology; Chemical Engineering	2	2%
Marine Drugs	Pharmacology, Toxicology and Pharmaceutics	2	2%
Microbial Cell Factories	Biochemistry, Genetics and Molecular Biology; Immunology and Microbiology; Chemical Engineering	2	2%
Trends in Biotechnology	Biochemistry, Genetics and Molecular Biology; Chemical Engineering	2	2%
Trends in Plant Science	Agricultural and Biological Sciences	2	2%
Others (the remaining 49 journals with one paper in each)	–	49	49%
Total	–	100	100%

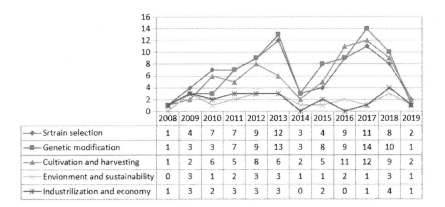

	2008	2009	2010	2011	2012	2013	2014	2015	2016	2017	2018	2019
Srtrain selection	1	4	7	7	9	12	3	4	9	11	8	2
Genetic modification	1	3	3	7	9	13	3	8	9	14	10	1
Cultivation and harvesting	1	2	6	5	8	6	2	5	11	12	9	2
Envionment and sustainability	0	3	1	2	3	3	1	1	2	1	3	1
Industrilization and economy	1	3	2	3	3	3	0	2	0	1	4	1

FIGURE 12.5 The distribution of papers based on topic and publication year.

role in the development of new and improved technologies through supports such as subsidiaries, finance, coordinates and legislature [174]. In the United Kingdom, for instance, nearly 66% of world-class research is supported by the government [175]. To this end, the universities actively involved in research on FGB were investigated based on the number of publications a year, in order to identify a possible correlation between the number of publications and the role countries have in the elaboration of the papers. The analysis results show that nearly one-third of the publications are carried out by universities or research centers in the United States. Figure 12.6 shows the number of the five countries that have the biggest contribution to the research in the field of FGB.

As stated in subsection 12.2.4, the contracts with industry and the grants from the government fuelled rapid progress in the research on algal biofuel and adoption of GM biomass in the United States in 2008–2009. Although the program yielded

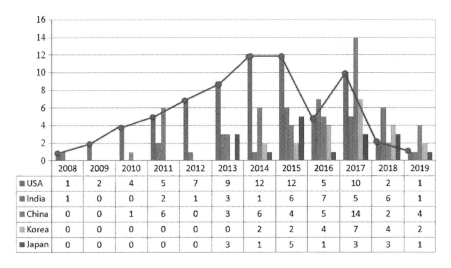

	2008	2009	2010	2011	2012	2013	2014	2015	2016	2017	2018	2019
USA	1	2	4	5	7	9	12	12	5	10	2	1
India	1	0	0	2	1	3	1	6	7	5	6	1
China	0	0	1	6	0	3	6	4	5	14	2	4
Korea	0	0	0	0	0	0	2	2	4	7	4	2
Japan	0	0	0	0	0	3	1	5	1	3	3	1

FIGURE 12.6 The distribution of papers based on topic and publication year.

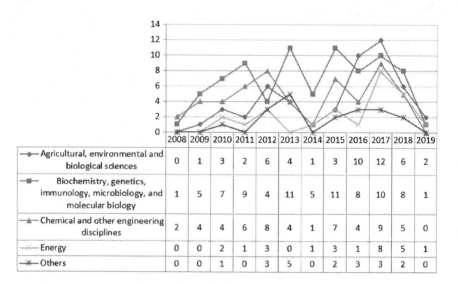

	2008	2009	2010	2011	2012	2013	2014	2015	2016	2017	2018	2019
Agricultural, environmental and biological sciences	0	1	3	2	6	4	1	3	10	12	6	2
Biochemistry, genetics, immunology, microbiology, and molecular biology	1	5	7	9	4	11	5	11	8	10	8	1
Chemical and other engineering disciplines	2	4	4	6	8	4	1	7	4	9	5	0
Energy	0	0	2	1	3	0	1	3	1	8	5	1
Others	0	0	1	0	3	5	0	2	3	3	2	0

FIGURE 12.7 The distribution of papers based on application areas and the publication year.

some success, none of the projects have proven to be economical in large-scale bio-fuel production. Hence, in the United States, support from the industry has declined as the proportion of research publications from other countries, such as China and India, has markedly increased in recent years. It is shown that for the period of 2008 to 2014 roughly coincided with the end of the stagnation period in countries such as China, India, Korea and Japan. The sharp decline in 2014 shown in Figure 12.6 is recovered in the following years by other countries, mainly China and India. After 2014, the number of publications with US contribution started to decline gradually, except for 2017. The distribution of papers based on application areas and publication years are shown in Figure 12.7.

The results shown in Figure 12.7 reveal that the number of publications on FGB in the field of "biochemistry, genetics, immunology, microbiology, and molecular biology" was the most in the two decades periods. It is nothworthy that the 100 selected papers have the least contribution to "energy" researches.

12.4 FUTURE WORKS

The results for the meta-analysis show that "strain selection" and "genetic modification" have the most number of studies, whereas the "environmental and sustainability" and "industrialization and economy" have the least. A great proportion of the selected articles deal with the algal species modified with genome editing. CRISPR genome editing techniques play an important role in the recent expansion of the studies on gene editing. It has enhanced efficiency, simplicity and performance of the process, and there is hope for further improvement in developing more safe methods for genome editing in the future.

Environmental and sustainability, as well as legitimate FGB, are topics that are covered in a handful of reviews, but specifically highlighted in two review studies

[21, 104] that investigated the environmental and health risks and mitigation strategy of FGB production. On the other hand, success in the commercialization of FGB lies in resolving the problems dealing with market uncertainties. The uncertainty and diversity in regulation discourage scientists and biotechnology companies securing their research money. Factors such as legitimacy concerns, health and environmental risks and the cost of biomass purchasing are the important problems that require a strong international commitment to reach a global consensus in providing a clear roadmap on deployments of alternative fuels and, more specifically, on the role of GM organisms in the future transition from fossil to biofuels. Safe deployment and use of GM products is a topic that caught the world's countries in a polarized and endless debate that needs to be delegated based on constraints and opportunities.

Release of GM algae into the environment is the main concern in the cultivation and harvesting steps of the FGB. The majority of literature on the environmental and health risks of GM organisms are related to terrestrial plants, and only a handful of studies have investigated the hazards presented by the discharge of GM algae from an inland open-pond cultivation medium. The literature review shows that there is no research in the field of FGB to study the potential hazards caused by the dispersion of the GM algae into open waters, such as rivers, canals and seas or underground sources. Furthermore, the literature lacks the practical strategies to monitor or control the unintended (or intended) release of GM algae into aquatic resources. Hence, further studies are required in this field to clarify the missing details and to establish a reliable risk-scoring system for algae to ensure that the exploitation in coastal, or within aquatic resources, will not harm ecosystems or human beings.

12.5 CONCLUSION

This chapter presented a goal-directed analysis of the bibliographic information on FGB production as one of the recent and reliable sources of biofuel. The FGB has great advantages compared to other biofuel generations that have been extracted from non-marinal sources and also from the third generation of biofuel. This chapter covered nearly two decades of research on FGB from 2008 to 2019, with a focus on aspects other than microbiological ones. A systematic literature search was carried out and the selected papers were included into a meta-analysis. These selected papers were divided into five categories that included strain selection, genetic modification, cultivation and harvesting, environmental and sustainability industrialization and economy. To this end, 100 studies were carefully selected about FGB production.

Most of the selected studies were published in 2013, 2016 and 2017. Based on the Scopus database, the selected papers were from 15 different application areas, which were shortlisted based on their criteria into five main groupings. In this regard, "biochemistry, genetics, immunology, microbiology and molecular biology" was the most important application area with 35 papers. Additionally, 65 international journals were considered in the current review chapter. "Renewable and sustainable energy reviews" and "bioresource technology" were ranked first among the considered journals in terms of studies on the production of the FGB.

In selecting a suitable strain, we need to consider various parameters, including genetic modification, cultivation and harvesting in order to achieve our ultimate goals. A great number of papers have studied the important parameters in selecting a specific strain for biofuel production. The availability of the genomic information and the establishing of the genetic model make some microalgae strains a more favorable target for genetic engineering. The main consideration in strain selection is the required composition for biofuel production, including the molecules of lipids, proteins, carbohydrates and nucleic acids that determine the whole process from cultivation up to the energy extraction stage.

Algae are one of the most environmentally friendly options for biofuel production. The CO_2 sequestration of microalgae is higher than many terrestrial plants. Algae eliminate organic and inorganic contaminants from the medium during growth, which is then introduced for the treatment of municipal, agricultural and industrial wastewater. Furthermore, water consumption in the production of the microalgae is almost a quarter of the average green water footprint in three terrestrial plants. Due to the shorter life cycle of the algal feedstock, microalgae is considered as a more sustainable choice for biofuel production. In this review paper, several successful studies on environmental benefits and sustainability of the produced FGB were discussed. The potential health and environmental risks in mass production of microalgae, including competition with native species, change in natural habitat, toxicity and horizontal gene transfer are the most important concerns. It was found that parameters such as wind were important factors in spreading genes into the surrounding area. Since biofuel production does not economically compete with fossil fuel, cost reduction is an important step toward an economically sustainable product. Culturing fast-growing microalgae on the largest possible production scale, combining algae cultivation with wastewater treatment, and the genetic modification of micro algae are the choices that make FGB more economically feasible.

This review chapter has some limitations, which can be considered as the objective of future studies. First, this review is focused on the production phase of the genetically modified biofuel, and the shared area with the third generation of biofuel such as harvesting and energy extraction is not detailed in this study. Secondly, due to the excellent ability of microalgae to yield a large amount of product, the focus of the reviewed studies was on biofuel production from microalgae compared to macroalgae. Hence, in this work, it was attempted to deal mainly with the production phase of microalgae. Third, since the number of papers on the environmental and health risks of genetically modified microalgae is so few, and although there was an obvious increase in the number of papers in strain selection, genetic modification and exploitation of microalgae, the number of papers in this area is constantly low from 2008 to 2019. Furthermore, the number of papers on the economy and industrialization of FGB are quite few and need to have studies conducted to investigate the solutions to commercialize the production of FGB. The interaction of the environmental and economic aspects of the FGB production is another topic that needs further investigation for the industrial-scale production of FGB.

Articles published at the end of 2019 (if any) have not been included in the present chapter because of the limited reporting time. The present review can be expanded for future studies. Another limitation is that the data were collected from journals, while the examined documents did not include conference papers, textbooks, doctoral and master's theses and unpublished papers on the FGB. Therefore, in future study, the data can be collected from these sources, and the obtained results can be compared with the data obtained and reported in this study. One more limitation is that all the papers were extracted from the journals written in English, which implies that the scientific journals in other languages were not involved in the review. However, the researchers believe that this chapter comprehensively reviewed most of the papers published by international journals. Hence, the current review chapter can provide future academic scholars with a better understanding of the production aspects of the FGB. This study can be used by academics and managers as a basis for further research. It can also help practitioners make more appropriate decisions using these approaches and be a guide to scholars, improving the discussed methodologies. The authors of this chapter carefully selected and summarized the available papers of several publishers in the Scopus database. However, a number of relevant outlets remained beyond the scope of the current study. Therefore, future researchers will be able to review the papers that were not considered in the current review. Another limitation is associated with the fact that the chapter presents a review of numerous works on the problem of using the recent studies on FGB published in various journals. The extensive diversity of the studies in terms of scope, discipline and research area, as well as the limited space of the present review obstructs the authors from including all references used in the meta-analysis.

REFERENCES

1. S. Amin, "Review on biofuel oil and gas production processes from microalgae", *Energy Conversion and Management*, vol. 50, pp. 1834
2. R. A. Lee, and J.-M. Lavoie, "From first-to third-generation biofuels: Challenges of producing a commodity from a biomass of increasing complexity", *Animal Frontiers*, vol. 3, pp. 6–11, 2013.
3. K. Ullah, V. K. Sharma, M. Ahmad, P. Lv, J. Krahl, and Z. Wang, "The insight views of advanced technologies and its application in bio-origin fuel synthesis from lignocellulose biomasses waste, a review", *Renewable and Sustainable Energy Reviews*, vol. 82, pp. 3992–4008, 2017.
4. P. Borse, and A. Sheth, "Technological and commercial update for first-and second-generation ethanol production in India", In: Chandel, A. K., Sukumaran, R. K. (Eds), *Sustainable Biofuels Development in India*. Springer, pp. 279–297, 2017.
5. D. L. Aguilar et al., "Operational strategies for enzymatic hydrolysis in a biorefinery", Kumar S., Sani R. (Eds), In: *Biorefining of Biomass to Biofuels*. Springer, pp. 223–248, 2018.
6. L. Allen, "Is Indonesia's fire crisis connected to the palm oil in our snack food?", *The Guardian*, 2015.
7. P. Jacobson, "SE Asian governments dismiss finding that 2015 haze killed 100,300", *Mongabay*, 2016.
8. F. Darvishi, Z. Fathi, M. Ariana, and H. Moradi, "Yarrowia lipolytica as a workhorse for biofuel production", *Biochemical Engineering Journal*, vol. 127, pp. 87–96, 2017.
9. S. Behera, R. Singh, R. Arora, N. K. Sharma, M. Shukla, and S. Kumar, "Scope of algae as third generation biofuels", *Frontiers in Bioengineering and Biotechnology*, vol. 2, p. 90, 2015.
10. D. Verma, A. Singla, B. Lal, and P. M. Sarma, "Conversion of biomass-generated syngas into next-generation liquid transport fuels through microbial intervention: Potential and current status", *Current Science*, vol. 110, pp. 329–336, 2016.

11. S. Chinnasamy, P. H. Rao, S. Bhaskar, R. Rengasamy, and M. Singh, "Algae: A novel biomass feedstock for biofuels", In: R. Arora, ed. *Microbial Biotechnology*: Energy *and* Environment. CAB International: Wallingford, UK, pp. 224–239, 2012.

12. L. Brennan, and P. Owende, "Biofuels from microalgae—A review of technologies for production, processing, and extractions of biofuels and co-products", *Renewable and Sustainable Energy Reviews*, vol. 14, pp. 557–577, 2010.

13. Z. Shokravi, H. Shokravi, M. A. Aziz, and H. Shokravi, "Algal biofuel: A promising alternative for fossil fuel", In: M. M. B. A. Aziz, ed. Fossil Free *Fuels*. Taylor and Francis, 2019.

14. B. Sajjadi, W.-Y. Chen, A. A. A. Raman, and S. Ibrahim, "Microalgae lipid and biomass for biofuel production: A comprehensive review on lipid enhancement strategies and their effects on fatty acid composition", *Renewable and Sustainable Energy Reviews*, vol. 97, pp. 200–232, 2018.

15. M. M. A. Aziz et al., "Two-stage cultivation strategy for simultaneous increases in growth rate and lipid content of microalgae: A review", *Renewable and Sustainable Energy Reviews*, 2019 (Under rev).

16. M. Aouida et al., "Growth dependent silencing and resetting of DGA1 transgene in Nannochloropsis salina", *Renewable and Sustainable Energy Reviews*, vol. 7, pp. 1–13, 2017.

17. J. W. Chen et al., "Identification of a malonyl CoA-acyl carrier protein transacylase and its regulatory role in fatty acid biosynthesis in oleaginous microalga *Nannochloropsis oceanica*", *Biotechnology and Applied Biochemistry*, vol. 64, no. 5, pp. 620–626, 2017.

18. B. Bharathiraja et al., "Aquatic biomass (algae) as a future feed stock for bio-refineries: A review on cultivation, processing and products", *Renewable and Sustainable Energy Reviews*, vol. 47, pp. 634–653, 2015.

19. X. B. Tan, M. K. Lam, Y. Uemura, J. W. Lim, C. Y. Wong, and K. T. Lee, "Cultivation of microalgae for biodiesel production: A review on upstream and downstream processing", *Chinese Journal of Chemical Engineering*, vol. 26, pp. 17–30, 2018.

20. S. Aikawa, S. Ho, A. Nakanishi, J. Chang, T. Hasunuma, and A. Kondo, "Improving polyglucan production in cyanobacteria and microalgae via cultivation design and metabolic engineering", *Biotechnology Journal*, vol. 10, pp. 886–898, 2015.

21. B. Abdullah et al., "Fourth generation biofuel: A review on risks and mitigation strategies", *Renewable Sustainable Energy Reviews*, vol. 107, pp.37–50, 2019.

22. X. Li et al., "Heterogeneous sulfur-free hydrodeoxygenation catalysts for selectively upgrading the renewable bio-oils to second generation biofuels", *Renewable and Sustainable Energy Reviews*, vol. 82, pp. 3762–3797, 2018.

23. W.-H. Leong, J.-W. Lim, M.-K. Lam, Y. Uemura, and Y.-C. Ho, "Third generation biofuels: A nutritional perspective in enhancing microbial lipid production", *Renewable and Sustainable Energy Reviews*, vol. 91, pp. 950–961, 2018.

24. D. F. Correa, H. L. Beyer, H. P. Possingham, S. R. Thomas-Hall, and P. M. Schenk, "Biodiversity impacts of bioenergy production: Microalgae vs. first generation biofuels", *Renewable and Sustainable Energy Reviews*, vol. 74, pp. 1131–1146, 2017.

25. B. P. J. Andrée, V. Diogo, and E. Koomen, "Efficiency of second-generation biofuel crop subsidy schemes: Spatial heterogeneity and policy design", *Renewable and Sustainable Energy Reviews*, vol. 67, pp. 848–862, 2017.

26. S. K. Bhatia, S. H. Kim, J. J. Yoon, and Y. H. Yang, "Current status and strategies for second generation biofuel production using microbial systems", *Energy Conversion and Management*, vol. 148, pp. 1142–1156, 2017.

27. F. Saladini, N. Patrizi, F. M. Pulselli, N. Marchettini, and S. Bastianoni, "Guidelines for emergy evaluation of first, second and third generation biofuels", *Renewable and Sustainable Energy Reviews*, vol. 66, pp. 221–227, 2016.

28. Z. U. Islam, Y. Zhisheng, E. B. Hassan, C. Dongdong, and Z. Hongxun, "Microbial conversion of pyrolytic products to biofuels: A novel and sustainable approach toward second-generation biofuels", *Journal of Industrial Microbiology and Biotechnology*, vol. 42, pp. 1557–1579, 2015.

29. C. J. Hayes, D. R. Burgess, and J. A. Manion, "Combustion pathways of biofuel model compounds: A review of recent research and current challenges pertaining to first-, second-, and third-generation biofuels", *Advances in Physical Organic Chemistry*, vol. 49. pp. 103–187, 2015.

30. E. Smeets, A. Tabeau, S. Van Berkum, J. Moorad, H. Van Meijl, and G. Woltjer, "The impact of the rebound effect of the use of first generation biofuels in the EU on greenhouse gas emissions: A critical review", *Renewable and Sustainable Energy Reviews*, vol. 38, pp. 393–403, 2014.

31. T. Stankus, "Reviews of science for science librarians: Second-generation biofuels feedstock: Crop wastes, energy grasses, and forest byproducts", *Science and Technology Libraries*, vol. 33, pp. 143–164, 2014.

32. T. R. Brown, "Technoeconomic assessment of second-generation biofuel pathways: Challenges and solutions", *Biofuels*, vol. 4, pp. 351–353, 2013.

33. W. Thompson, and S. Meyer, "Second generation biofuels and food crops: Co-products or competitors?", *Global Food Security*, vol. 2, pp. 89–96, 2013.

34. Y. Li-Beisson, and G. Peltier, "Third-generation biofuels: Current and future research on microalgal lipid biotechnology", *OCL – Oilseeds and Fats, Crops and Lipids*, vol. 20, 2013.

35. J. S. Rowbotham, P. W. Dyer, H. C. Greenwell, and M. K. Theodorou, "Thermochemical processing of macroalgae: A late bloomer in the development of third-generation biofuels?", *Biofuels*, vol. 3, pp. 441–461, 2012.

36. T. Damartzis, and A. Zabaniotou, "Thermochemical conversion of biomass to second generation biofuels through integrated process design—A review", *Renewable and Sustainable Energy Reviews*, vol. 15, pp. 366–378, 2011.

37. M. Mohammadi, G. D. Najafpour, H. Younesi, P. Lahijani, M. H. Uzir, and A. R. Mohamed, "Bioconversion of synthesis gas to second generation biofuels: A review", *Renewable and Sustainable Energy Reviews*, vol. 15, pp. 4255–4273, 2011.

38. J. Lü, C. Sheahan, and P. Fu, "Metabolic engineering of algae for fourth generation biofuels production", *Energy and Environmental Science*, vol. 4, pp. 2451–2466, 2011.

39. M. A. Martin, "First generation biofuels compete", *New Biotechnology*, vol. 27, pp. 596–608, 2010.

40. J. Hromádko, J. Hromádko, P. Miler, V. Hönig, and M. Cindr, "Technologies in second-generation biofuel production", *Chemicke Listy*, vol. 104, pp. 784–790, 2010.

41. S. N. Naik, V. V Goud, P. K. Rout, and A. K. Dalai, "Production of first and second generation biofuels: A comprehensive review", *Renewable and Sustainable Energy Reviews*, vol. 14, pp. 578–597, 2010.

42. R. E. H. Sims, W. Mabee, J. N. Saddler, and M. Taylor, "An overview of second generation biofuel technologies", *Bioresource Technology*, vol. 101, pp. 1570–1580, 2010.

43. C. R. Carere, R. Sparling, N. Cicek, and D. B. Levin, "Third generation biofuels via direct cellulose fermentation", *International Journal of Molecular Sciences*, vol. 9, pp. 1342–1360, 2008.

44. Ł. Jęczmionek, S. Oleksiak, I. Skręt, and A. Marchut, "The second-generation biofuels", *Przemysl Chemiczny*, vol. 85, pp. 1570–1574, 2006.

45. M. Levine-Clark, and E. L. Gil, "A comparative citation analysis of Web of Science, Scopus, and Google Scholar", *Journal of Business & Finance Librarianship*, vol. 14, pp. 32–46, 2008.

46. G. Ennas, B. Biggio, and M. C. Di Guardo, "Data-driven journal meta-ranking in business and management", *Scientometrics*, vol. 105, pp. 1911–1929, 2015.

47. T. E. Carter, T. E. Smith, and P. J. Osteen, "Gender comparisons of social work faculty using H-Index scores", *Scientometrics*, vol. 111, pp. 1547–1557, 2017.
48. A. Martín-Martín, E. Orduna-Malea, and E. D. López-Cózar, "Coverage of highly-cited documents in Google Scholar, Web of Science, and Scopus: A multidisciplinary comparison", *Scientometrics*, vol. 116, pp. 2175–2188, 2018.
49. M. Gusenbauer, "Google Scholar to overshadow them all? Comparing the sizes of 12 academic search engines and bibliographic databases", *Scientometrics*, vol. 118, pp. 177–214, 2019.
50. A. Aghaei Chadegani et al., "A comparison between two main academic literature collections: Web of Science and Scopus databases", *Asian Social Science*, vol. 9, pp. 18–26, 2013.
51. C. López-Illescas, F. de Moya-Anegón, and H. F. Moed, "Coverage and citation impact of oncological journals in the Web of Science and Scopus", *Journal of Informetrics*, vol. 2, pp. 304–316, 2008.
52. A. Demirbas, and M. F. Demirbas, "Importance of algae oil as a source of biodiesel", *Energy Conversion and Management*, vol. 52, pp. 163–170, 2011.
53. E. Nwokoagbara, A. K. Olaleye, and M. Wang, "Biodiesel from microalgae: The use of multi-criteria decision analysis for strain selection", *Fuel*, vol. 159, pp. 241–249, 2015.
54. N. M. D. Courchesne, A. Parisien, B. Wang, and C. Q. Lan, "Enhancement of lipid production using biochemical, genetic and transcription factor engineering approaches", *Journal of Biotechnology*, vol. 141, pp. 31–41, 2009.
55. B. Chen, C. Wan, M. A. Mehmood, J. S. Chang, F. Bai, and X. Zhao, "Manipulating environmental stresses and stress tolerance of microalgae for enhanced production of lipids and value-added products—A review", *Bioresource Technology*, vol. 244, pp. 1198–1206, 2017.
56. A. Banerjee, C. Banerjee, S. Negi, J. S. Chang, and P. Shukla, "Improvements in algal lipid production: A systems biology and gene editing approach", *Critical Reviews in Biotechnology*, vol. 38, pp. 369–385, 2018.
57. T. Capell, and P. Christou, "Progress in plant metabolic engineering", *Current Opinion in Biotechnology*, vol. 15, pp. 148–154, 2004.
58. Y. S. Shin, H. I. Choi, J. W. Choi, J. S. Lee, Y. J. Sung, and S. J. Sim, "Multilateral approach on enhancing economic viability of lipid production from microalgae: A review", *Bioresource Technology*, vol. 258, pp. 335–344, 2018.
59. A. Robles-Medina, P. A. González-Moreno, L. Esteban-Cerdán, and E. Molina-Grima, "Biocatalysis: Towards ever greener biodiesel production", *Biotechnology Advances*, vol. 27, pp. 398–408, 2009.
60. M. Lapuerta, J. Barba, A. D. Sediako, M. R. Kholghy, and M. J. Thomson, "Morphological analysis of soot agglomerates from biodiesel surrogates in a coflow burner", *Journal of Aerosol Science*, vol. 111, pp. 65–74, 2017.
61. Z. Li, L. Qiu, X. Cheng, Y. Li, and H. Wu, "The evolution of soot morphology and nanostructure in laminar diffusion flame of surrogate fuels for diesel", *Fuel*, vol. 211, pp. 517–528, 2018.
62. B. Ravindran, M. B. Kurade, A. N. Kabra, B.-H. Jeon, and S. K. Gupta, "Recent advances and future prospects of microalgal lipid biotechnology", In: S. Gupta, A. Malik, F. Bux, eds. *Algal Biofuels*. Springer, pp. 1–37, 2017.
63. C.-Y. Chen et al., "Microalgae-based carbohydrates for biofuel production", *Biochemical Engineering Journal*, vol. 78, pp. 1–10, 2013.
64. R. Radakovits, R. E. Jinkerson, A. Darzins, and M. C. Posewitz, "Genetic engineering of algae for enhanced biofuel production", *Eukaryotic Cell*, vol. 9, pp. 486–501, 2010.
65. H. C. Greenwell, L. M. L. Laurens, R. J. Shields, R. W. Lovitt, and K. J. Flynn, "Placing microalgae on the biofuels priority list: A review of the technological challenges", *Journal of the Royal Society Interface*, vol. 7, pp. 703–726, 2010.

66. I. M. Remmers, R. H. Wijffels, M. J. Barbosa, and P. P. Lamers, "Can we approach theoretical lipid yields in microalgae?", *Trends in Biotechnology*, vol. 36, pp. 265–276, 2018.
67. E. S. Shuba, and D. Kifle, "Microalgae to biofuels: 'Promising' alternative and renewable energy, review", *Renewable and Sustainable Energy Reviews*, vol. 81, pp. 743–755, 2018.
68. N. K. Sharma, S. P. Tiwari, K. Tripathi, and A. K. Rai, "Sustainability and cyanobacteria (blue-green algae): Facts and challenges", *Journal of Applied Phycology*, vol. 23, pp. 1059–1081, 2011.
69. D. Grizeau, L. A. Bui, C. Dupré, and J. Legrand, "Ammonium photo-production by heterocytous cyanobacteria: Potentials and constraints", *Critical Reviews in Biotechnology*, vol. 36, pp. 607–618, 2016.
70. M. Xie, W. Wang, W. Zhang, L. Chen, and X. Lu, "Versatility of hydrocarbon production in cyanobacteria", *Applied Microbiology and Biotechnology*, vol. 101, pp. 905–919, 2017.
71. M. Kamalanathan, P. Chaisutyakorn, R. Gleadow, and J. Beardall, "A comparison of photoautotrophic, heterotrophic, and mixotrophic growth for biomass production by the green alga Scenedesmus sp. (Chlorophyceae)", *Phycologia*, vol. 57, pp. 309–317, 2017.
72. A. Ghosh et al., "Progress toward isolation of strains and genetically engineered strains of microalgae for production of biofuel and other value added chemicals: A review", *Energy Conversion and Management*, vol. 113, pp. 104–118, 2016.
73. Y. J. Tamayo-Ordóñez et al., "Advances in culture and genetic modification approaches to lipid biosynthesis for biofuel production and in silico analysis of enzymatic dominions in proteins related to lipid biosynthesis in algae", *Phycological Research*, vol. 65, pp. 14–28, 2017.
74. A. Tabernero, E. M. M. del Valle, and M. A. Galán, "Evaluating the industrial potential of biodiesel from a microalgae heterotrophic culture: Scale-up and economics", *Biochemical Engineering Journal*, vol. 63, pp. 104–115, 2012.
75. J. Lowrey, M. S. Brooks, and P. J. McGinn, "Heterotrophic and mixotrophic cultivation of microalgae for biodiesel production in agricultural wastewaters and associated challenges—A critical review", *Journal of Applied Phycology*, vol. 27, pp. 1485–1498, 2015.
76. M. Isleten-Hosoglu, I. Gultepe, and M. Elibol, "Optimization of carbon and nitrogen sources for biomass and lipid production by *Chlorella saccharophila* under heterotrophic conditions and development of Nile red fluorescence based method for quantification of its neutral lipid content", *Biochemical Engineering Journal*, vol. 61, pp. 11–19, 2012.
77. J. Wang, H. Yang, and F. Wang, "Mixotrophic cultivation of microalgae for biodiesel production: Status and prospects", *Applied Biochemistry and Biotechnology*, vol. 172, pp. 3307–3329, 2014.
78. N. Arora, A. Patel, P. A. Pruthi, and V. Pruthi, "Synergistic dynamics of nitrogen and phosphorous influences lipid productivity in *Chlorella minutissima* for biodiesel production", *Bioresource Technology*, vol. 213, pp. 79–87, 2016.
79. P. Přibyl, V. Cepák, and V. Zachleder, "Production of lipids in 10 strains of Chlorella and Parachlorella, and enhanced lipid productivity in *Chlorella vulgaris*", *Applied Microbiology and Biotechnology*, vol. 94, pp. 549–561, 2012.
80. J. Fan, Y. Cui, M. Wan, W. Wang, and Y. Li, "Lipid accumulation and biosynthesis genes response of the oleaginous *Chlorella pyrenoidosa* under three nutrition stressors", *Biotechnology for Biofuels*, vol. 7, p. 17, 2014.
81. T. M. Mata, R. Almeidab, and N. S. Caetanoa, "Effect of the culture nutrients on the biomass and lipid productivities of microalgae *Dunaliella tertiolecta*", *Chemical Engineering*, vol. 32, pp. 973, 2013.

82. C. Yeesang, and B. Cheirsilp, "Effect of nitrogen, salt, and iron content in the growth medium and light intensity on lipid production by microalgae isolated from freshwater sources in Thailand", *Bioresource Technology*, vol. 102, pp. 3034–3040, 2011.

83. M. Y. Roleda, S. P. Slocombe, R. J. G. Leakey, J. G. Day, E. M. Bell, and M. S. Stanley, "Effects of temperature and nutrient regimes on biomass and lipid production by six oleaginous microalgae in batch culture employing a two-phase cultivation strategy", *Bioresource Technology*, vol. 129, pp. 439–449, 2013.

84. L. Xin, H. Hong-ying, G. Ke, and S. Ying-xue, "Effects of different nitrogen and phosphorus concentrations on the growth, nutrient uptake, and lipid accumulation of a freshwater microalga *Scenedesmus* sp.", *Bioresource Technology*, vol. 101, pp. 5494–5500, 2010.

85. M. Chen, H. Tang, H. Ma, T. C. Holland, K. Y. S. Ng, and S. O. Salley, "Effect of nutrients on growth and lipid accumulation in the green algae *Dunaliella tertiolecta*", *Bioresource Technology*, vol. 102, pp. 1649–1655, 2011.

86. M. I. Khan, J. H. Shin, and J. D. Kim, "The promising future of microalgae: Current status, challenges, and optimization of a sustainable and renewable industry for biofuels, feed, and other products", *Microbial Cell Factories*, vol. 17, 2018.

87. Q. Hu et al., "Microalgal triacylglycerols as feedstocks for biofuel production: Perspectives and advances", *Plant Journal*, vol. 54, pp. 621–639, 2008.

88. R. Radakovits, R. E. Jinkerson, A. Darzins, and M. C. Posewitz, "Genetic engineering of algae for enhanced biofuel production", *Eukaryotic Cell*, vol. 9, pp. 486–501, 2010.

89. Z. Y. Du et al., "Enhancing oil production and harvest by combining the marine alga *Nannochloropsis oceanica* and the oleaginous fungus *Mortierella elongata*", *Biotechnology for Biofuels*, vol. 11, 2018.

90. S. Y. Choi et al., "Transcriptome landscape of *Synechococcus elongatus* PCC 7942 for nitrogen starvation responses using RNA-seq", *Scientific Reports*, vol. 6, p. 30584, 2016.

91. J. Xue, Y. F. Niu, T. Huang, W. D. Yang, J. S. Liu, and H. Y. Li, "Genetic improvement of the microalga *Phaeodactylum tricornutum* for boosting neutral lipid accumulation", *Metabolic Engineering*, vol. 27, pp. 1–9, 2015.

92. C. J. Salas-Montantes et al., "Lipid accumulation during nitrogen and sulfur starvation in *Chlamydomonas reinhardtii* overexpressing a transcription factor", *Journal of Applied Phycology*, vol. 30, no. 3, pp. 1–13, 2018.

93. N. N. Zulu, J. Popko, K. Zienkiewicz, P. Tarazona, C. Herrfurth, and I. Feussner, "Heterologous co-expression of a yeast diacylglycerol acyltransferase (ScDGA1) and a plant oleosin (AtOLEO3) as an efficient tool for enhancing triacylglycerol accumulation in the marine diatom *Phaeodactylum tricornutum*", *Biotechnology for Biofuels*, vol. 10, Article Number: 187, 2017.

94. V. da Silva Ferreira, and C. Sant'Anna, "Impact of culture conditions on the chlorophyll content of microalgae for biotechnological applications", *World Journal of Microbiology and Biotechnology*, vol. 33, pp. 20, 2017.

95. M. Atta, A. Idris, A. Bukhari, and S. Wahidin, "Intensity of blue LED light: A potential stimulus for biomass and lipid content in fresh water microalgae *Chlorella vulgaris*", *Bioresource Technology*, vol. 148, pp. 373–378, 2013.

96. C. L. Teo, M. Atta, A. Bukhari, M. Taisir, A. M. Yusuf, and A. Idris, "Enhancing growth and lipid production of marine microalgae for biodiesel production via the use of different LED wavelengths", *Bioresource Technology*, vol. 162, pp. 38–44, 2014.

97. P. M. Schenk et al., "Second generation biofuels: High-efficiency microalgae for biodiesel production", *Bioenergy Research*, vol. 1, pp. 20–43, 2008.

98. S. R. Medipally, F. M. Yusoff, S. Banerjee, and M. Shariff, "Microalgae as sustainable renewable energy feedstock for biofuel production", *BioMed Research International*, vol. 2015, Article Number: 519513, 2015.

99. R. Sharon-Gojman, S. Leu, and A. Zarka, "Antenna size reduction and altered division cycles in self-cloned, marker-free genetically modified strains of *Haematococcus pluvialis*", *Algal Research*, vol. 28, pp. 172–183, 2017.
100. S. L. Pahl, A. K. Lee, T. Kalaitzidis, P. J. Ashman, S. Sathe, and D. M. Lewis, "Harvesting, thickening and dewatering microalgae biomass", In: M. Borowitzka, N. Moheimani, eds. *Algae for Biofuels and Energy*. Springer, pp. 165–185, 2013.
101. G. Singh, and S. K. Patidar, "Microalgae harvesting techniques: A review", *Journal of Environmental Management*, vol. 217, pp. 499–508, 2018.
102. T. Mathimani, and N. Mallick, "A comprehensive review on harvesting of microalgae for biodiesel – Key challenges and future directions", *Renewable and Sustainable Energy Reviews*, vol. 91, pp. 1103–1120, 2018.
103. A. I. Barros, A. L. Gonçalves, M. Simões, and J. C. M. Pires, "Harvesting techniques applied to microalgae: A review", *Renewable and Sustainable Energy Reviews*, vol. 41, pp. 1489–1500, 2015.
104. T. A. Beacham, J. B. Sweet, and M. J. Allen, "Large scale cultivation of genetically modified microalgae: A new era for environmental risk assessment", *Algal Research*, vol. 25, pp. 90–100, 2017.
105. R. O'Driscoll, M. E. J. Stettler, N. Molden, T. Oxley, and H. M. ApSimon, "Real world CO_2 and NOx emissions from 149 Euro 5 and 6 diesel, gasoline and hybrid passenger cars", *Science of the Total Environment*, vol. 621, pp. 282–290, 2018.
106. B. Zhu, G. Chen, X. Cao, and D. Wei, "Molecular characterization of CO_2 sequestration and assimilation in microalgae and its biotechnological applications", *Bioresource Technology*, vol. 244, pp. 1207–1215, 2017.
107. N. K. Singh, A. K. Upadhyay, and U. N. Rai, "Algal technologies for wastewater treatment and biofuels production: An integrated approach for environmental management", In: S. Gupta, A. Malik F. Bux, eds. *Algal Biofuels*. Springer, pp. 97–107, 2017.
108. D. L. Sutherland, C. Howard-Williams, M. H. Turnbull, P. A. Broady, and R. J. Craggs, "Enhancing microalgal photosynthesis and productivity in wastewater treatment high rate algal ponds for biofuel production", *Bioresource Technology*, vol. 184, pp. 222–229, 2015.
109. D. De Francisci, Y. Su, A. Iital, and I. Angelidaki, "Evaluation of microalgae production coupled with wastewater treatment", *Environmental Technology*, vol. 39, pp. 581–592, 2018.
110. A. Guldhe et al., "Prospects, recent advancements and challenges of different wastewater streams for microalgal cultivation", *Journal of Environmental Management*, vol. 203, pp. 299–315, 2017.
111. O. Komolafe, S. B. Velasquez Orta, I. Monje-Ramirez, I. Y. Noguez, A. P. Harvey, and M. T. Orta Ledesma, "Biodiesel production from indigenous microalgae grown in wastewater", *Bioresource Technology*, vol. 154, pp. 297–304, 2014.
112. L. Wang et al., "Cultivation of green algae *Chlorella* sp. in different wastewaters from municipal wastewater treatment plant", *Applied Biochemistry and Biotechnology*, vol. 162, pp. 1174–1186, 2010.
113. A. A. H. Khalid, Z. Yaakob, S. R. S. Abdullah, and M. S. Takriff, "Enhanced growth and nutrients removal efficiency of *Characium* sp. cultured in agricultural wastewater via acclimatized inoculum and effluent recycling", *Journal of Environmental Chemical Engineering*, vol. 4, pp. 3426–3432, 2016.
114. H. Kamyab et al., "*Chlorella pyrenoidosa* mediated lipid production using Malaysian agricultural wastewater: Effects of photon and carbon", *Waste and Biomass Valorization*, vol. 7, pp. 779–788, 2016.
115. A. Polishchuk et al., "Cultivation of nannochloropsis for eicosapentaenoic acid production in wastewaters of pulp and paper industry", *Bioresource Technology*, vol. 193, pp. 469–476, 2015.

116. R. P. Singh, P. K. Singh, R. Gupta, and R. L. Singh, "Treatment and recycling of wastewater from textile industry", In: R. Singh, R. Singh, eds. *Advances in Biological Treatment of Industrial Waste Water and Their Recycling for a Sustainable Future.* Springer, pp. 225–266, 2019.

117. Y. Wang et al., "Perspectives on the feasibility of using microalgae for industrial wastewater treatment", *Bioresource Technology*, vol. 222, pp. 485–497, 2016.

118. I. Matsuwaki, S. Harayama, and M. Kato, "Assessment of the biological invasion risks associated with a massive outdoor cultivation of the green alga, Pseudochoricystis ellipsoidea", *Algal Research*, vol. 9, pp. 1–7, 2015.

119. J. P. Hewett, A. K. Wolfe, R. A. Bergmann, S. C. Stelling, and K. L. Davis, "Human health and environmental risks posed by Synthetic Biology R&D for energy applications: A literature analysis", *Applied Biosafety*, vol. 21, pp. 177–184, 2016.

120. S. J. Szyjka et al., "Evaluation of phenotype stability and ecological risk of a genetically engineered alga in open pond production", *Algal Research*, vol. 24, pp. 378–386, 2017.

121. D. A. Russo, A. P. Beckerman, and J. Pandhal, "Competitive growth experiments with a high-lipid *Chlamydomonas reinhardtii* mutant strain and its wild-type to predict industrial and ecological risks", *AMB Express*, vol. 7, p. 10, 2017.

122. S. Katsanevakis et al., "Impacts of invasive alien marine species on ecosystem services and biodiversity: A pan-European review", *Aquatic Invasions*, vol. 9, pp. 391–423, 2014.

123. N. Goldenfeld, and C. Woese, "Biology's next revolution", *Nature*, vol. 445, p. 369, 2007.

124. A. A. Snow, and V. H. Smith, "Genetically engineered algae for biofuels: A key role for ecologists", *BioScience*, vol. 62, pp. 765–768, 2012.

125. S. Mazard, A. Penesyan, M. Ostrowski, I. T. Paulsen, and S. Egan, "Tiny microbes with a big impact: The role of cyanobacteria and their metabolites in shaping our future", *Marine Drugs*, vol. 14, no. 5, p. 97, 2016.

126. R. H. Wijffels, O. Kruse, and K. J. Hellingwerf, "Potential of industrial biotechnology with cyanobacteria and eukaryotic microalgae", *Current Opinion in Biotechnology*, vol. 24, pp. 405–413, 2013.

127. L. Barsanti, and P. Gualtieri, "Is exploitation of microalgae economically and energetically sustainable?", *Algal Research*, vol. 31, pp. 107–115, 2018.

128. Y. Chisti, "Biodiesel from microalgae", *Biotechnology Advances*, vol. 25, pp. 294–306, 2007.

129. Y. Chisti, "Constraints to commercialization of algal fuels", *Journal of Biotechnology*, vol. 167, pp. 201–214, 2013.

130. T. Zhang, X. Xie, and Z. Huang, "Life cycle water footprints of nonfood biomass fuels in China", *Environmental Science and Technology*, vol. 48, pp. 4137–4144, 2014.

131. P.-Z. Feng, L.-D. Zhu, X.-X. Qin, and Z.-H. Li, "Water footprint of biodiesel production from microalgae cultivated in photobioreactors", *Journal of Environmental Engineering*, vol. 142, p. 4016067, 2016.

132. C. Crofcheck, and M. Crocker, "Application of recycled media and algae-based anaerobic digestate in *Scenedesmus* cultivation", *Journal of Renewable and Sustainable Energy*, vol. 8, p. 1, 2016.

133. J. Lowrey, M. S. Brooks, and R. E. Armenta, "Nutrient recycling of lipid-extracted waste in the production of an oleaginous thraustochytrid", *Applied Microbiology and Biotechnology*, vol. 100, pp. 4711–4721, 2016.

134. B. Sialve, N. Bernet, and O. Bernard, "Anaerobic digestion of microalgae as a necessary step to make microalgal biodiesel sustainable", *Biotechnology Advances*, vol. 27, pp. 409–416, 2009.

135. E. Barbera, E. Sforza, S. Kumar, T. Morosinotto, and A. Bertucco, "Cultivation of *Scenedesmus obliquus* in liquid hydrolysate from flash hydrolysis for nutrient recycling", *Bioresource Technology*, vol. 207, pp. 59–66, 2016.

136. Y. Zhang, A. Kendall, and J. Yuan, "A comparison of on-site nutrient and energy recycling technologies in algal oil production", *Resources, Conservation and Recycling*, vol. 88, pp. 13–20, 2014.

137. C. R. Talbot, "Comparing nutrient recovery via rapid (Flash Hydrolysis) and conventional hydrothermal liquefaction processes for *Microalgae* cultivation", Master of Science (MS), thesis, Civil/Environmental Engineering, Old Dominion University, 2015.

138. P. Kenny, and K. J. Flynn, "Physiology limits commercially viable photoautotrophic production of microalgal biofuels", *Journal of Applied Phycology*, vol. 29, pp. 2713–2727, 2017.

139. X. Tan, M. K. Lam, Y. Uemura, J. W. Lim, C. Y. Wong, and K. T. Lee, "Cultivation of microalgae for biodiesel production: A review on upstream and downstream processing", *Chinese Journal of Chemical Engineering*, vol. 26, no. 1, pp. 17–30, 2017.

140. S. K. Hess, B. Lepetit, P. G. Kroth, and S. Mecking, "Production of chemicals from microalgae lipids – status and perspectives", *European Journal of Lipid Science and Technology*, vol. 120, 1700152, 2018.

141. I. A. Nascimento et al., "Screening microalgae strains for biodiesel production: Lipid productivity and estimation of fuel quality based on fatty acids profiles as selective criteria", *Bioenergy Research*, vol. 6, pp. 1–13, 2013.

142. S. Leu, and S. Boussiba, "Advances in the production of high-value products by microalgae", *Industrial Biotechnology*, vol. 10, pp. 169–183, 2014.

143. R. H. Wijffels, and M. J. Barbosa, "An outlook on microalgal biofuels", *Science*, vol. 329, pp. 796–799, 2010.

144. Y. Maeda, T. Yoshino, T. Matsunaga, M. Matsumoto, and T. Tanaka, "Marine microalgae for production of biofuels and chemicals", *Current Opinion in Biotechnology*, vol. 50, pp. 111–120, 2018.

145. R. Davis, A. Aden, and P. T. Pienkos, "Techno-economic analysis of autotrophic microalgae for fuel production", *Applied Energy*, vol. 88, pp. 3524–3531, 2011.

146. S. Nagarajan, S. K. Chou, S. Cao, C. Wu, and Z. Zhou, "An updated comprehensive techno-economic analysis of algae biodiesel", *Bioresource Technology*, vol. 145, pp. 150–156, 2013.

147. O. Konur, "Current state of research on algal biohydrogen", In: S. K. Kim, C. G. Lee, eds. *Marine Bioenergy: Trends and Developments*. Boca Raton, FL: CRC Press, pp. 393–422, 2015.

148. K. Y. Show, Y. Yan, M. Ling, G. Ye, T. Li, and D. J. Lee, "Hydrogen production from algal biomass – Advances, challenges and prospects", *Bioresource Technology*, vol. 257, pp. 290–300, 2018.

149. K. Y. Show, D. J. Lee, J. H. Tay, C. Y. Lin, and J. S. Chang, "Biohydrogen production: Current perspectives and the way forward", *International Journal of Hydrogen Energy*, vol. 37, pp. 15616–15631, 2012.

150. C. E. de Farias Silva, and A. Bertucco, "Bioethanol from microalgae and cyanobacteria: A review and technological outlook", *Process Biochemistry*, vol. 51, pp. 1833–1842, 2016.

151. R. P. Rastogi, A. Pandey, C. Larroche, and D. Madamwar, "Algal green energy – R&D and technological perspectives for biodiesel production", *Renewable and Sustainable Energy Reviews*, vol. 82, pp. 2946–2969, 2018.

152. J. Spruijt, R. Schipperus, A. M. J. Kootstra, and C. L. M. de Visser, *AlgaeEconomics: Bio-Economic Production Models of Micro-Algae and Downstream Processing to Produce Bio Energy Carriers*. EnAlgae Swansea University, 2015.

153. E. S. Salama et al., "Recent progress in microalgal biomass production coupled with wastewater treatment for biofuel generation", *Renewable and Sustainable Energy Reviews*, vol. 79, pp. 1189–1211, 2017.

154. Y. Gong, and M. Jiang, "Biodiesel production with microalgae as feedstock: From strains to biodiesel", *Biotechnology Letters*, vol. 33, pp. 1269–1284, 2011.

155. I. Sizova, A. Greiner, M. Awasthi, S. Kateriya, and P. Hegemann, "Nuclear gene targeting in Chlamydomonas using engineered zinc-finger nucleases", *The Plant Journal*, vol. 73, pp. 873–882, 2013.

156. H. Li, C. R. Shen, C.-H. Huang, L.-Y. Sung, M.-Y. Wu, and Y.-C. Hu, "CRISPR-Cas9 for the genome engineering of cyanobacteria and succinate production", *Metabolic Engineering*, vol. 38, pp. 293–302, 2016.

157. M. Serif, B. Lepetit, K. Weißert, P. G. Kroth, and C. Rio Bartulos, "A fast and reliable strategy to generate TALEN-mediated gene knockouts in the diatom *Phaeodactylum tricornutum*", *Algal Research*, vol. 23, pp. 186–195, 2017.

158. D. Stukenberg, S. Zauner, G. Dell'Aquila, and U. G. Maier, "Optimizing CRISPR/cas9 for the diatom *Phaeodactylum tricornutum*", *Frontiers in Plant Science*, vol. 9, p. 740, 2018.

159. Q. Wang, Y. Lu, Y. Xin, L. Wei, S. Huang, and J. Xu, "Genome editing of model oleaginous microalgae *Nannochloropsis* spp. by CRISPR/Cas9", *The Plant Journal*, vol. 88, pp. 1071–1081, November 2016.

160. K. Takahashi et al., "Lipid productivity in TALEN-induced starchless mutants of the unicellular green alga *Coccomyxa* sp. strain Obi", *Algal Research*, vol. 32, pp. 300–307, 2018.

161. Y. S. Shin, J. Jeong, T. H. T. Nguyen, J. Y. H. Kim, E. Jin, and S. J. Sim, "Targeted knockout of phospholipase A_2 to increase lipid productivity in *Chlamydomonas reinhardtii* for biodiesel production", *Bioresource Technology*, vol. 271, pp. 368–374, 2019.

162. W. L. Chu, "Strategies to enhance production of microalgal biomass and lipids for biofuel feedstock", *European Journal of Phycology*, vol. 52, pp. 419–437, 2017.

163. S. Jeon et al., "Current status and perspectives of genome editing technology for microalgae", *Biotechnology for Biofuels*, vol. 10, Article Number: 267, 2017.

164. A. M. Ruffing, "Engineered cyanobacteria: Teaching an old bug new tricks", *Bioengineered Bugs*, vol. 2, pp. 136–149, 2011.

165. R. L. Kitchener, and A. M. Grunden, "Methods for enhancing cyanobacterial stress tolerance to enable improved production of biofuels and industrially relevant chemicals", *Applied Microbiology and Biotechnology*, vol. 102, pp. 1617–1628, 2018.

166. C. Banerjee, K. K. Dubey, and P. Shukla, "Metabolic engineering of microalgal based biofuel production: Prospects and challenges", *Frontiers in Microbiology*, vol. 7, p. 432, 2016.

167. W. Jiang, A. J. Brueggeman, K. M. Horken, T. M. Plucinak, and D. P. Weeks, "Successful transient expression of Cas9 and single guide RNA genes in *Chlamydomonas reinhardtii*", *Eukaryotic Cell*, vol. 13, pp. 1465–1469, 2014.

168. M. Nymark, A. K. Sharma, T. Sparstad, A. M. Bones, and P. Winge, "A CRISPR/Cas9 system adapted for gene editing in marine algae", *Scientific Reports*, vol. 6, p. 24951, 2016.

169. K. Baek et al., "DNA-free two-gene knockout in *Chlamydomonas reinhardtii* via CRISPR-Cas9 ribonucleoproteins", *Scientific Reports*, vol. 6, p. 30620, 2016.

170. P.-H. Kao, and I.-S. Ng, "CRISPRi mediated phosphoenolpyruvate carboxylase regulation to enhance the production of lipid in *Chlamydomonas reinhardtii*", *Bioresource Technology*, vol. 245, pp. 1527–1537, 2017.

171. F. Daboussi et al., "Genome engineering empowers the diatom *Phaeodactylum tricornutum* for biotechnology", *Nature Communications*, vol. 5, Article Number: 3831, 2014.

172. I. Ajjawi et al., "Lipid production in *Nannochloropsis gaditana* is doubled by decreasing expression of a single transcriptional regulator", *Nature Biotechnology*, vol. 35, p. 647, 2017.

173. K. E. Wendt, J. Ungerer, R. E. Cobb, H. Zhao, and H. B. Pakrasi, "CRISPR/Cas9 mediated targeted mutagenesis of the fast growing cyanobacterium Synechococcus elongatus UTEX 2973", *Microbial Cell Factories*, vol. 15, p. 115, 2016.
174. E. Lauridsen, "The role of government agencies in supporting research on fertility regulation", vol. 2. Background documents. Copenhagen, Denmark, Scriptor, 901–10,1983.
175. Universities UK, *University Funding Explained*. Universities UK. (accessed June), vol. 22, p. 2018, 2016.

Index

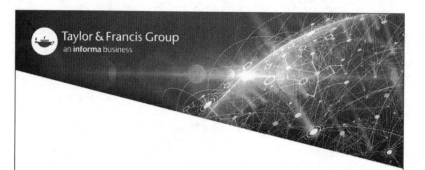